'This book presents an excellent account of the socio-regulatory environment in seven shale gas holding countries, highlighting the intractable tension between securing energy supply and a sustainable habitat. More importantly even, Madeline Taylor and Tina Hunter have written an absolutely timely book on a crucial aspect of the unconventional gas conundrum: how agricultural landowners can be empowered vis-à-vis the gas industry, and how the regulatory governance makes a difference. This book will be an invaluable resource for both scholars and practitioners alike.'

— *Andreas Goldthau, Royal Holloway University of London, UK*

Agricultural Land Use and Natural Gas Extraction Conflicts

Onshore unconventional gas operations, in most jurisdictions, operate on the legal principle that all activities during exploration and extraction are 'temporary' in nature. The concept that the onshore unconventional gas industry has a temporary effect on the land on which it operates creates a regulatory paradox. On one hand, unconventional gas activities create energy security, national wealth and a bourgeoning export industry. On the other, agricultural land and agriculturalists may be significantly disadvantaged by unconventional gas activities potentially producing permanent damage to non-renewable fertile soils and spoiling the underground water tables. Thus, threatening future food security and food sovereignty.

This book explores the socio-regulatory dimensions of coexistence between agricultural and onshore unconventional gas land uses in the jurisdictions with the highest concentration of proven unconventional gas reserves – Australia, Canada, the USA, the UK, France, Poland and China. In exploring the differing regulatory standpoints of unconventional gas land uses on productive farming land in the chosen jurisdictions, this book provides an original three-part categorisation of regulatory approaches addressing the coexistence of agricultural land and unconventional gas namely: adaptive management, precautionary and, finally, statism. It offers a timely and topical approach to socio-legal natural resource governance theory based on the participation, transparency and empowerment for agricultural landholders, examining how differing frameworks such as the collective bargaining framework can create equitable and sustainable contractual arrangements with unconventional gas companies.

Madeline Taylor is currently an Academic Fellow at The University of Sydney Law School, Australia.

Tina Hunter is the Director of the Aberdeen University Centre for Energy Law (AUCEL), UK, and the Professor in Petroleum Law at the University of Aberdeen.

Earthscan Studies in Natural Resource Management

For more information on books in the Earthscan Studies in Natural Resource Management series, please visit the series page on the Routledge website: http://www.routledge.com/books/series/ECNRM/

Agricultural Land Use and Natural Gas Extraction Conflicts

A Global Socio-Legal Perspective

Madeline Taylor and Tina Hunter

Routledge
Taylor & Francis Group

LONDON AND NEW YORK

from Routledge

First published 2019
by Routledge
2 Park Square, Milton Park, Abingdon, Oxon OX14 4RN

and by Routledge
52 Vanderbilt Avenue, New York, NY 10017

First issued in paperback 2020

Routledge is an imprint of the Taylor & Francis Group, an informa business

British Library Cataloguing in Publication Data
A catalogue record for this book is available from the British Library

Library of Congress Cataloging-in-Publication Data
Names: Taylor, Madeline, author. | Hunter, Tina, author.
Title: Agricultural land use and natural gas extraction conflicts : a global
socio-legal perspective / Madeline Taylor and Tina Hunter.
Description: New York : Routledge, 2018. | Series: Earthscan studies in
natural resource management | Includes bibliographical references and index.
Identifiers: LCCN 2018031149 (print) | LCCN 2018033662 (ebook) |
ISBN 9780203702178 (eBook) | ISBN 9781138572232 (hardback) | ISBN
9780203702178 (ebk)
Subjects: LCSH: Land use. | Natural gas.
Classification: LCC HD111 (ebook) | LCC HD111 .T39 2018 (print) |
DDC 333.76--dc23
LC record available at https://lccn.loc.gov/2018031149

ISBN 13: 978-0-367-58298-2 (pbk)
ISBN 13: 978-1-138-57223-2 (hbk)

Typeset in Bembo
by Taylor & Francis Books

Contents

Illustrations

Figures

Table

Foreword

The extent to which the development of shale gas (and shale oil) over the past decade has changed the global energy game is difficult to overstate. It has already had significant implications for geopolitics as well as for energy markets, and experts are not necessarily in agreement regarding the impact it will have in the coming decades, except that it will not be negligible. Macro-level considerations of energy security, climate change and the like can, however, mask the fact that the micro-level reality of the individual drill pad is not infrequently a story of contestation, if not indeed conflict, as the imperatives of the unconventional hydrocarbon industry come into contact with those of agriculture. Whether it is a matter of competition for water resources, fears of contamination of aquifers or the general disruption of the countryside, the arrival of shale operations in long-established rural settings raises the issue of potentially conflicting normative orders. The precise manner in which those conflicts are resolved says much about a jurisdiction's view of macro-level considerations as well as having implications for higher-level potential conflicts, for example, between food security and energy security. The extent to which any jurisdiction is successful in achieving a balance in the context of this multi-level and dynamic problematic is a matter of considerable interest not only to immediate stakeholders, but also to policymakers, legislators, regulators and interested parties in other countries also. This timely book offers the first comprehensive review of these issues in a number of key nations and will no doubt find a ready readership among the many who must grapple with these challenges in the years ahead.

John Paterson
Professor of Law
University of Aberdeen

Preface and acknowledgements

Globally, nations are searching for a sustainable alternative to traditional fossil fuels. Unconventional gas has arisen as a potential 'stepping stone' fuel before production of renewable energies. Onshore unconventional gas operations, in most jurisdictions, operate on the legal principle that all activities during exploration and extraction are 'temporary' in nature. The concept that the onshore unconventional gas industry has a temporary effect on the land on which it operates creates a regulatory paradox. On one hand, unconventional gas activities create energy security, national wealth and a bourgeoning export industry. On the other, agricultural land may be significantly disadvantaged by unconventional gas activities potentially producing permanent damage to non-renewable fertile soils and spoiling the underground water tables. Thus, threatening future food security.

This book explores the socio-regulatory dimensions of coexistence between agricultural and onshore unconventional gas land uses in the jurisdictions with the highest concentration of proven unconventional gas reserves – Canada, Australia, the USA, the UK, France, Poland and China. In exploring the differing policy standpoints of unconventional gas land uses on productive farming land in the chosen jurisdictions, this book provides an original three-part categorisation of socio-regulatory approaches addressing the coexistence of agricultural land and unconventional gas namely: adaptive management, precautionary and finally, statist based governance. It offers a timely and topical approach to socio-legal natural resource governance theory based on the participation, transparency and empowerment for agricultural landholders, examining how differing frameworks such as the collective bargaining framework can create equitable and sustainable contractual arrangements with unconventional gas companies.

The research and writings within this book is dedicated to the land and the brave farmers who have lost their lives trying to safeguard it.

This book is the resulting work of Madeline Taylor and Tina Hunter, who have collaborated with equal contribution.

Acronyms and abbreviations

12th FYP	12th National Five Year Plan of 2011–2015
13th FYP	13th National Five Year Plan of 2016–2020
ac	acres
ACCC	Australian Competition and Consumer Commission
ADR	alternative dispute resolution
ALC	Agricultural Land Commission
ALCA	*Agricultural Land Commission Act*, SBC 2002, c 36
ALR	Agricultural Land Reserve
AMEC	Association of Mining and Exploration Companies
APC	Australian Productivity Commission
APLNG	Australia Pacific Liquefied Natural Gas
APPEA	Australian Petroleum Production and Exploration Association
Bcf	billion cubic feet
Bcm	billion cubic metres
BFPR	*Basic Farmland Protection Regulation 1994* (PRC)
BGS	British Geological Survey
bpd	barrels per day
CAD	Canadian dollar
CAPL	Canadian Association of Petroleum Landmen
CBNRM	Community-based Natural Resource Management
CCA	Conduct and Compensation Agreement
CESCR	Committee on Economic, Social and Cultural Rights
Charter	Community Engagement Charter of the United Kingdom Onshore Operators Group
CNPC	China National Petroleum Corporation
CO$_2$	carbon dioxide
COAG	Council of Australian Governments
CSG	coal seam gas
CSO	civil society organisation
DEC	Department of Environmental Conservation
DECC	Department of Energy and Climate Change
DILGP	Department of Infrastructure, Local Government and Planning
DOH	Department of Health

EA	Environmental Authority
EC	European Commmission
ECL	*Environmental Conservation Law* 3 NY ENV Law Consol § 23–0301 (McKinney, 2014)
EEA	European Economic Area
EEC	European Economic Community
EIA	Environmental Impact Assessment
EIS	Environmental Impact Statement
EPA	*Environmental Protection Act 1994* (Qld)
EPBCA	*Environment Protection and Biodiversity Conservation Act 1999* (Cth)
EPMR	Environmental Protection and Management Regulation
EU	European Union
FAO	Food and Agriculture Organization of the United Nations
FIRB	Foreign Investment Review Board
FSRU	floating storage and regasification unit
GAB	Great Artesian Basin
GC	Gasfields Commission
GCA	*Gasfields Commission Act 2013* (Qld)
GDP	gross domestic product
GHG	greenhouse gas
GMO	genetically modified organism
Green and Styles Report	Professor Peter Styles and Dr Christopher Green's independent report entitled *Preese Hall Shale Gas Fracturing: Review and Recommendations for Induced Seismic Mitigation*
ha	hectares
HF	hydraulic fracturing
HSE	Health and Safety Executive of the United Kingdom
HVHF	high-volume hydraulic fracturing
Hydrocarbon Directive	*Directive 94/22/EC of the European Parliament and of the Council of 20 May 1994 on the Conditions for Granting and Using Authorizations for the Prospection, Exploration and Production of Hydrocarbons*
ICESCR	*International Covenant on Economic, Social and Cultural Rights*
ICPPC	International Coalition to Protect the Polish Countryside
IEA	International Energy Agency
ILO	International Labour Organization
ILUA	Indigenous Land Use Agreement
IMF	International Monetary Fund
Independent Review	*Independent Review of the Gasfields Commission Queensland and Associated Matters*
IPC	International NGO/CSO Planning Committee for Food Sovereignty

JLCNY	Joint Landholders Coalition Of New York
LAC	*Land Access Code 2016* (Qld)
LAF	Land Access Framework
LAL	*Land Administration Law 1999* (PRC)
LNG	liquefied natural gas
LPA	Local Planning Authority
LVC	La Vía Campesina
MERCPA	*Mineral and Energy Resources (Common Provisions) Act 2014* (Qld)
MHR	Municipal Home Rule
MLR	Ministry of Land Resources
MLUF	Multiple Land Use Framework
MNR	Ministry of Natural Resources
MOU	Memorandum of Understanding
MRA	*Mineral Resources Act 1989* (Qld)
MRL	*Mineral Resources Law* (PRC)
NEB	National Energy Board
Network	European Science and Technology Network on Unconventional Hydrocarbon Extraction
NFF	National Farmers Federation
NGO	non-government organisation
NLP	National Landcare Program
NPPF	National Planning Policy Framework
NPPG	National Planning Practice Guideline
NSW	New South Wales
OCP	Official Community Plan
OECD	Organisation for Economic Co-operation and Development
OGAA	*Oil and Gas Activities Act*, SBC 2008, c 36
OGC	Oil and Gas Commission
OGIA	Officer of Groundwater Impact Assessment
OGSML	Oil, Gas and Solution Mining Law
PA Act	*Petroleum Act 1934* (Qld)
PAA	Priority Agricultural Area
PEDL	Petroleum Exploration and Development License
PGPSA	*Petroleum and Gas (Production and Safety) Act 2004* (Qld)
PLA	Priority Living Area
PNGA	*Petroleum and Natural Gas Act*, RSBC 1996, c 361
PRC	People's Republic of China
PRRD	Peace River Regional District
RIDA	Regional Interests Development Approvals
RIMRL	Rules for the implementation of the *Mineral Resource Law* (PRC)
Rio Declaration	*Rio Declaration on Environment and Development*
Royal Society Report	Royal Society and Royal Academy of Engineering report entitled *Shale Gas Extraction in the UK: A Review of Hydraulic Fracturing*

RPI Reg	*Regional Planning Interests Regulation 2014* (Qld)
RPIA	*Regional Planning Interests Act 2014* (Qld)
SCA	Strategic Cropping Area
SCL Act	*Strategic Cropping Land Act (2011)* (Qld)
SCLC	Steuben County Landowners Coalition
SEPA	Scottish Environment Protection Agency
SEQRA	State Environmental Quality Review Act
SGEIS	Supplemental Generic Environmental Impact Statement
SRB	Surface Rights Board
SWDA	*Safe Water Drinking Act 1974*
Tcf	trillion cubic feet
TCLG	Tioga County Landowners Group
TFEU	2010 Treaty on the Functioning of the European Union
TOC	Total Organic Carbon
UDHR	*Universal Declaration of Human Rights*
UG	unconventional gas
UGR	unconventional gas resource
UK	United Kingdom
UKCS	United Kingdom Continental Shelf
UKOOG	United Kingdom Onshore Operators Group
UN	United Nations
UNEP	United Nations Environmental Programme
UNHRC	United Nations Human Rights Council
USA	United States of America
WA	*Water Act 2000* (Qld)
WFS	1996 World Food Summit
WSCC	West Sussex County Council
WTO	World Trade Organization

1 Introduction

Introduction and background

If the twentieth century was the century of oil, then the twenty-first century will be known as the century of gas during the transition to renewable energy. Since the early 2000s, there has been a marked shift in the value and importance of gas as an energy resource. A mere 20 years ago, gas was seen as an inconvenience in the production of oil – to be burnt or drained off and seen as the scourge of an oilfield. That position has changed dramatically. Today, the importance of gas has risen until it possibly exceeds oil in its status as a source of hydrocarbon fuel. To date the extraction of gas has come from conventional oil fields, and concomitant with the extraction of it was the extraction of oil, seen as the real prize.[1] However, as climate change science has evolved, and there has been recognition of the impact of the use of coal as a primary energy resource, other forms of hydrocarbons have been examined and their potential as primary sources of energy. The hydrocarbon that has been assessed as being of the most value for the generation of electricity is that of gas. Many see gas as the transition fuel.[2] That is the bridge between the use of hydrocarbons, particularly coal, and the move to renewable sources of energy. Indeed, some say that it will be necessary to have gas as a continued primary source of energy to augment and supplement a renewable energy source, particularly wind, until battery technology is such that electricity can be stored at a mass scale. This is because gas provides the ability to generate electricity relatively quickly and can be effectively used to supplement renewable sources of energy in cogeneration. The value of gas is such that Yergin has described it as the new prize and energy security as the new quest.[3]

The early 2000s saw an energy revolution in the USA. At the start of the century, low oil and gas production saw a decline in the US economy and a reliance on imports. After many years of wars in the Middle East, a stagnating economy after the Global Financial Crisis and low economic growth, the 'shale gas revolution' brought about access to 'sustainable' cheap gas in the USA and sparked economic growth on the back of it. This has enabled the USA to become a net oil and gas exporter for the first time since 1975. Such bounty in the USA from shale gas, combined with the assessment and discovery that other jurisdictions have access to such wealth in their own unconventional gas deposits,[4] has encouraged many other States to try and emulate such success.

The reason for the focus on hydraulic fracturing in shale gas production is two-fold. Firstly, is the fear (real or imagined) of contamination of both surface and groundwater resources as a result of hydraulic fracturing. Secondly, is the use and disposal of water that is used within the hydraulic fracturing process. Both of these concerns point to a focus to varying extents within this book, due to the nexus between water and agriculture. Water is essential for the production of agricultural products; although farmers often do not need access to water of high drinking standard, it must be potable water that does not cause harm to the environment. Furthermore, these water resources must be permanent not ephemeral. Hydraulic fracturing requires up to 1.5–16 million gallons of water per hydraulic fracture (each well can require anywhere between 5–15 fractures). Of this water use, approximately 50 per cent will be returned to the surface contaminated with the chemicals that are added during the hydraulic fracturing processes as well as naturally occurring contaminants that return to the surface. This then raises another challenge, that of the disposal of contaminated water often on agricultural land. It is this interaction between hydraulic fracturing, unconventional gas production, agricultural activities and the broader context that is further explored here.

An added dimension to the extraction of shale gas is the value of agricultural land. Such value may be both objective and subjective, clearly agricultural land provides food security for a nation. However, many States are also concerned with the issue of energy security. For some States, one type of security may trump the other. For example, for Poland, a nation dependent upon Russia for gas and therefore energy security, and the country that bore the brunt of the impact and effects of the Ukrainian Gas Crisis, energy security is paramount. Although recently, with a change in political party, Poland has a renewed focus on its food security and agricultural land protection. For other countries, such as France, the value of its food both in economic and cultural terms is much higher than that of energy security. This is because France has both a strong relationship with the provenance of its food and also has high energy security due to its use of nuclear power as an energy source. This book explores these complex relationships between energy security and food security within the context of unconventional gas extraction. Primarily focusing on shale gas, it examines the extraction of unconventional gas within a socio-legal context.

Scope of the book

This book explores the socio-regulatory dimensions of coexistence between agricultural and onshore unconventional gas land uses in the jurisdictions with some of the highest concentration of proven unconventional gas reserves – Queensland, Australia; British Columbia, Canada; The UK; the state of New York, USA; France; Poland and China. In exploring the differing policy standpoints of unconventional gas exploitation on productive farming land in the chosen jurisdictions, this book provides an original analysis of socio-regulatory approaches to the coexistence of agricultural land and unconventional gas:

adaptive management, the precautionary principle and, finally, a statist approach. It offers a timely and topical consideration to socio-legal natural resource governance theory based on the participation, transparency and empowerment for agricultural landholders.

This book is novel and contemporary. Globally, nations are searching for sustainable alternatives to traditional fossil fuels. Unconventional gas has arisen as a potential 'stepping stone' fuel before production of renewable energies. For example, in some Australian states, coal seam gas has been touted as the country's energy future as a result of a political environment currently hostile to coal production and in the developmental stages of large scale renewable energy projects. Significant empirical research exists relating to the sociological and scientific impacts of unconventional gas and agricultural coexistence. However, to date, there has been little comparative socio-legal research of agricultural protection and onshore unconventional gas coexistence between civil and common law based systems in relation to agricultural landholder rights, arable land protection and unconventional gas activities. This book develops a richer understanding of the operation of unconventional gas sectors within rural communities globally.

The second part of the book focuses on comparative jurisdictional case studies of differing legal systems and their regulatory treatment of agricultural land, and farmers, during and after unconventional gas activities. Drawing on this international commentary and context, the third part of this book looks to socio-regulatory solutions for these conflicting land uses by attaining community-based natural resource governance. The concluding chapters then examine the proposed solution of collective bargaining for agricultural landholders in producing equitable solutions of coexistence with unconventional gas activities. This book also acknowledges the broader socio-legal context for the discussion of agricultural land protection regulatory frameworks beyond that of recent work focusing on environmental justice. It focuses on contributing to the growing research work in collectivism within rural communities affected by unconventional gas activities. In an original approach, this book parallels the alternative coexistence of socio-legal frameworks to govern agricultural land use and unconventional gas activities evident in the seven chosen jurisdictions.

The insights provided by this book have potential relevance to future policy and legal practice in the agricultural and energy sectors throughout the chosen jurisdictions and globally. This work argues that coexistence requires a new approach to governance, reflected in collectivism and collective bargaining, in order to recognise the interests of private agricultural landholders in a sustainable and equitable way during and after unconventional gas activities. Arguably, the most pressing challenge facing resource-rich nations is how to plan for a future that is economically productive, preserves agricultural land and its future food production. This, in turn, equates to the need for regulation that reflects this recent shift in priorities from mining to preserve agriculture sustainably. This book explores the creation and adoption of a novel legal vehicle, collective bargaining, to increase the bargaining power of rural landholders during unconventional gas activities. Currently, no global socio-legal academic study has performed a legal analysis of the potential of collective bargaining for agricultural landholders and unconventional gas titleholders. Consequently, this book will fill a current gap in the literature by focusing on the application of new theories in an unconventional gas context: adaptive management, precautionary principle and a statist approach.

The importance of coexistence and contestation in the context of unconventional gas

As jurisdictions increase drilling for unconventional gas, the contestation of agricultural land and unconventional gas will become increasingly apparent. Considering these multiple and often contrasting demands on the land and the complexities they generate, agriculture is a highly politicised regulatory space.[5] It is also a multi-actor sector, with decision-makers ranging from individual farmers to national law-makers, each with their own incentives, perspectives and priorities. Coexistence seeks to recognise and equitably manage the interests of different stakeholders, regulatory bodies and private companies. This ensures effective regulation of competing land uses, rather than one sector being privileged to the disadvantage of another. The regulatory aim of States seeking to develop their unconventional gas sector is that of coexistence. For example, in Queensland, Australia, the stated purpose of its unconventional gas regulation is to 'achieve coexistence of landholders, regional communities and the onshore gas industry in Queensland'.[6] Consequently, coexistence is fundamental to achieving effective regulation of unconventional gas exploration and extraction with one specific stakeholder group, agricultural landholders, as 'prime agricultural and quality water resources must not be compromised for future generations'.[7]

Coexistence, the distribution of benefits from the unconventional gas industry and its effects on agricultural land is therefore the decisive problem facing States. In contrast, the concept of contestation is underpinned by objections to norms, principles and regulations, by rejecting or refusing to implement them as a mode of critique and opening up discourse of potential reform.[8] Therefore, contestation is often the starting point for regulatory reform and, if successfully implemented, provides a landscape of coexistence between land uses. As a socio-legal practice, contestation gives rise to the need for States and companies to attain a social license, in addition to a legal State license, to operate.

Compliance with regulation is but one dimension to facilitating coexistence, while attaining a social license to operate for unconventional gas companies can unify opposing social and political interests. It is in the interests of the State, unconventional gas titleholders and agricultural landholders that they must coexist, given that the lifespan of unconventional gas production activities in most jurisdictions is up to 50–60 years. The complex relationship between landholders, companies and the State characterises the unconventional gas sector. The economic significance of the sector often influences political and regulatory decision-making, leading to intense social scrutiny, as is the case in New York State, USA. At the heart of the problem, is the absence of a 'one size fits all' regulatory benchmark to determine an equitable distribution between protecting farmland and reaping extractive revenues and energy security from unconventional gas sources.

The access, use and management of agricultural land have been drastically altered in States with a newfound interest in unconventional gas extraction.[9] This has led to the contestation and modification of the relationship between rural communities, the State and gas companies. In particular, agricultural landholders often hold a strong sense of value in protecting its finite nature, a concept shared by academics, economists and philosophers alike and as

> Land draws special attention because of its well-known special attributes: land is essential for physical human survival, it provides the essential context for social and political life, it often embeds symbolic or religious values, it is largely irreplaceable, each location is unique, and the supply of land is almost finite.[10]

It is the role of the State to shape the legal conditions for unconventional gas development. In the context of the future sustainability of unconventional gas as an energy source is the challenge of striking the appropriate balance between arable land protection and energy security. At one end of the spectrum, is the emphasis of food security and ensuring a State is self-sufficient in its yields and preserves its soil security in order to feed its citizens. At the other end of the spectrum, is the desire of the State to achieve energy policy targets for delivering energy independence and often securing the economic benefits of developing unconventional gas, whatever the cost. Regulating for unconventional gas must consequently find a balance between contestation and coexistence. This book seeks to establish that land contestation creates conditions ripe for *ex ante* regulatory assessment to evaluate whether the chosen regulatory approach of adaptive management, the precautionary principle or a statist approach will effectively promote land use coexistence. This book performs a comparative examination of the different models of agricultural land use zonings and authorisations for unconventional gas activities in a number of differing jurisdictions, noting their similarities and differences before leading into potential recommendations to guide future regulation.

Unconventional gas resources and their development

Whereas the production of gas has hitherto been from conventional reservoirs, increasingly unconventional reservoirs are being utilised for the extraction of oil and gas. In particular, shale formations, the source formation for conventional reservoirs, are now viewed as reservoirs within themselves for the production of gas. This is because the use of existing technologies, promulgated by George Mitchell in Texas in the 2000s, has enabled these reservoirs to be exploited for their gas potential, bypassing the need to wait millions of years for gas to be liberated into the conventional reservoirs above. As a consequence, since 1975 the USA has moved from being a net energy importer to a major producer of oil and gas from such reservoirs and is expected to be the world's largest producer of both oil and gas by 2019, surpassing even that of Saudi Arabia (oil) and Russia (gas).[11] Indeed, Yergin has described the advent of hydraulic fracturing as 'the most important energy innovation so far in the 21st century.'[12]

Although the production of gas from such unconventional reservoirs has created great economic advantages and potential for the USA, something many countries wish to emulate, it has also created conflict. Such conflict has occurred particularly in new jurisdictions, where unconventional gas exploration has commenced.[13] This potential conflict arises because shale reservoirs often underlie agricultural land. The reason for this phenomenon is the geological circumstances in which shale reservoirs are formed. Shale deposits occur in three environments: marine, fluvial and lacustrine. Each of these depositional environments contains high organics and fine particles resulting in the formation of unconventional gas millions of years later as a result of continued deposition, heating and pressure. In today's environment, those same depositional landscapes produce agricultural land, indeed often prime agricultural land with high soil quality, that supports cropping and other such intensive agricultural activities. Therein lies the problem. Shale reservoirs underlie existing agricultural land. Such agricultural land includes both high-value cropping land and pasture land.

Some of the best shale gas fields, for example the Barnett, Eagleford and Marcellus in the USA, contain high organics, greater than or equal to 5 per cent, producing high yields of gas. These may or may not be overlaid by productive agricultural land such as in the USA. In other jurisdictions, for instance the UK, Poland and France, existing shale formations are overlaid by arable agricultural land. It is this scenario that creates a three-dimensional conundrum. In order to liberate the gas that lies underneath the land it is necessary to undertake drilling and production activities on the surface. Yet, it is this surface land that is often prime agricultural land. Therefore, in order to produce unconventional gas, it is necessary to utilise agricultural land. For some countries, the choice between agricultural land and shale gas is easy. For others, there is a need to have both. But this raises another more important question: Should a country need to choose between agricultural land and unconventional gas production or should the two activities be able to exist alongside each other? It is this question that forms the focus of the book.

In most jurisdictions, onshore unconventional gas operations operate on the legal principle that all activities during exploration and extraction are 'temporary' in nature. The concept that the onshore unconventional gas industry has a temporary effect on the land on which it operates creates a regulatory paradox.[14] On one hand, unconventional gas activities create energy security, national wealth and a bourgeoning export industry. On the other, agricultural land and agriculturalists may be significantly disadvantaged by 'temporary' unconventional gas activities potentially producing permanent damage to non-renewable fertile soils and spoiling the underground water table, thus threatening future food security. The challenge for regulators is to balance the economic and energy needs related to unconventional gas extraction with protecting agricultural land for its use.

Hydraulic fracturing is a process developed in Texas in the late 1990s by George Mitchell. It was developed as an attempt to create the necessary porosity and permeability in the tight, small-grained shale rocks to enable the shale gas to migrate freely and be recovered at the surface. Fractures are created in rocks through the application of pressurised fluid to the target geology to increase its porosity and permeability.[15] The use of water as a fracturing agent (known as 'slickwater fracturing') was developed for shale rocks as a method for extracting the gas known to be inside the rock but that was previously inaccessible. This new generation of hydraulic fracturing is characterised by the use of high volumes of water, and often several million gallons of water are used for each fracture. The environmental impact of hydraulic fracturing is the high volume of water used for each hydraulic fracture, and the impact is particularly prominent in areas where there is competition for water, especially in agricultural areas where there is low water availability.[16]

In the USA, the shift from conventional oil and gas production to unconventional oil and gas production was relatively seamless. Commencing in Texas, production of shale gas raised a few eyebrows in a state that has been used to the production of oil and gas for over a hundred years. However, in some areas of the USA, particularly the Marcellus area of the northeastern USA, the commencement of the extraction of shale gas caused both physical issues as well as social implications arising from environmental impacts.[17] Community contestation and concern was highlighted in Josh Fox's 2010 documentary *Gasland* which identified shale gas extraction as a cause of contamination of aquifers and broader environmental harm. The impact of *Gasland*, both within the USA, and more importantly, outwith, was enormous. Numerous jurisdictions moved from considering unconventional gas extraction as a logical progression from the development of conventional oil and gas, to an activity that was to be feared and reviled. Since that time, shale gas extraction has been the subject of countless studies. These studies have ranged from individual studies undertaken by academics to EU-wide studies.[18]

To take one jurisdiction, Australia has seriously contemplated the extraction of unconventional gas in all six states and the Northern Territory. Most of these jurisdictions have ended up undertaking studies regarding the environmental safety and impact of the activity. These studies have varied in quality and outcomes. It should be noted that perhaps the best example of a comprehensive study in relation to shale gas extraction is that of the *Scientific Inquiry into Hydraulic Fracturing in the Northern Territory* (The Pepper Report).[19] However, it is important to note that the vast majority of reports worldwide focus on the activity of hydraulic fracturing, and it is this that creates the most contestation for opponents of unconventional gas.

However, hydraulic fracturing is merely a well stimulation technique to create permeability and porosity within shale gas reserves in order to enable the hydrocarbons to flow. In the extraction of shale gas from shale reservoirs hydraulic fracturing is always required. In the extraction of other forms of unconventional gas resources, especially tight gas and coal seam gas (CSG), hydraulic fracturing is not always required. Indeed, for the extraction for CSG to occur, hydraulic fracturing is used in less than 10 per cent of all wells. Rather, CSG is produced by dewatering the seam; removing water from coal cleats and thereby liberating the gas. Production of CSG itself has other environmental impacts and effects and is considered in Chapter 5 since it is commercially undertaken in Queensland, Australia. In most other jurisdictions, the extraction of unconventional gas centres around shale gas, which is the focus of much of this book.

Overview of the book

This book is comprised of 13 chapters arranged in three parts. This first chapter provides an introduction and contextual background to the phenomenon of agricultural land and unconventional gas regulatory contestation in a global political context.

Chapter 2 explores natural resource governance concepts relating to both unconventional gas and agricultural lands within both legal and social literature. Natural resource governance refers to the ways in which society manages the supply of, or access to, the natural resources upon which it relies for survival and development. It examines not only the ownership of gas resources, but also how the extraction of a non-renewable resource from below land creates a new regulatory paradigm – that of prioritising or reconciling competing resource needs in an increasingly populous world.

Chapter 3 analyses the scholarly debate and literature relating to the human right to adequate food and national food security obligations according to the *International Covenant on Economic, Social and Cultural Rights.* This chapter then explores food sovereignty as the defining concept at the heart of a growing socio-political food movement in Canada, France and the USA. The narrative of food sovereignty collectivises the rights of local farmers, growers and consumers as part of the demand for a sustainable food system in the face of growing oil and gas operations. This 'bottom up' rights-based approach has created community supported agriculture protecting farmers' rights during energy activities and extraction globally.

Chapter 4 outlines and examines the three differing socio-legal governance approaches demonstrated in the chosen case study jurisdictions. The first, adaptive management, is based on 'learning through doing' – the legislative response to multiple land uses through regulation. The second, the precautionary principle, examines the reactive stance which favours one land use over another. Lastly, the statist approach, which prioritises political will and energy security to the exclusion of other land uses.

Chapter 5 will analyse the current agricultural land protection and petroleum regulatory frameworks of Queensland, Australia – representing a regulatory example of adaptive management, with less protective measures for agricultural land and its farmers during CSG activities. The attempts at natural resource governance provided by the Gasfields Commission and Land Access Code will be analysed against the socio-political context of farmer land access and compensation agreements in Queensland.

Chapter 6 provides an examination of the Agricultural Land Reserve Regime of British Columbia, Canada as an archetype of adaptive management by examining the role of the Agricultural Land Commission and Oil and Gas Commission within the province. The legislative regime promotes coexistence of reserved agricultural land, particularly in the Peace-River Region.

Chapter 7 will analyse the conflict and legal framework in relation to shale gas extraction and agriculture. The energy security and food security nexus in the UK is one of the most marked in the unconventional gas extraction debate. Opposition to shale gas extraction in the UK is fierce, with division evident along party and class lines. The chapter will focus on the current regulatory framework, analysing the interplay between the planning system, environmental protection and agricultural activities in the UK. The recent permanent moratorium on shale gas extraction to preserve agriculture will also be analysed as will its implications for land access, agricultural production and compensation to land users.

Chapter 8 provides a socio-legal examination of an alternative 'bottom-up' regulatory approach of landholder coalitions in New York State, USA, by empowering agricultural landholders during and after shale gas operations. The regulatory moratorium on hydraulic fracturing and shale gas activities within New York State is also discussed as another example of precautionary regulation.

Chapter 9 provides an examination of France; the second largest holder of shale gas resources in Europe and the first country to enact a ban via legislation on all hydraulic fracturing of shale deposits from 2011. This chapter observes the application of the precautionary principle by the French legislature coupled with an analysis of the Ardèche movement.

Chapter 10 analyses the social, political and legal issues relating to shale gas extraction in a jurisdiction that was overwhelmingly pro-extraction, but where poor results and a change in government have altered the perception of the value and use of shale gas exploitation on agricultural land. Poland provides a unique study since it seeks to exploit its shale gas resources as a means of securing energy independence from Russia, but also wants agricultural reform to enhance its basic agricultural sector. This study of Poland will provide a novel analysis of a country where energy security threatened to hold primacy over food security.

Chapter 11 analyses the current statist-based regulations in China governing shale gas extraction, pursuant to the Technical Policy for the Pollution Prevention and Control of Oil and Gas Exploitation. Further, the chapter examines the *Basic Farmland Protection Regulation of 1994* as the regulatory basis for conversion of land within farmland districts that is deemed 'unavoidable to build national projects', including national shale gas activities. Consequently, Chapter 11 illustrates the socio-legal approach by Chinese environmental regulations relying upon political processes, rather than scientific advice, as a risk-accepting regulatory approach.

Chapter 12 provides analysis of the theoretical constructs of social capital and the concept of collective action as a form of natural resource governance. The proposal of a collective bargaining vehicle for agricultural landholders, within this chapter, in improving their negotiation position in unconventional gas agreements is both timely and contemporary given the political interest in ameliorating the position of agricultural landowners and land access arrangements in the chosen jurisdictions.

Chapter 13 concludes by exploring solutions founded in collective bargaining proposed to aid in a number of legal frameworks in the empowerment and protection of arable land and its farmers in the face of a growing unconventional industry globally.

Book contribution

The contribution of this book is in the application of a comparison of alternative regulatory approaches to managing coexistence of unconventional gas in a global survey. With unconventional gas to be developed for LNG export and to ease gas shortages across the world, an examination of the legislative frameworks of some of the most important gas producing States is timely. It is intended this book will also contribute to the wider study of regulation in considering how land has developed and exists in jurisdictions in response to legal issues related to unconventional gas development; namely, the impact of the energy source on agricultural land and concerns over land use, access and compensation.

Although significant empirical research exists in relation to the sociological and scientific impacts of unconventional gas and agricultural coexistence, there has been little comparative socio-legal research concerning unconventional gas at both land use and land access levels. Therefore, this book is the first in providing an in-depth functional analysis relating to coexistence and land use.

As evident in the numerous jurisdictions analysed within this book, regulatory concern about the development of unconventional gas relating to land and incorporated damage to productive farming land, land access and compensation are not unique to any State. It is intended that the insights provided by this book may hold relevance for future regulatory directions in managing conflicting interests and avoiding regulatory burdens in jurisdictions seeking to create, or maintain, their unconventional gas industry, as it provides an examination of varying styles of regulation for these activities. For any State wishing to sustain a position of global competitiveness in unconventional gas, there is a pressing need to create a cross-government regulatory system that encourages industry development, improves compliance and will not result in a diminution of landholder rights. This book provides an analysis of unconventional gas frameworks to identify any regulatory gaps or successes and considers alternative regulatory tools to achieve the balancing of competing interests – those of farms and unconventional gas titleholders.

Notes

1 As stated by DeGoyler, observing about the Middle East, 'that oil in this region is the single greatest prize in all history'. This concept of 'the prize' became the title of Daniel Yergin's comprehensive volume on the history of oil. De Goyler in Daniel Yergin, *The Prize: The Epic Quest for Oil, Money and Power* (Simon & Schuster, 1991) 358.

2 For example, the Finkel Review into renewable energy in Australia notes that gas could, and should, play a role in low carbon electricity generation. See Alan Finkel, *Independent Review into the Future Security of the National Electricity Market – Blueprint for the Future* (2017) <https://www.energy.gov.au/publications/independent-review-future-security-national-electricity-market-blueprint-future>.

3 Daniel Yergin, *The Quest: Energy, Security and the Remaking of the Modern World* (2012).

4 US Energy Information Administration, *Technically Recoverable Shale Oil and Shale Gas Resources: An Assessment of 137 Shale Formations in 41 Countries Outside the United States* (2013).

5 Claire A. Dunlop and Claudio M. Radaelli, *Handbook of Regulatory Impact Assessment* (Edward Elgar, 2016).

6 *Gasfields Commission Act 2013* (Qld) s 3. *Regional Planning Interests Act 2014* (Qld) s 3 (c): One of the purposes of this Act is to achieve 'coexistence, in areas of regional interest, of resource activities and other regulated activities with other activities, including, for example, highly productive agricultural activities'.

7 Office of the Chief Economist, *Review of the Socio-economic Impacts of Coal Seam Gas in Queensland* (2015). For purposes of clarity, it is noted that the effect of the unconventional gas industry on Aboriginal treaty rights and lands and the ability of indigenous stakeholders to coexist harmoniously with the unconventional gas industry is not addressed within this book.

8 Antje Wiener, '*A Theory of Contestation* – A Concise Summary of Its Argument and Concepts' 49(1) *Polity* 110.

9 It is noted that the use of the term coexistence in this book does not extend to the argument of environmentalists relating to soil, water and land degradation as a result of hydraulic fracturing. This is primarily a scientific and highly technical discussion outside the scope of this book.

10 Rachelle Alterman, 'Land-use Regulations and Property Values: The "Windfalls Capture" Idea Revisited' in Nancy Brooks, Kieran Donaghy and Gerrit Jan Knaap (eds), *The Oxford Handbook of Urban Economics and Planning* (ebook, Oxford University Press, 2011).

11 Faith Birol, 'US to Overtake Russia as Top Oil Producer by 2019 at Latest: IEA' (2018) *Reuters* 27 February 2018 <https://www.reuters.com/article/us-energy-iea/u-s-to-overtake-russia-as-top-oil-producer-by-2019-at-latest-iea-idUSKCN1GB0C6>.

12 Daniel Yergen, 'Yergin on the Next Energy Revolution' (2014) *McKinsey Quarterly*, April 2014, <https://www.mckinsey.com/business-functions/sustainability-and-resource-productivity/our-insights/daniel-yergin-on-the-next-energy-revolution>.

13 Especially in Australia and France.

14 This temporary nature is that the activity will be there for a period, and then will no longer have an impact on the land. This 'temporary' status can range from a few months, such as in an exploration well being drilled or a production well prepared, to many years, such as for a production well.

15 Tina Hunter, 'All Hydraulic Fracturing is Equal, But Some is More Equal Than Others: An Overview of the Types of Hydraulic Fracturing and the Environmental Impacts' (2014) 29(3) *Australian Environment Review* 66, 66.

16 Ibid.

17 New York State Department of Health, *A Public Health Review of High Volume Hydraulic Fracturing for Shale Gas Development* (2014); United States Environmental Protection Agency, *Hydraulic Fracturing for Oil and Gas: Impacts from the Hydraulic Fracturing Water Cycle on Drinking Water Resources in the United States* (2016).
18 European Commission, *The Shale Gas 'Revolution': Challenges and Implications for the EU* (2013).
19 Rachael Pepper et al., *Scientific Inquiry into Hydraulic Fracturing of Onshore Unconventional Reservoirs in the Northern Territory* (2018) <https://frackinginquiry.nt.gov.au/>.

Part I

Socio-regulatory theories related to unconventional gas extraction and agricultural activities

2 Natural resource governance and land use conflict

Introduction

For any consideration of the interplay between unconventional gas extraction and agricultural activities, it is essential to consider the nature of the socio-legal framework within which these activities occur. This framework is natural resource governance, which refers to the way in which society manages the supply of, and access to, the natural resources upon which it relies for both its survival and development. There are many aspects of natural resource governance, with much of the literature and research relating to resource extraction, particularly mining in developing countries.[1] However, in this chapter, the analysis is confined to natural resource governance, specifically in its application with respect to the use and enjoyment of agricultural land where unconventional gas extraction occurs.

The governance of natural resources necessarily requires a contemplation of the use of the land and will not only incorporate the regulation of how the land is used, but also how the differing land uses interact with each other. It is naïve to think that there is only one use of land. Rather, as demonstrated by the extraction of shale gas on agricultural land, there are often multiple and competing uses of land, by numerous and differing stakeholders. Such competing uses are not always (although often are) associated with a two-dimensional paradigm. For example, the competing uses of the land taking place on the surface of the land only. Rather, these competing interests are associated with the three-dimensional use of the land. This three-dimensional analysis of the land includes an examination of activities at the surface competing with the underground uses of the land. This three-dimensional approach is needed when considering the natural resource governance of agricultural activities and unconventional gas extraction. Therefore, there is a need to examine concepts of land and resource ownership, as well as a consideration of the fundamental aspects of property as a 'bundle of rights'.

While examining issues of unconventional gas resource ownership, this chapter also explores the new regulatory paradigm that arises from extracting non-renewable resources beneath land surfaces which, themselves, are the site of production, such as those developed in agriculture. This paradigm is concerned with reconciling or prioritising competing resource needs in an increasingly populous world. This chapter firstly explores the concept of ownership and resource development, as well as how the extraction of a subsurface natural resource reflects within the concept of property rights. As part of this consideration of ownership, concepts of dominion and estates are considered. The chapter then considers the theoretical foundations of natural resource governance and its application in shale gas extraction. It then focuses on the collective governance of natural resources, which arises from individualistic responses to Hardin's concept of *The Tragedy of the Commons*.[2] A related concept considered in this chapter is that of Community-Based Natural Resource Management. Finally, this chapter addresses how natural resource governance conflict can be managed within the unconventional gas context through a consideration of land access regulation such as multiple land use frameworks.

Ownership, natural resources and the governance paradigm

Land ownership systems have arisen out of two fundamental concepts. Firstly, the common law and its relationship with the Crown and, secondly, that of civil law jurisdictions and private ownership. Essentially, the paradigm difference between such legal systems is that of reservation to the Crown. Under common law, the reservation of some minerals for the exclusive use of the Sovereign (Crown Prerogative of Royal Metals) was legitimised in 1568, when the *Case of Mines* recognised the Crown's entitlement to gold and silver.[3] With the transplant and colonisation of the common law to other jurisdictions during the eighteenth and nineteenth centuries, respectively, as occupation occurred, the concept of Crown Prerogative of Royal Metals was adopted in these jurisdictions also. The concept of Crown Prerogative, which today has the broader scope of Crown Reservation, is central to the common law Doctrine of Tenure and the Doctrine of Estates,[4] and has been embedded in the legal system of almost all common law countries.[5] As such, reservation of some minerals to the Crown has always prevailed in common law jurisdictions. This Crown Prerogative has been increasingly extended through statute in common law countries, and now encompasses the reservation of petroleum. The reservation of petroleum has extended to unconventional shale gas in common law jurisdictions (for example, the UK, Canada and Australia), particularly in the twentieth century.[6]

In most common law jurisdictions, Crown (or State) Reservation remains, having been incorporated into law by statute rather than common law. The legal status of ownership in common law countries is that of the relationship between a landowner and their land. Under such a relationship, the landowner has title to his land in fee simple,[7] subject to Crown reservation.[8] This creates two primary forms of relationships – that of the fee simple absolute and the lesser interest of a leasehold estate. This view of the relationship between a landowner and his property was expanded by Lord Coke:

property in land is not merely to allege some causal physical affinity with a particular piece of land, but rather to stake out some sort of claim to the legitimacy of one's personal space in this land. It is to assert that the land is 'proper' to one; that one has some significant self-constituting, self-realising, self-identifying connection with the land; that the land is, in some measure, an embodiment of one's personality and autonomy. The claim 'property' in land is to arrogate at least a limited form of sovereignty over the land and to allege that one has some emotion or investment–backed security in it. To have 'property' in land connotes, ultimately, a deeply instinctive self-affirming sense of belonging and control; and it is precisely this sense of possessory control which identifies the two proprietary estates acknowledged today in English law, the fee simple absolute (or freehold estate) and the term of years absolute (or leasehold estate).[9]

In contrast, this relationship, and the concept of Crown Reservation has not been applied and incorporated into the land law framework of the USA. Rather, allodial ownership of land prevails in the USA, so land ownership is absolute. This is in contrast with other legal systems, where there is conditional ownership of land in fee simple, a hallmark of the doctrine of tenure.

A fundamental aspect of real property ownership is 'the ability to control access to the land. It is fundamental to the rights of a freeholder and leaseholder that their legal rights include the ability to exclude access to the land'.[10] Under allodial title, the landowner has absolute ownership of their land and is not subject to any reservation of the State or federal government. This creates private ownership of mineral rights, such as the private ownership of shale gas by surface estate landholders in the USA. This form of ownership is unique in that it grants mineral ownership rights to private landholders which, in other common law jurisdictions, are reserved only for the Crown. However, it is important to note that subsurface mineral ownership rights can be alienated or severed and transferred to third parties from the prevailing surface rights. Such an example is in Texas, USA where severance of mineral ownership from the land grants title to subsurface minerals, typically to private oil and gas companies, via a mineral lease to extract unconventional gas.[11] Associated with such private rights is the *Rule of Capture*, which enables the owner of the mineral rights to take from the land any oil or gas captured within their leasehold area.

Crown Reservation, and the severance of surface and subsurface estates under the allodial ownership system, demonstrate the complex relationship of property. This means that property is essentially a 'bundle of rights' that can be split between either private individuals, or between private individuals and the State where public interest prevails. Such splitting of the bundle of rights necessarily may create conflict. This conflict relates to the fundamental concept of the legal right to exclude others from property, as iterated by Brandeis, J in *International News Service*:

An essential element of individual property is the legal right to exclude others from enjoying it. If the property is private, the right of exclusion may be absolute; if the property is affected with a public interest, the right of exclusion is qualified.[12]

Such exclusion is fundamental in natural resource governance. Where landholders'[13] land is subject to other interests (such as subsurface interests), their capacity to exclude others is diminished or removed. It is this loss of capacity to exclude that can create conflict, particularly when competing land uses exist. Such potential conflict is demonstrated on land where a three-dimensional conflict occurs. This conflict relates to the use of the surface of the land for agricultural activities, particularly in relation to intensive farming such as cropping, and the extraction of unconventional gas from beneath the land for either private or public benefit. For example, in common law jurisdictions with Crown or statutory reservation, the right to explore for and recover unconventional gas can be granted to third parties, such as a petroleum company, by the State via a licensing system. This grant of the right to explore and take unconventional gas trumps the ownership of land in fee simple or a leasehold estate, and therefore the right to exclusive possession of the land is compromised. This means that the landholder holds their land subject to the grant of the interest by the Crown. Naturally, the extraction of the resource will require access to land and create some fundamental changes to that land. However, it also creates tension, since the landholder no longer has control over access to his or her land.

The nature of unconventional shale gas extraction means that the use of the surface for subsurface activities and the interaction between the subsurface title-holder and the landholder is more intense. In addition, the impact on the land is also increased because a greater area of land is required to extract shale gas.[14] Fundamentally, this creates conflict between the surface landholder and the interests of the subsurface resource titleholder. Existing governance frameworks need to address this in order to manage the conflict that arises and to enable these activities to coexist. This task is particularly important in jurisdictions where energy security and food security are of equal importance. Such management may diminish in importance where energy security trumps food security, or in jurisdictions where food security dominates over energy security.[15] The relationships between food security and energy security will be explored in the individual comparative country studies in Part II of this book.

In order to understand the relationship between these fundamental property concepts, it is important to explore the theory of natural resource governance and how it relates to unconventional gas governance and land coexistence.

Theoretical foundations of natural resource governance

Governance is defined by Salih as, 'the exercise of power in an institutional context with the main aim of directing, controlling, and regulating activities concerned with the public interest'.[16] In a normative sense, a governance framework also allows for a discussion of the need for legitimacy, accountability, transparency and discourse.[17] Natural resource management is defined by Baromey as, 'Actual decisions and actions concerning policy and practice regarding how resources and the environment are appraised, protected, allocated, developed, used, rehabilitated, remediated and restored, monitored and evaluated'.[18]

Since the inception of natural resource management, a shift is evident by law-makers and commentators to natural resource governance terminology, as opposed to natural resource management. This is due to the fact that the term management indicates a 'dependence on the technical knowhow and expertise of state agencies, a bureaucratic and monopolised environmental approach'.[19] From the 1980s onwards, governance frameworks have shifted the emphasis towards deliberative democratic process and collaboration as part of a co-management process. This shift is highlighted by Singleton: 'governance systems that combine state control with local, decentralised decision making and accountability and which, ideally, combine the strengths and mitigate the weaknesses of each'.[20]

According to Martin, Williams and Kennedy, natural resource governance must include policies that are:

1 Economically efficient;
2 Institutionally coordinated;
3 Behaviourally effective; and
4 Trans-tenure.

In order to optimise:

1 Economic production from natural resources;
2 Within the constraints of water and land capacity for extraction;
3 Fairness in the allocation of costs and benefits; and
4 Restoration of landscapes to produce public goods.[21]

Natural resource governance theory, therefore, is drawn from environmental impact assessments that give provenance to natural, social and environmental resources.[22] Natural resource governance refers to the ways in which society manages the supply of or access to the natural resources upon which they rely for their survival and development.[23] As human society is fundamentally dependent on natural resources, ensuring the ongoing access to, or a steady provision of natural resources is central to governance, and thus legal regulation.[24] Historically, this access has been organised through a range of schemes varying in degrees of formality and involvement, such as the National Landcare Program ('NLP') in Australia.[25] Since the early 1990s, the NLP has been the centrepiece of natural resource governance in Australia, providing funding to Landcare groups and attracting increasing farmer participation.[26] Nevertheless, the effectiveness of the NLP has increasingly come into question. For example, farmers are expected, on the one hand, to put community interests before their own by internalising the costs of providing off-site and long-term social and environmental benefits. On the other hand, farmers are often asked to become more 'entrepreneurial' and productive in order to compete in export markets in an environment of agricultural deregulation.[27]

Natural resource governance goes to the heart of regulatory governance. Specific governance issues include establishing hierarchies between the different resources or deciding which ones are strategic and need to be secured as a priority, as in the case of agricultural land and unconventional gas. A 'natural' resource by definition is one that is apparent and in existence within the need for human intervention. Hence, the arable lands or the gas within them, rather than the crop that grows on them, comprise a country's natural resources by which some form of benefit can be derived immaterially or materially.[28]

Collective governance of natural resources

Collective governance of natural resources by resource users emerged from the individual responses to Hardin's essay on the *Tragedy of the Commons*,[29] which is described as:

> the philosophy that there is a conflict between the existence of a world with finite dimensions and resources, and the actions of the rational human being in maximising his or her short term interests. The 'commons' refers to the earth's natural resources which are regarded as belonging to no individual or state (for example, the oceans or the atmosphere). The tragedy refers to the inevitability that the commons will be degraded and ultimately exhausted as a result of the actions of the human race.[30]

In response to the *Tragedy of the Commons*, Gilles et al. identify examples where communities and stakeholders have avoided the *Tragedy of the Commons* by the effective management of grazing lands or pastures.[31] The case studies presented by Gilles et al. demonstrate that, contrary to theoretical predictions, collective action can ensure resource governance of agricultural land. Further, Laerhoven argues that in some cases collective governance of natural resources by a community of organised users achieves more sustainable outcomes than either State management or privatisation.[32]

Since the publication of the Brundtland Report in 1987,[33] community participation natural resource governance has been given more attention by researchers and policymakers.[34] Moreover, the intellectual movement towards community ownership of resources, rather than government or private ownership, has driven greater community participation in resource management.[35] At the same time, deep-rooted relationships between decentralisation, collective action and property rights have caused many governments to recognise the important role of communities and local landholders in natural resource governance. As noted by Rahman, Hickey and Sarker, the 'establishment of accountability, transparency, participation and equity in informal institutions' enhances the collective action of a community.[36] Ostrom summarised that collective action drives the community to formulate the rules in use, or *de facto* rules, to sustain resource use under a decentralised property rights regime.[37]

One of the major challenges in the field of natural resource governance arises during conflicts between stakeholders over the valuation and utilisation of natural resources. One such example appears in Australia, where the Queensland Government adopted the concept of natural resource management, rather than natural resource governance, defining the concept as:

> Managing Queensland's natural resources in a responsible way involves the long-term care and use of our soil, water and vegetation, and the ecosystems they form, in a way that also supports the economic and social needs of the community. In Queensland, this is achieved through a collaborative approach involving partners across all levels of government, industry and the broader community.[38]

Regionalism, as part of collective governance in agricultural communities, is essential to the development of the natural resources of the community, as stakeholders in these communities experience a strong trust and collaborative dependency on their regional leaders.[39] The practical implementation of State and regional based natural resource governance is often a source of policy and legal contest, such as in the mining regions of Queensland. Previously, many of these regions have been predominantly economically tied to agriculture. However, the recent development of significant unconventional gas resources has impacted on resource governance. According to Mayere and Donehue:

> Regional councils in Queensland have limited say in planning and decision making with regards to resource-related activities, despite having significant responsibilities in terms of providing essential infrastructure. This leaves them in a difficult situation, with little direct control over a new and rapidly growing economic force.[40]

A complex legislative relationship, governed by the *Mineral Resources Act 1989* (Qld) (MRA), the *Petroleum and Gas (Production and Safety) Act 2004* (Qld) (PGPSA) and the *Sustainable Planning Act 2016* (Qld), has arisen due to contestation over the regulation of activities, impacting both regional and local planning in Queensland. This relationship was further hampered by the collection of resource rent by the State Government from activities arising in regional areas. The policy response to this rent-seeking behaviour was the *Royalties for Regions Program*, established in 2012. The program aimed at reinvesting resource rent back into producing regions through a system whereby regions may apply for funding.[41] The program was concluded in 2016.

Community-based natural resource governance

Community-Based Natural Resource Management ('CBNRM') has also emerged in socio-legal literature, as the credibility and capacity of communities to manage their own natural resources has increased, leading toward the greater devolution of state power to the local level.[42] CBNRM involves the joint consideration of environmental as well as socioeconomic objectives and outcomes within a communal social network. CBNRM emphasises local knowledge and tradition, as expressed in local property rights and traditional values.[43] The success or failure of CBNRM is outlined by Kellert et al. as:

> The equitable distribution of benefits across a wide range of community members and the empowerment of community members (including the ability to effectively engage in conflict resolution and increased production and widespread distribution of knowledge and more sustainable use of resources).[44]

Kellert et al. argue that a lack of procedural justice may lead people to contest externally-controlled management decisions, turning to more CBNRM.[45] Their analysis outlined five shared characteristics of CBNRM programs, which are 'involving local institutions in resource management; empowering local authority; reconciling socio-economic development and environmental conservation; legitimising local land rights; and including traditional knowledge into resource management'.[46] Importantly, CBNRM can be utilised by rural-landholder collective bargaining vehicles, which is considered in Chapter 12 of this book.

For example, the Queensland Competition Authority recognises that 'the regulatory regime for CSG has developed incrementally, in response to the challenges of a newly emerging industry – an industry for which many of the impacts were largely unknown and untested'.[47] Therefore, the opportunity to create a system which reflects CBNRM more adequately, and the pivotal role of agricultural landholders in the CSG extraction process, is now a pressing and vital issue in Queensland.

Managing natural resource governance conflicts

Where conflict over natural resource use and management occurs, it is in the interest of all of the stakeholders to reach and maintain a position where all of the parties can live with the activities being undertaken. Coexistence seeks to recognise and equitably manage the interests of different stakeholders, including landholders, regulatory bodies and private companies. This ensures effective regulation of competing land uses, rather than one sector being privileged to the disadvantage of another.

According to Everingham et al., coexistence in the context of CSG and community sustainability, conservation of biodiversity and integrated approaches to land use planning in Australia includes, 'effective management by resource companies, and by regulators, of operating practices, off-site impacts, and the distribution of benefits'.[48] Coexistence and the distribution of benefits from the industry as it affects agricultural landowners and the unconventional gas industries in Australia is therefore the decisive intention of policymakers.

The Australian national *Multiple Land Use Framework* (MLUF), released in 2013, defines coexistence as a:

> Principle that acknowledges and respects the rights of all land users and the potential of all regulated land uses, while ensuring that regulated land is not restricted to a sole use without considering the implications or consequences for other potential land uses and the broader benefits to all Australians.[49]

This definition of coexistence, at the national level, is applicable as a benchmark in managing competing interests during unconventional gas activities. It is in the interests of the State, petroleum titleholders and landholders to coexist given that the average project lifecycle of unconventional gas activities can span up to three decades.

Conclusion

This chapter has examined the importance of natural resource governance in relation to conflict between agricultural activities and shale gas extraction. This necessarily includes an examination of the regulatory paradigms of land ownership, three-dimensional property rights and the governance of conflicting resource uses between multiple stakeholders. Therefore, an understanding of natural resource governance is essential to the understanding of either prioritising or reconciling competing resource needs in all jurisdictions surveyed within this book. The concepts surrounding natural resource governance aid in discerning the regulatory approaches of States in regulating for or prohibiting unconventional gas activities on agricultural land.

Notes

1 For instance, refer to the Natural Resource Governance Institute, *Extractive Industries Transparency International* (2018) <https://eiti.org/>.
2 Garrett Hardin, 'The Tragedy of the Commons' (1968) 162(3859) *Science* 1243.
3 *Case of Mines* (1568) 1 Plowd 310. This case also conferred the right to annex non-precious metals in the surrounding areas.
4 The Doctrine of Tenure/Doctrine of Estates refers to the way in which rights over land are held and distributed and trace its origins to the Norman Conquest in 1066. Since that time, land has been distributed to the Lords by the Monarch under a system where the lords did not receive ownership of their land, but rather a lesser bundle of rights, being 'tenants' of the land, holding an 'estate' in land 'of' the Crown, with the 'estate in land' being the right to possess the land for a period of time, subject to certain conditions. See, also, Brendan Edgeworth, *Butt's Land Law* (Thompsons Reuters, 7th edn, 2017).

5 The exception to this is the United States, which has allodial title.
6 For example, s 2 of the *Petroleum Act 1934* (UK) vests ownership of petroleum in the Crown. Similarly, ownership of petroleum is vested in the State in the following examples: s 6 of the *Petroleum (Onshore) Act 1991* (NSW); s 26 of the *Petroleum and Gas (Production and Safety) Act 2004* (Qld); Pt 9 of the *Petroleum and Natural Gas Act*, RSBC 1996, c 361; and s 50 of the *Land Act* RSBC 1996 c 245.
7 Fee simple is recognized by Kirby J as the 'local equivalent of full ownership' in *Wik Peoples v Queensland* (1996) 187 CLR 1, 250.
8 See *AG v Bronw* (1947) Legge 312, where this premise is confirmed.
9 Kevin Gray and Susan Francis Gray, 'The Idea of Property in Land' in Susan Bright and John K. Dewar (eds), *Land Law: Themes and Perspectives* (Oxford University Press, 1998) 15, 4.
10 Michael Weir and Tina Hunter, *Property Rights, and Coal Seam Gas Extraction: The Modern Property Right Conundrum* (2014) 2 *Property Law Review* 71, 74.
11 See *Benge v Scharbauer* (1953) 259 S W 2d 166 (Tx); *Acker v Guinn* (1971) 464 S W 2d 348 (Tx); and *Amarillo Oil Co v Energy-Agri Products Inc* (1990) 794 S W 2d 20 (Tx).
12 *International News Service v Associated Press* (1918) 248 U.S. 215, 248.
13 The term landholder in this context refers to any holder of the surface interest of the land. It specifically includes either the landowner (fee simple or allodial) or the holder of a surface leasehold interest, such as a pastoral lease or an agricultural lease.
14 Because of the nature of the geology of the shale and the need for hydraulic fracturing, unconventional shale gas extraction requires the drilling of a great number of wells. This in turn means that there is a need for a greater number of well pads, although the drilling of multiple wells from a single pad can significantly reduce this. See George E. King, *Hydraulic Fracturing 101: What Every Representative, Environmentalist, Regulator, Reporter, Investor, University Researcher, Neighbor and Engineer Should Know about Estimating Frac Risk and Improving Frac Performance in Unconventional Gas and Oil Wells* (SPE International, 2012).
15 For example, in France in Chapter 9.
16 M.A. Mohamed Salih, 'Governance of Food Security in the 21ˢᵗ Century' in Hans Günter Brauch et al. (eds), *Facing Global Environmental Change: Environmental, Human, Energy, Food, Health and Water Security Concepts* (Springer, 2009) 34, 34.
17 Hemant R. Ojha, Andy Hall and Rasheed Sulaiman, *Adaptive Collaborative Approaches in Natural Resource Governance: Rethinking Participation, Learning and Innovation* (Routledge, 2012).
18 Neth Baromey, *Ecotourism as a Tool for Sustainable Rural Community Development and Natural Resources Management in the Tonle Sap Biosphere Reserve* (Kassel University Press, 2008) 19.
19 Seema Arora-Jonsson, *Gender, Development and Environmental Governance: Theorizing Connections* (Routledge, 2013) 6.
20 Sara Singleton, *Constructing Cooperation: The Evolution of Institutions of Co-management* (University of Michigan Press, 1998) 7.
21 Paul Martin, Jacqueline Williams and Amanda Kennedy, 'Creating the Next Generation Rural Landscape Governance: The Challenge for Environmental Law Scholarship' in Paul Martin, Li Zhiping, Qin Tianbao, Anel Du Plessis and Yves Le Bouthillier (eds), *Environmental Governance and Sustainability* (IUCN Academy of Environmental Law, 2012) 50, 57.
22 An Environmental Impact Assessment (EIA) is defined as 'An assessment of the possible impact of a proposed action undertaken to enable environment and heritage protection and biodiversity conservation'. See Australian Government Department of Environment and Energy, *Annual Report 2016–2017* (2017) 350.
23 Nonna Moartinov-Bennie and Angela Hecimovic, 'Assurance of Australian Natural Resource Management' (2010) 12(4) *Public Management Review* 549.

24 Michael Lockwood, Julie Davidson, Allan Curtis, Elaine Stratford and Rod Griffith, 'Governance Principles for Natural Resource Management' (2010) 23(10) *Society and Natural Resources* 986.

25 Australian Government, *National Landcare Program* (Australian Government Department of Environment and Energy and Department of Agriculture and Water Resources, 2016) <http://www.nrm.gov.au/>.

26 Andrew Hodges and Tim Goesch, 'Natural Resource Management: Results for 2004–05 from an ABARE Survey of Australian Farmers' (2006) 13(3) *Australian Commodities* 569.

27 Stewart Lockie and Vaughan Higgins, 'Roll-out Neoliberalism and Hybrid Practices of Regulation in Australian Agri-environmental Governance' (2007) 23 *Journal of Rural Studies* 1.

28 John Dryzek and David Schlosberg, *Debating the Earth: The Environmental Politics Reader* (Oxford University Press, 2005).

29 Garrett Hardin, 'The Tragedy of the Commons' (1968) 162(3859) *Science* 1243, 1245.

30 Peter Butt and Peter Nygh, *Encyclopaedic Australian Legal Dictionary* (LexisNexis, 2016) 200.

31 Jere Lee Gilles and Keith Jamtgaard, 'The Commons Reconsidered' (1982) 4(2) *Rangelands* 51.

32 Frank van Laerhove, 'Traditions and Trends in the Study of the Commons' (2007) *International Journal of the Commons* 13.

33 Alex T. Davidson, Michael Dence and The Royal Society of Canada, Institute for Research on Public Policy, *The Brundtland Challenge and the Cost of Inaction* (IRPP, 1988).

34 Giles Atkinson, Simon Dietz and Eric Neumayer, *Hand Book of Sustainable Development* (Edward Elgar Publishing, 2007) 2.

35 Elinor Ostrom, 'Self-organization and Social Capital' (1995) 4(1) *Industrial and Corporate Change* 131.

36 H.M. Tuihedur Rahman, Gordon M. Hickey and Swapan Kumar Sarker, 'A Framework for Evaluating Collective Action and Informal Institutional Dynamics Under a Resource Management Policy of Decentralization' (2012) 83 *Ecological Economics* 32, 32.

37 Elinor Ostrom, Roy Gardner, and James Walker, *Rules, Games, and Common-Pool Resources* (The University of Michigan Press, 1994).

38 Queensland Government, *Annual Outlook 2015 Queensland Regional Natural Resource Management Investment Program 2013–2018* (Department of Natural Resources and Mines, 2016) <https://www.dnrm.qld.gov.au/__data/assets/pdf_file/0006/281355/nrm-annual-outlook.pdf>.

39 J.A. Bellamy, *Federalism and Regionalism in Australia: New Approaches, New Institutions?* (ANU ePress, 2007).

40 Severine Mayere and Paul A. Donehue, 'Perceptions of Land-use Uncertainty in Queensland's Resource-based Regions' (2014) 51(3) *Australian Planner* 212, 212.

41 Queensland Government Department of State Development, *Royalties for the Regions* (2015) <http://www.drd.wa.gov.au/rfr/whatisrfr/Pages/default.aspx>.

42 Mark Baker and Jonathan Kusel, *Community Forestry in the United States: Past Practice, Crafting the Future* (Island Press, 2003).

43 Carol Griffin, 'Watershed Councils: An Emerging Form of Public Participation in Natural Resource Management' (1999) 35(3) *Journal of the American Water Resource Association* 505.

44 Stephen R. Kellert, Jai N. Mehta, Syma A. Ebbin and Laly L. Lichtenfeld, 'Community Natural Resource Management: Promise, Rhetoric, and Reality' (2000) 13 (8) *Society and Natural Resources* 705, 706.

45 Ibid, 705.

46 Ibid, 709.

47 Queensland Competition Authority, *Coal Seam Gas Review* (2014) <http://www.qca.org.au/getattachment/aaaeab4b-519f-4a95-8a65-911bc46cc1d3/CSG-investigation.aspx>.

48 Jo-Anne Everingham, Nina Collins, Will Rifkin, Daniel Rodriguez, Thomas Baumgartl, Jim Cavaye and Sue Vink, 'How Farmers, Graziers, Miners and Other Gas-Industry Personnel See Their Potential for Coexistence in Rural Queensland' (2014) 6(2) *SPE Economics and Management* 122, 122; Tina Hunter, Submission No 9 to the Productivity Commission, *Regulatory Burden on the Upstream Petroleum (Oil and Gas) Sector*, August 2008.
49 The *Multiple Land Use Framework* was developed by the Energy Council of the Council of Australian Governments (COAG) in recognition of the conflict arising over land use and access associated with the development of unconventional gas resources on agricultural land. This was especially important because of the multiple and sequential land use issues, challenges and opportunities that prevail. See, Standing Council on Energy and Resources, *Multiple Land Use Framework* (Council of Australian Governments, 2013) <http://www.coagenergycouncil.gov.au/sites/prod.energycouncil/files/publications/documents/Multiple%20Land%20Use%20Framework%20-%20Dec%202013.pdf>.

3 The value of agricultural land

The right to food, food security and food sovereignty

Introduction

The jurisdictions examined in this book face multiple challenges – to build economic development based on population growth whilst maintaining protection of arable lands and ensure energy security.[1] In recognising this conflict, Williams, Milligan and Stubbs state

> it is a defensible proposition that the only development activities that should be acceptable in a region are those that allow the landscape to maintain its function indefinitely. It would be folly to secure one natural resource while putting at risk renewable long-term resource use.[2]

In addition, Williams, Milligan and Stubbs describe 'maintain(ing) the landscape function indefinitely is a thorny policy issue'.[3] This is not, least of all, due to some jurisdictions giving primacy to the attainment of energy security over other land use objectives. This is the case, even though achievement of land use objectives underpins economic growth. For example, in the headlong rush to industrialisation, China has fundamentally and permanently altered its landscape, with economic wealth seemingly gained at the expense of the natural environment: From 1997 to 2015 China has witnessed a decrease in farmland of 8.2 million hectares. According to the Food and Agricultural Organisation of the United Nations (FAO), China's per capita arable land availability is 0.092 hectares, equating to −40 per cent of the world average.[4] In contrast, in France, a country with a strong cultural attachment to the rural landscape and a profitable food industry, the strong farming lobby *Mutualité Sociale Agricole* has spearheaded a campaign that has led to the protection and maintenance of farmland.

The example of France illustrates another dimension to the shale gas and energy security debate; that is, food security and the use of land to secure and maintain future food 'accessibility and availability'. Internationally, this right to food is enshrined in several instruments, including the *Universal Declaration of Human Rights* (UDHR)[5] and the *International Covenant on Economic, Social and Cultural Rights* (ICESCR).[6] This customary international law, on its own, is a mere guideline point, requiring national States to enact domestic legislation in the form of a National Food Plan, or similar guide-line to create food security regulation. However, these too are in a sense non-binding, leaving the interpretation of food security to the particular cultural, economic and political goals of each State. Although the jurisdictions examined in this book are party to the ICESCR, the interpretation of the basic human right to food differs.[7] This demonstrates that food security may be a secondary consideration when a State attempts to attain the goal of energy security, of which the development of unconventional gas is a critical component.

This chapter will analyse the ICESCR right to adequate food and food security in the context of the four elements as outlined by the FAO: 'availability, access, utilisation and stability'.[8] This is followed by a discussion of the broader and more holistic concept of food sovereignty, which encapsulates the right of agricultural landholders to agrarian citizenship in a food production economy. This chapter concludes with an examination of the fundamental policy dilemma of food security versus energy security as States continue to increase development of unconventional gas, often at the expense of rural landscapes and the agricultural economy.

The human right to adequate food

The ICESCR requires signatory States to fulfil the right to adequate food, sustainably and without discrimination as a basic human right since the right to food is an 'inclusive right inherent to all people, and is a right to all nutritional elements that a person needs to live a healthy and active life including the means to access them'.[9] Article 11(6) contains the normative content of the human right that 'the right to adequate food is realised when every man, woman and child, alone or in a community with others, has physical and economic access at all times to adequate food or means for its procurement'.[10]

When the ICESCR entered into force in 1966,[11] the right to adequate food[12] became legally binding and codified, with 160 Member States recognising:

> The right of everyone to an adequate standard of living for himself and his family, including *adequate food*, clothing and housing, and to the continuous improvement of living conditions. The States Parties will take appropriate steps to ensure the realization of this right ....[13]

The right to food is fundamentally linked to the right to have an adequate standard of living.[14] Under Article 11(2) of ICESCR, the right to be free from hunger requires States to:

> [Recognise] the fundamental right of everyone to be free from hunger, [and] shall take, individually and through international co-operation, the measures, including specific programmes, which are needed:

(a) To improve methods of production, conservation and distribution of food by making full use of technical and scientific knowledge, by disseminating knowledge of the principles of nutrition and by developing or reforming agrarian systems in such a way as to achieve the most efficient development and utilization of natural resources; and

(b) Taking into account the problems of both food-importing and food-exporting countries, to ensure an equitable distribution of world food supplies in relation to need.[15]

Under Article 11(1) of ICESCR, the right to adequate food[16] is a 'relative standard', whereas Article 11(2) of ICESCR establishes the right to be free from hunger as an 'absolute standard'.[17] ICESCR provides for progressive realisation for signatories, recognised as meaning States should progress as 'expeditiously as possible towards to goal of full implementation of the right to food'.[18] These requirements are considered the 'core content' of the right to food, the digression from which is a breach of human rights.[19] Adequacy is strongly linked with the notion of sustainability in stating food must be 'accessible for both present and future generations'.[20]

Eide states that sustainability 'implies that the *physical … environment* in which food is procured must be … protected from erosion of distortion'.[21] In light of this, States must maintain a level of agricultural protection sufficient to ensure the long-term viability of food for their population.[22] Therefore, preserving and protecting the physical environment in which food is produced is key to sustainably adhering to principles of food security, as enshrined within the right to adequate food.

The right to adequate food was reaffirmed in the 1996 World Food Summit (WFS) and consequential *Rome Declaration on World Food Security*.[23] The WFS committed to 'clarify the content of the right to adequate food and the fundamental right of everyone to be free from hunger'.[24] This resulted in the adoption of The Committee on Economic, Social and Cultural Rights' (CESCR) General Comment No. 12. General Comment No. 12 is instrumental in providing an imperative, guiding interpretation for effective implementation of the right into States' domestic policy and lawmaking.[25] General Comment No. 12 provides the 'core content'[26] of the right to adequate food as:

The availability of food in a quantity and quality sufficient to satisfy the dietary needs of individuals, free from adverse substances, and acceptable within a given culture; and

The accessibility of such food in ways that are sustainable and that do not interfere with the enjoyment of other human rights.[27]

'Availability', as referred to in General Comment No. 12, provides for safeguarding food by 'feeding oneself directly from *productive land* or other natural resources, or having means for the procurement of food through well-functioning distribution, processing, and market systems'.[28] The scope of 'accessibility' encompasses 'economic accessibility in relation to personal and household costs associated with acquiring food for an adequate diet without compromising other basic needs'.[29] The 'accessibility' of information implies the 'right to seek, receive, and impart information concerning food and nutrition issues'.[30] This form of accessibility is important in relation to consumer knowledge of products, as well as for farmers' knowledge of the food system. Importantly, 'sustainable access' of food incorporates the notion of long-term availability and accessibility.[31] Therefore, sustainable food and its availability is tangentially linked to a State protecting its agricultural lands to ensure adequate access to food.[32] As examined in Chapter 4, the statist approach to food security serves this need. Notably, a statist approach is characterised by a high level of regulatory intervention by the State, including control over State-owned agricultural land.

Food security

The 1996 World Trade Organization (WTO) World Food Summit defined food security as 'existing when all people at all times have access to sufficient, safe, nutritious food to maintain a healthy and active life'.[33] The FAO refined the definition of food security to include broader socio-legal rights, stating that food security is 'a situation that exists when all people at all times have physical, social and economic access to sufficient, safe and nutritious food that meets their dietary needs for an active and healthy life'.[34] Over time, the definition of food security has been redefined to be tangibly linked to the right to food, which is realised 'when every man, woman and child, alone or in community with others, has physical and economic access at all times to adequate food or means for its procurement'.[35] Therefore, as summarised by Lawrence, the four dimensions of food security are evident according to the FAO:

> food availability (sufficient quantities of food for distribution); food access (ability to obtain foods that are available locally and in markets); utilisation (the capacity to prepare foods in a way that supports nutritional well-being and energy needs); and, stability (adequate access at all times).[36]

These dimensions make explicit reference to the rights of the individual, as well as the community. They extend the concept of food security beyond the economic aspects of the food supply chain to the protection of agricultural land as the source of all food stuffs. Initially, at the time of its initial adoption, the right to food was seen as a symbol and aspirational.[37] Now, the right to food is an operational tool widely recognised as a key to the success of sustainable food security strategies. A major threat to food security is the loss of agricultural land and 'land grabs'. This phenomenon represents the shift occurring in the jurisdictions examined in this book, from largely State-owned agriculture to private, agribusiness, export-orientated cropping. Such developments continue to create major concerns regarding food security.[38]

'Land grabbing' is defined as

> land acquisitions ... [that are] ... underhanded or unfair; when few bene-
> fits flow to the local economy; when foreign land purchase is accompanied
> by local dispossession; where there is minimal or no consideration for the
> environment; and where the decision to sell land to foreigners is made by
> political elites largely independently of the interests of local farmers and
> communities.[39]

It demonstrates the extreme scenario whereby a State fails to protect its agricultural lands by permitting large-scale land acquisitions, thereby failing to implement food security. This phenomenon is most evident in developing nations.[40] However, the leniency towards foreign acquisition of agricultural lands in Australia is also evident, typified by the sale of Cubbie Station in Queensland, one of the state's largest rural properties, spanning 93,000 hectares over the Murray-Darling Basin, to a Chinese government-owned corporation.[41] While the Australian government deemed that this did not contravene food security and Foreign Investment guidelines, there was community unease relating to national sovereignty over productive lands. The acquisition of Australian assets by foreign owners is assessed and managed by the Foreign Investment Review Board (FIRB).[42] FIRB reviews any proposed invest-ment of individuals with agricultural land holdings exceeding $15 million, or any proposed investment in agricultural land of a foreign government, with an exception for review of proposed investments applying to Australia's trade agreement partners.[43]

Food sovereignty

The concept of food sovereignty refers to the right of communities, peoples and States to independently determine their own food and agricultural policies. The human right to adequate food was encapsulated in the concept of food sovereignty, advocated for by the La Via Campesina (LVC) movement. The LVC stated at the Non-Governmental Organisation/Civil Society Organisation (NGO/CSO) forum of the 1996 World Food Summit that, '[f]ood is a basic human right. This right can only be realised in a system where food sovereignty is guaranteed'.[44] The food sovereignty definition provided in the Declaration of Nyéléni can be taken to be the most representative explanation of food sovereignty, as more than 500 representatives of NGOs and more than 80 countries agreed on its definition as:

> The right of peoples to healthy and culturally appropriate food produced
> through ecologically sound and sustainable methods, and their right to
> define their own food and agriculture systems.[45]

This relatively recent emergence of food sovereignty as a subject of national policy reflects a growing public concern to secure domestic food supply by preserving agricultural lands. It also reflects concerns about the right of peoples to define, protect and regulate domestic agricultural production and land policies that promote safe, healthy and ecologically sustainable as well as culturally appropriate food production.[46] Maintaining productive farmland and sustainable farms as well as providing adequate volumes of foodstuffs have led to measures aimed at protecting farmland and farm activities in many countries. For example, in Canada, *The People's Food Plan* argues that protection of agricultural land is needed to achieve food sovereignty to 'increase protection for agriculture lands … rural and remote communities must be recognised as a key priority that guides all land use policy decisions'.[47] The movement's assumption is that food production should support and respect biodiversity, the productive capacity of the land, preserve resources and cultural values.[48] This view calls for protection of local markets against foreign imports, regulation of production to prevent against surpluses and the abolition of export aids.

Proponents of food sovereignty view the failure of agricultural policy and regulation as a problem rooted in decades of neoliberal globalisation, forced upon developing countries undemocratically by national political leaders in collaboration with the International Monetary Fund (IMF), World Bank and WTO.[49] In their view, these organisations have enabled increased trade liberalisation, deregulation and, therefore, powerful food oligopolies to form controlling agricultural prices and practices. Food sovereignty as a concept questions the perspectives and institutions that favour a neoliberal approach to agriculture and development, creating the appearance that the market is uniformly liberalised.[50]

In their policy statement *An Answer to the Global Food Crisis: Peasants and Small Farmers can Feed the World*, LVC describes several changes national governments can make to improve the human right to food. One of LVC's arguments against neoliberalism is that food is too important to be left under the auspices of corporations and a deregulated market. This argument is in direct contrast to the 30 years of the neoliberal approach to agriculture for developed nations. Consequently, it advocates increasing agricultural protectionist measures and multilateral support for the right to food. LVC's resistance efforts illustrate backlash against neoliberal globalisation and corporate approaches to agriculture, which has evidently shifted cultural perceptions and influenced policy.[51]

Research on implementing food sovereignty in legal instruments remains scarce, focussing mainly on the concept and its role within civil society movements in developing nations' constitutional instruments.[52] However, recent literature has identified factors obstructing the institutionalisation of food sovereignty into legal policies. Windfuhr and Jonsén[53] identify critical factors obstructing the successful implementation of food sovereignty into national policy, including:

> the dominant neoliberal development paradigm; the market-based, production approach to food security that privileges modern technological advancements (biotechnology) over traditional knowledge; the coherence and timeliness of proposed instruments to support food sovereignty; the

privileging of international trade policies over policies to promote food security and rural livelihoods; and the narrowing of access to community resources through expropriation, patenting, and privatisation.[54]

Further, in order to enforce the legality of food sovereignty in international law, López argues a new branch of international law (International Law for the Rural World) as an actionable right and legal concept must be created and enforced:

> But if that which is desired is to achieve a ruling as detailed and broad as the program that defends Food Sovereignty, it is worthwhile to propose the creation of a branch within international law, which could be denoted as International Law for the Rural World.[55]

This aspirational concept could bring about the necessary legal enforceability of food sovereignty as a State's right, a people's right and a right of local communities.

While the original definition of food sovereignty has evolved since the movement's inception, core elements have remained. According to Schanbacher, 'we can now juxtapose current conceptions of food security and a purely economic understanding of globalization with a more culturally and politically attuned understanding of globalization, global hunger, and poverty'.[56] Attendant to these policies is the consolidation of small-scale, family and self-sufficient farms into large-scale mechanised farms that practice mono-cropping, employ capital intensive methods of production, produce only for export and harm biodiversity.[57] Challenging the premise that market-led reforms produce more efficient farms, food sovereignty recognises the total benefits of small-scale agriculture by focusing not only on economic gains, but also on how agriculture promotes biodiversity, connects farmers and families to the land. This provides an intimate link between farmers, their crops and the foods that they produce and consume.

The right to adequate food and the concept of food sovereignty as a basis for international development policy thus share some common ground and differences in scope and nature. As stated by Edelman:

> Advocates typically suggest that 'food sovereignty' is diametrically opposed to 'food security', but historically there actually has been considerable slippage and overlap between these concepts. Food sovereignty theory has usually failed to indicate whether the 'sovereign' is the nation, region or locality, or 'the people'. This lack of specificity about the sovereign feeds a reluctance to think concretely about the regulatory mechanisms necessary to consolidate and enforce food sovereignty, particularly limitations on long-distance and international trade and on firm and farm size.[58]

Food security and the concept of food sovereignty both stress the importance of sustainability, agrarian reform, State sovereignty and international responsibility. According to Beuchelt and Virchow, 'While the right to food does not specifically addresses [sic] the agricultural production model, agricultural trade and international markets, all these aspects need to be aligned within the spirit of the right to food'.[59] The final difference is in the legal validity of the concept. While the right to adequate food is an international human right, and thus binding, the concept of food sovereignty is only part of national law in few countries within the Global South.

However, as Schanbacher argues, 'While UN organizations such as the FAO and [the International Fund for Agricultural Development] recognise the shortcomings of World Bank, IMF, and WTO policies, the current definition of food security still falls short of outlining the demands voiced by food sovereignty'.[60] Policy proposals aimed at implementing the concept of food sovereignty at an international level include: a ban on food storage and hoarding by transnational corporations and the private sector; international treaties and competition laws aimed at limiting the concentration and market power of major agri-food corporations and international commodity agreements which regulate the total food output on world markets.[61]

The increasing approval of the concept of food sovereignty is beginning to find its way into the debates and policies of United Nations (UN) organisations and national governments in both developing and developed countries.[62] In addition to the LVC, at a regional, non-governmental level, there is the example of the European Platform for Food Sovereignty, founded in 2003. However, the International Labour Organization (ILO), the United Nations Environmental Programme (UNEP) and the World Bank have not adopted the concept of food sovereignty. The FAO appears to deal more frequently with the concept of food sovereignty, but no official FAO document to date contains the concept. The FAO has a cooperation agreement with the International NGO/CSO Planning Committee for Food Sovereignty (IPC), and it included in its *Glossary on Right to Food* a definition of food sovereignty.[63] However, the United Nations Human Rights Council (UNHRC) is currently the only UN body which has intensively discussed the concept of food sovereignty within its documents and meetings. Several regional agreements have incorporated the concept of food sovereignty into agricultural policies. The impetus for the establishment of international-level policies has diminished due to debates within the broader movement as to whether such strategies were effective, given a concern that such legal strategies would have a 'demobilising effect'.[64] Eight countries to date have integrated the concept of food sovereignty into their national legislation – usually combined with the right to adequate food.[65]

The weakness of the food sovereignty concept is that it is not yet a comprehensive concept which addresses hunger in both urban and rural populations. The exclusive focus on the issue of access to productive resources and national, or local, self-determination of agricultural and food policies fails to take account of the need for adequate and functioning systems of local and national institutions, legal frameworks, education, health care and hard infrastructure. The legal approach to the right to adequate food is internationally agreed upon and has been ratified by nearly every country in the world. Haugen concludes that the human right to food is more precise, has stronger support among nations and is closer to legally binding obligations than the food sovereignty concept.[66] Likewise, Beuchelt and Virchow opine that 'the legal approach to adequate food appears to be the more promising way to reduce global hunger as it applies to all human beings, including small food producers'.[67] Therefore, they recommend 'the continued reliance on the right to adequate food rather than introducing the concept of food sovereignty in national and international policy making'.[68]

Concerns over food security and food sovereignty have led to various policy initiatives to secure arable land. For example, land use and land access regulatory regimes may be used to protect high quality farmland from development. Land use planning has been considered a major tool in ensuring the future viability of agricultural land, through legislation to protect farmland via agricultural zoning. Thus, land use and land access regulatory regimes can enable food security and food sovereignty for the State. However, the application of these concepts is uneven across jurisdictions, not least because there is no uniform concept of food security. Less politicised is the Right to Food, which is supported and enacted by ICESCR and has specific legal measures to ensure signatory States uphold the right to food for their citizens.

Unconventional gas presents a new hurdle for States seeking to safeguard both food security and food sovereignty. In all jurisdictions surveyed within this book, unconventional gas is located beneath agricultural land and within rural communities. Therefore, States must balance the interests of the unconventional gas sector and other industries. In particular, the safeguarding of agricultural land, equitable land access for landholders and ensuring no long-term impacts to groundwater are essential to the functioning of food security. Thus, unconventional gas creates a new challenge for States to ensure it does not affect the viability of existing agricultural land and impact the securitisation of food. As recognised by Larson, 'the pursuit of energy security can often frustrate efforts to achieve food security'.[69] As nations seek to realise the security of both energy and food, the physical and economic access to sufficient food is often threatened and in conflict with the need to develop unconventional gas reserves on agricultural land.

The concept of food sovereignty emerged from a broader social debate about the growth of the industrialised food economy and its effect on localised food production. The Green Revolution in the 1980s demonstrated the rise of global food agribusinesses which created mono-cropping and diminished the rights and economic stability of small scale farmers. This was particularly evident in developing countries, displacing small agricultural landowners in favour of mega agricultural production. In Venezuela, the response from agricultural landowners was to collectivise and form a coalition to protect the rights and autonomy of small scale cropping agricultural producers. In so doing, it demonstrated the principles of food sovereignty, ecological sustainability and accountability to protect and secure the rights of agricultural producers in the face of global corporate entities. Similarly, the rise of unconventional gas again creates a new pressure on agricultural landholders. The original, 1996 definition of food sovereignty did not anticipate the rise of the unconventional gas industry, but there are similarities with the protection of arable lands on that the basis of food production is being threatened. The lack of specific protection measures and accountability of either agribusiness or unconventional gas companies can create diminished rights of agricultural landholders.

In Australia, for example, a coalition of landowners in Queensland have formed an opposition group to unconventional gas activities on agricultural land, based on the uncertainty of the environmental effects on the water table, the productivity of agricultural land and the rights of landowners in negotiating land access agreement with gas companies. Conflict between agriculture and CSG production has become apparent in some of Australia's most productive farming areas, including the Darling Downs in Queensland,[70] as well as threatening the exceptionally fertile Liverpool Plains in New South Wales. The Darling Downs region, situated over the large reserves of CSG in the Surat Basin, is 'one of Queensland's most important agricultural assets'.[71] It comprises 11 per cent of Queensland's area and produces around a quarter of Queensland's agricultural output.[72] The Darling Downs region as an agricultural area in Queensland is unparalleled. It produces 88.1 per cent of Queensland's eggs, 65 per cent of total cotton value, 95.7 per cent of stone fruit value, 66.6 per cent of grain sorghum production, 48.2 per cent of wheat production and 64.5 per cent of pig production.[73]

Thus, the relationship between CSG and agricultural activity in Queensland and the possibility of further development in other states and territories has raised concerns about the impact of CSG, particularly, and unconventional gas development on food security and food sovereignty generally.[74] The main concerns include the impact on food production, the use and contamination of underground and surface water resources, agricultural landholders' rights over land, the effects on the socioeconomic environment in the affected regions and the future viability of farming in these areas.

Hydraulic fracturing has become synonymous with a threat to valuable farming land, leading to permanent bans on the technology in some jurisdictions such as France as examined in Chapter 9. The threat of hydraulic fracturing to agricultural land, groundwater and future food security is due to water use and possible subsequent depletion of aquifers, and water contamination (both surface water and ground water) arising from both shale gas and CSG activities. A further threat arises from the dewatering of coal seams, necessary for the production of CSG, and the briny water produced during the dewatering process. Consequently, public sentiment and opposition to unconventional gas is highly divisive in regional areas, and the industry has been vilified for its practices in developing unconventional gas on farming lands.

Conclusion

Food security and food sovereignty serve to demonstrate another dimension to the energy–food nexus, requiring States to balance different land uses; in this instance agricultural land and the rise of unconventional gas as a valuable and profitable energy resource. The political debate has tended towards a reactive approach to recognising and acknowledging the effects of unconventional gas extraction on sustainable and productive agricultural land. Similarly, the concept of food sovereignty demonstrates agricultural landowners' concern that their rights and land tenure may be eroded in favour of inequitable land access agreements that favour those companies undertaking unconventional gas extraction.

The USA, Canada, Australia, France and the UK have all explored the concept of food security or food sovereignty.[75] As signatories of ICESCR, all jurisdictions examined within this book hold the duty to uphold the right to food for its citizens. However, as demonstrated in this chapter this application is fragmentary and unevenly applied, hence food security and food sovereignty remain as outliers in the energy security debate, yet to be promulgated or fully explored by States in the development of unconventional gas. Seen through the lens of food security, unconventional gas in each of these jurisdictions has the potential to pose a threat to the viability of a secure food supply, particularly if land is degraded or unavailable for food production. In addition, food sovereignty has also become a key concern of agricultural landowners, fearful for their continued viability of agricultural land if unconventional gas activities occur unfettered. A broader policy debate, capturing a discussion on sustainable agricultural and land use landowner interests is currently a missing element in the energy security versus food security challenge.

Notes

1 Ian Gray and Geoffrey Lawrence, *A Future for Regional Australia: Escaping Global Misfortune* (Cambridge University Press, 2001).
2 John Williams, Ann Milligan and Tim Stubbs, 'Whole of Landscape Assessment and Planning in the Management of Unconventional Gas Exploration and Production in Australia' in R. Quentin Grafton, Ian G. Cronshaw and Michal C. Moore (eds), *Risks, Rewards and Regulation of Unconventional Gas: A Global Perspective* (Cambridge University Press, 2016) 427, 427.
3 Ibid.
4 Food and Agricultural Organisation, *Country Programming Framework 2012–2015 for People's Republic of China* (UN FAO, 2015) <http://www.fao.org/3/a-ax529e.pdf>.
5 *Universal Declaration of Human Rights*, GA Res 217A, 3rd sess, 183rd plen mtg, UN Doc A/810 at 71 (1948) (*Universal Declaration of Human Rights*) art 25(1).
6 *International Covenant on Economic, Social and Cultural Rights*, opened for signature 16 December 1966, 993 UNTS 3 (entered into force 3 November 1976).
7 Ibid.
8 Food and Agricultural Organisation, *The State of Food Security and Nutrition in the World 2017 Building Resilience for Peace and Food Security* (UN FAO, 2017).
9 The UN states 'it is not simply a right to a minimum ration of calories, proteins and other specific nutrients.' See United Nations, *The Right to Adequate Food Fact Sheet: 34* (OHCHR, 2000) <http://www.ohchr.org/Documents/Publications/FactSheet34en.pdf>.
10 *International Covenant on Economic, Social and Cultural Rights*, adopted and opened for signature, ratification and accession by General Assembly resolution 2200A (XXI) of 16 December 1966 entry into force 3 January 1976, in accordance with article 27. art 11(6).
11 The Right to Food was first recognised in article 25 of the *Universal Declaration of Human Rights* and later reaffirmed in article 11 of the *International Covenant on Economic, Social and Cultural Rights* (Article 11), and through the right to life, in the *International Covenant on Economic, Social and Cultural Rights*, adopted and opened for signature, ratification and accession by General Assembly resolution 2200A (XXI) of 16 December 1966 entry into force 3 January 1976, in accordance with article 27.

12 Flavio Luiz Schieck Valente and Ana Maria Suarez Franco, 'Human Rights and the Struggle Against Hunger: Laws, Institutions and Instruments in the Fight to Realise the Right to Adequate Food' (2010) 2 *Yale Human Rights and Development Law Journal* 422.

13 *International Covenant on Economic, Social and Cultural Rights*, opened for signature 16 December 1966, 993 UNTS 3 (entered into force 3 November 1976) art 11(1).

14 'For a person to have an adequate standard of living they must have as a minimum; easy access to nutritious, quality and affordable food, easy access to sufficient, safe and affordable water for personal and domestic use, adequate housing for themselves and their family and a healthy environment'. *International Covenant on Economic, Social and Cultural Rights*, opened for signature 16 December 1966, 993 UNTS 3 (entered into force 3 November 1976) art 11.

15 Ibid, art 11(2).

16 'Adequate food' is not defined within General Comment No. 12; however, it is understood the Committee interprets the term as being free from adverse substances, culturally acceptable and be physically and economically available to all parties. See, also, United Nations, *General Comment 12: The Right to Adequate Food: U.N. Doc. E/C.12/1999/5* (CESCR, 1999).

17 Smita Narula, 'Reclaiming the Right to Food as a Normative Response to the Global Food Crisis' (2011) 13 *Yale Human Rights and Development Law Journal* 405.

18 United Nations, *International Covenant on Economic, Social and Cultural Rights General Comment No 12* (CESCR, 2009) <http://www.unhcr.org/refworld/docid/4a60961f2.html>.

19 Dietary needs are extrinsically linked to the right to work, education, health, water and life and are consequently an important foundation for all human rights.

20 United Nations, *International Covenant on Economic, Social and Cultural Rights General Comment No 12* (CESCR, 2009) <http://www.unhcr.org/refworld/docid/4a60961f2.html>.

21 Special Rapporteur on the Right to Food, *The New International Economic Order and the Promotion of Human Rights* (ECOSOC, 1987), U.N. Doc. E/CN.4/Sub.2/1987/23.

22 Further, according to General Comment No. 12, States 'shall take measures to improve methods of production, conservation and distribution of food ... by reforming agrarian systems'.

23 *Rome Declaration on World Food Security*, opened for signature 13 November 1996, E/CN.4/RES/1998/23.

24 Ibid, 61, objective 7.4.

25 Comment No. 12 was drafted to accelerate and clarify the promotion of food security and 'became an authoritative interpretation of a piece of binding law when it was adopted by the UN Economic and Social Council in 1999', see Otto Hospes, *Overcoming Barriers to the Implementation of the Right to Food* (HALSHA, 2010) <https://hal.archives-ouvertes.fr/hal-00650148/document>.

26 It is noted that General Comment No. 12 presents recommendations towards the realisation of the right to food as non-binding soft law, orientating governments on the implementation of the binding provision of Article 11 ICESCR.

27 United Nations, *General Comment 12: The Right to Adequate Food: U.N. Doc. E/C.12/1999/5* (CESCR, 1999).

28 Ibid, 12.

29 'Accessibility' also refers to the realisation of physically accessible food that is accessible to everyone including physically vulnerable individuals, free from discrimination. See, also, Kerstin Mechlem, 'Food Security and the Right to Food in the Discourse of the United Nations' (2004) 10 *European Law Journal* 4.

30 United Nations, *General Comment 12: The Right to Adequate Food: U.N. Doc. E/C.12/1999/5* (CESCR, 1999).

31 Sigrun Skogly, 'Right to Adequate Food: National Implementation and Extra-territorial Obligations' (2007) 11 *Max Planck Yearbook of United Nations Law* 340.

32 *International Covenant on Economic, Social and Cultural Rights*, opened for signature 16 December 1966, 993 UNTS 3 (entered into force 3 November 1976) art 1(2); United Nations, *General Comment 15: The Right to Water: U.N. Doc. E/C.12/2002/ 11* (CESCR, 2002).

33 World Trade Organization, *Food Security* (WTO, 2010) <http://www.who.int/ trade/glossary/story028/en/>.

34 Food and Agricultural Organization, *Chapter 2: Food Security: Concepts and Measurement* (FAO, 2006) <http://www.fao.org/docrep/005/y4671e/y4671e06.htm>.

35 Olivier de Schutter, *A Rights Revolution: Implementing the Right to Food in Latin America and the Caribbean* (UN Special Rapporteur to the Right to Food, 2012) <http://www.srfood.org/index.php/en/right-to-food>.

36 Geoffrey Lawrence, 'Re-evaluating Food Systems and Food Security: A Global Perspective' (2017) 53(4) *Journal of Sociology* 774, 780.

37 Olivier de Schutter, *Interim Report of the Special Rapporteur on the Right to Food: A/ 68/288* (UN Special Rapporteur on the Right to Food, 2013) 3.

38 Mohamed Behnassi, Shabbir A. Shahid and Joyce D'Silva, *Sustainable Agricultural Development: Recent Approaches in Resources Management and Environmentally-balanced Production Enhancement* (Springer, 2011).

39 Geoffrey Lawrence, 'Re-evaluating Food Systems and Food Security: A Global Perspective' (2017) 53(4) *Journal of Sociology* 774, 784.

40 John Anthony Allan, Martin Keulertz, Suvi Sojamo and Jeroen Warner (eds), *Handbook of Land and Water Grabs in Africa: Foreign Direct Investment and Food and Water Security* (Routledge, 2013).

41 Stephen Long, *Deal Finalised for Sale of Cubbie Station* (Australian Broadcasting Corporation, 2012) <http://www.abc.net.au/news/2012-10-12/sale-of-cubbie-station-confirmed/4310256>.

42 As enacted by the *Foreign Acquisitions and Takeovers Act 1975* (Cth); *Foreign Acquisitions and Takeovers Fees Imposition Act 2015* (Cth); and the *Register of Foreign Ownership of Water or Agricultural Land Act 2015* (Cth).

43 Jurisdictions falling within this exception include Singapore, Japan, USA, Republic of Korea and Malaysia. Thresholds for privately owned investors from free trade agreement partner countries that have the higher threshold for agricultural land acquisitions include: for Chile, New Zealand and United States, $1.134 billion, see *Foreign Acquisitions and Takeovers Act 2015* (Cth) for further detail.

44 Tina Beuchelt and Detlef Virchow, 'Food Sovereignty or the Human Right to Adequate Food: Which Concept Serves Better as International Development Policy for Global Hunger and Poverty Reduction?' (2012) 29(2) *Agriculture and Human Values* 259, 270.

45 See Declaration of Nyéléni, *Food Sovereignty Framework: Concept and Historical Context* (Nyéléni, 2008) <https://nyeleni.org/IMG/pdf/FoodSovereigntyFramework. pdf>.

46 Tina Beuchelt and Detlef Virchow, 'Food Sovereignty or the Human Right to Adequate Food: Which Concept Serves Better as International Development Policy for Global Hunger and Poverty Reduction?' (2012) 29(2) *Agriculture and Human Values* 259, 270.

47 Food Secure Canada, *Resetting the Table: A People's Food Policy for Canada* (People's Food Policy Project, 2015) 12 <http://foodsecurecanada.org/sites/default/files/fsc-resetting-2015_web.pdf>.

48 Ivette Perfecto, John H. Vandermeer and Angus Lindsay Wright, *Nature's Matrix: Linking Agriculture, Conservation and Food Sovereignty* (Earthscan, 2009).

49 Annette Aurelie Desmarais, Hannah Wittman and Nettie Wiebe, *Food Sovereignty: Reconnecting Food, Nature & Community* (Fernwood, 2010).

50 Amy Trauger, *Food Sovereignty in International Context: Discourse, Politics and Practice of Place* (Routledge, 2015).

51 Marie-Josée Massicotte, 'La Vía Campesina, Brazilian Peasants, and the Agribusiness Model of Agriculture: Towards an Alternative Model of Agrarian Democratic Governance' (2010) 85 *Studies in Political Economy* 69.

52 Hans Morten Haugen, 'Food Sovereignty – An Appropriate Approach to Ensure the Right to Food?' (2009) 78(3) *Nordic Journal of International Law* 263.

53 Michael Windfuhr and Jennie Jonsén, *Food Sovereignty: Towards Democracy in Localized Food Systems* (ITDG Publications, 2005).

54 Ibid, 22.

55 Miguel Ángel Martin López, 'A Study of the Application of Food Sovereignty in International Law' (2016) 4(2) *Groningen Journal of International Law* 14, 34.

56 William Schanbacher, *The Politics of Food: The Global Conflict Between Food Security and Food Sovereignty* (ABC-CLIO, 2010) 54.

57 Eric Holt-Giménez, Raj Patel and Annie Shattuck, *Food Rebellions!: Crisis and the Hunger for Justice* (Pambazuka, 2009).

58 Marc Edelman, 'Food Sovereignty: Forgotten Genealogies and Future Regulatory Challenges' (2014) 41 *The Journal of Peasant studies* 959, 959.

59 Tina Beuchelt and Detlef Virchow, 'Food Sovereignty or the Human Right to Adequate Food: Which Concept Serves Better as International Development Policy for Global Hunger and Poverty Reduction?' (2012) 29(2) *Agriculture and Human Values* 259, 272.

60 William Schanbacher, *The Politics of Food: The Global Conflict Between Food Security and Food Sovereignty* (ABC-CLIO, 2010) 74.

61 Michel Pimbert, *Towards Food Sovereignty: Another World is Possible for Food and Agriculture* (IIED, 2008).

62 Annette Aurelie Desmarais, Hannah Wittman and Nettie Wiebe, *Food Sovereignty: Reconnecting Food, Nature & Community* (Fernwood, 2010).

63 Food and Agricultural Organisation of the United Nations, *Glossary on Right to Food* (UN FAO, 2018) <http://www.fao.org/right-to-food/resources/glossary/en/>.

64 Wendy Godek, 'The Complexity of Food Sovereignty Policymaking: The Case of Nicaragua's Law 693' (Paper presented at Food Sovereignty: A Critical Dialogue International Conference, Yale University, USA, September 14–15, 2013) 1.

65 These countries are Venezuela, Mali, Senegal, Nepal, Bolivia, Ecuador, Egypt and Nicaragua. The majority of initiatives to legally implement the concept of food sovereignty have occurred over the past 6 years. As a result, not all laws are ratified yet or implemented in the form of specific strategies.

66 Hans Morten Haugen, 'Food Sovereignty – An Appropriate Approach to Ensure the Right to Food?' (2009) 78(3) *Nordic Journal of International Law* 263.

67 Tina Beuchelt and Detlef Virchow, 'Food Sovereignty or the Human Right to Adequate Food: Which Concept Serves Better as International Development Policy for Global Hunger and Poverty reduction?' (2012) 29(2) *Agriculture and Human Values* 259, 272.

68 Ibid, 273.

69 Rhett B. Larson, 'Reconciling Energy and Food Security' (2014) 48 *University of Richmond Law Review* 929.

70 Office of the Chief Economist, *Review of the Socioeconomic Impacts of Coal Seam Gas in Queensland* (Commonwealth of Australia, 2015) <https://industry.gov.au/Office-of-the-Chief-Economist/Publications/Documents/coal-seam-gas/Socioeconomic-impacts-of-coal-seam-gas-in-Queensland.pdf>.

71 Queensland Government, *Darling Downs Regional Plan* (Queensland Department of State Development, Infrastructure and Planning, 2013) <http://www.statedevelopment.qld.gov.au/resources/plan/darling-downs/darling-downs-regional-plan.pdf>.

72 Ibid, 3.

73 Arrow Energy, *Analysis of Agricultural Production and Issues in the Darling Downs: Surat Gas Project Supplementary Report to the Environmental Impact Statement Report prepared for Arrow*

Energy Pty Ltd and Coffey Environments Australia Pty Ltd (AEC Group, 2013) <https://www.arrowenergy.com.au/__data/assets/pdf_file/0003/8670/Appendix_14.pdf>.

74 Amanda Kennedy, *Environmental Justice and Land Use Conflict: The Governance of Mineral and Gas Resource development* (Routledge, 2017).

75 In the USA, the National Family Farm Coalition is a member of La Via Campensia. In Canada, the People's Food Policy Project outlines seven pillars of Canadian Food Sovereignty: (1) food for people; (2) values food provides; (3) localises food systems; (4) puts control locally; (5) builds knowledge and skills; (6) works with nature; and (7) recognises that food is sacred. The Australian Food Sovereignty Alliance (AFSA) seeks to create a nourishing and ecologically sound food system in Australia. The French Peasant Seed Network has adopted the concept of food sovereignty in conserving local seeds and supporting local agriculture. The UK People's Food Policy holds the vision of a food system where 'Land is recognised and valued as an essential resource for food and shelter and the basis for numerous social, cultural and spiritual practices. Land is no longer treated and traded as a commodity; instead, it is understood as a common good of the people'. Dee Butterly and Dr. Ian Fitzpatrick, *A People's Food Plan* (2017) 13.

4 Theories of adaptive management, precautionary principle and the statist approach

Introduction

The State requires a regulatory framework that balances the interests of unconventional gas and agricultural land if the policy objectives of that State are to develop both sectors. For example, State policy can be at odds with those of private gas companies, as market forces and profit incentives may not align with the broader public policy objectives of the State.[1] Consequently, the State must play a role in balancing the competing interests of stakeholders across the policy spectrum.

As defined by international law,[2] the State has a custodian role to develop its natural resources in the public interest as part of the basic constituent of the right to self-determination. At the core of a State's interests is energy security.[3] Energy security is inherently linked to national security, reliability and security of supply despite changing market conditions. The partnership between State and energy companies is more complex, as private gas companies are predisposed to be driven by profit maximisation for shareholders via exploration and development. The commercial ambitions of private companies are framed within the State regulatory approach, which provides the policy and regulation framework within which unconventional gas companies operate.

In assigning proprietary rights to the private sector for resource exploration, development and production in return for royalties as capital to be expended for the public good, the State owes a 'duty of development' in the public interest.[4] The core objective of a public resource system is for the State to encourage stability of supply and, in so doing, minimise conflicting interests and satisfy public interest duties. Under these circumstances, the challenge for the State as regulator is allocating and managing resources in the public interest for financial and economic returns whilst protecting natural resources on an ongoing basis.

The purpose of this chapter is to examine differing socio-legal regulatory theories fundamental to the administration of unconventional gas activities by the State. This provides a contextual framework for examining issues arising from unconventional gas activities on agricultural land. This chapter will first explore adaptive management; secondly it will explore the precautionary principle; and finally examine statist approaches to regulation. These are differing approaches to regulation that a State may employ in regulating its respective unconventional gas sector while managing a multitude of legal interests.

Policy

Klein and Marmor define policy as 'what governments do and neglect to do'.[5] Policy is the term used to describe specific choices by a State, but the notion also embraces general directions and philosophies. Given the wide array of definitions for policy, policy is described in three different but compatible ways by Althaus, Bridgman and Davis:

> First, policy can be an authoritative choice of a government. Second, policy is a hypothesis, an expression of theories about cause and effect. Finally, policy is explored as the objective of governmental action.[6]

A State's petroleum policy represents the current position and focus of the relevant government in developing its petroleum resource, including political, fiscal and economic policies. Consequently, petroleum policy is developed through a complex network of factors including the geolocation of a country, petroleum potential, political development, infrastructure and its regulation. The basis of a successful exhaustible natural resources policy is premised upon four cornerstones according to de Sa:

1 a competitive, stable and fair fiscal regime;
2 a transparent legal and regulatory framework;
3 strong institutions to implement them; and
4 sound environmental management systems.[7]

Dworkin's theory of policy taxonomy also comprises the legal theory of principle, policy and rules related to the development of natural resource policy comprising:

- principles, the norms protecting individual rights;
- policies, the norms promoting collective goals; and
- rules, implementing the principles and policies.[8]

In particular, Dworkin sees policy as a kind of standard that sets out a goal to be reached, generally an improvement in some economic, political or social feature of the community in that they stipulate that some present feature is to be protected from adverse change.[9] Therefore, both policy and principles are set apart from rules, which are the legislative instruments that outline what the law is.[10] As each State has the sovereign right to exploit its natural resources, its policy and tools to develop its corresponding regulation will serve to create economic diversification of its natural resource sector.[11]

Regulation based on petroleum policy that governs petroleum development utilises Dworkin's taxonomy, as summarised by Hunter and Chandler:

> Principles are the overarching values that exist, allowing a sovereign state to exploit its resources. Policies are generally the current position or focus of a government in developing a natural resource, and will encompass political and fiscal policies ... rules are the actual legislation established by the parliament.[12]

An effective petroleum policy, which by extension gives rise to effective regulation as explored throughout this chapter, must balance the interdependence between the State as owner of petroleum resources and private natural resource companies as petroleum explorers and exploiters.[13] Therefore, a petroleum policy must secure possession of, and access to, its petroleum resources effectively in allowing the development of its resources by private actors while ensuring long-term sustainability of land use and land access in the exploited area.

The theory of regulation

Regulation is intended to influence the behaviour of individuals, organisations and governmental actors to promote specific community objectives, including those relating to social and environmental concerns.[14] Historically, some economists have asserted that regulation is efficient when it maximises wealth and economic outcome.[15] However, this view fails to provide an ethical basis for action and cannot justify the distinction of rights in society. Thus, it cannot be used as a yardstick to measure regulatory decisions affecting society. Legitimate or 'good' regulation and regulatory regimes involve reference to two or more of the following five key inquiries:

1   Is the action or regime supported by legislative authority?
2   Is there an appropriate schema of accountability?
3   Are procedures fair, accessible, and open?
4   Is the regulator acting with sufficient expertise?
5   Is the action or regime efficient?[16]

The mechanisms utilised to regulate behaviour include rules, expectations and codes of conduct. Regulations are often administered by an authoritative administrative body that promotes and monitors compliance through the imposition of sanctions or other penal options. Regulation is devised and recognised in three distinct yet overlapping categories, including economic regulation relating to market decisions to promote efficiency, social regulation promoting the public interest where the economic implications are likely to be secondary, and administrative regulations overseen by public agencies requiring, distributing and generating information.[17]

Baldwin, Cave and Lodge make the additional distinction of regulation comprising of a specific set of commands where regulation involves the promulgation of a binding set of rules to be applied by a body devoted to this purpose and includes:

> A deliberate state influence – where regulation has a more broad sense and covers all state actions that are designed to influence business or social behaviour; and all forms of social or economic influence – where all mechanisms affecting behaviour, whether these are state-based or from other sources, such as market forces, are deemed regulation.[18]

Effective regulation is defined by the Council of Australian Governments (COAG) as 'standardising the exercise of bureaucratic discretion, so as to reduce discrepancies between government regulators, reducing uncertainty and lowering compliance costs. Regulatory measures should contain compliance strategies that ensure the greatest degree of compliance at the lowest cost to all parties'.[19]

Unconventional gas is classified as both administrative and economic in the regulatory context, since the promotion of resource activities and exploration promotes market and economic efficiency for the State and its citizens. Administrative bodies and tribunals act as direct monitors of unconventional gas activities via approvals and decisions as administrative policy bodies. It is recognised that regulation is required to safeguard the interests of parties with unequal bargaining power so they may negotiate effectively to protect their interests.

The regulatory role of the State in implementing its unconventional gas policy has many facets, ranging from exploration, production, and developing suitable legal regimes to managing resource revenue in the form of royalties. Without legal institutions to develop these legal regimes and roles of the State effectively, the State may risk becoming bound to private oil and gas companies and their standards to exploit unconventional gas.

Better regulation as effective regulation

Regulation, as defined by Selznick, concerns the 'sustained and focused control exercised by a public agency over activities that are valued by the community'.[20] The important elements of sustained and focused regulatory action implied in this definition illustrate intention taken with an explicit regulatory aim. As stated by Freiberg:

> The action is taken after a process, in which the goal, the regulatory action, and the consequences of taking the action, are related in meaningful ways. In other words, regulation is purposive and instrumental. It is intended to solve problems.[21]

Expanding on Selznick's explanation of regulation, Black defines regulation as 'the sustained and focused attempt to alter the behaviour of others according to defined standards and purposes with the intention of producing a broadly identified outcome or outcomes'.[22] Regulatory intention, as defined by Freiberg, is considered a primary purpose of regulation.[23]

Managing the uncertainty of risk surrounding a legal activity is the point at which regulatory intervention must occur to respond to this uncertainty. Regulation, therefore, necessarily affects the socio-legal landscape of a particular jurisdiction in which it operates and is enforced by the State. When, and how, a State responds to an uncertain legal landscape requires estimation of risks and weighing up benefits based on technological change. Without such regulatory regimes, a State may lose autonomy and control of its natural resource exploitation, resulting in less effective resource extraction and limited benefits for its citizens. In regulation, Baldwin et al. recognise that prevention is better than a cure, where waiting to regulate, or ineffective regulation, would create a high amount of injury or damage suffered.[24] An assessment of regulatory approaches requires an examination of the costs and benefits applicable to differing stakeholders and indeed, regulatees, in order to discern whether a regulatory system is operating effectively.

Regulation is distinguished from the broader concept of public policy, which consists of government processes that identify and respond to social problems.[25] Therefore, regulation occupies a narrower scope than governance. As discussed by Braithwaite, Coglianese and Levi-Faur, governance differs from regulation:

> Government and governance are about providing, distributing, and regulating. Regulation can be conceived as that large subset to government that is about steering the flow of events and behaviour, as opposed to providing and distributing. Of course, when regulators regulate, they often steer the providing and distributing that regulated actors undertake as well … it often makes sense to regard regulation as more narrow than governance.[26]

A core concept defining contemporary regulation is the management of conflicting interests intended to manage disparate stakeholders and 'communities', as suggested by Selznick's regulatory definition.[27]

Snir describes the regulatory task of the State as increasingly 'orchestrated' in pursuing 'public goals by promoting and empowering a network of public, private sector, and civil society actors and institutions, all of which are encouraged to engage in various "regulatory" (including self-regulatory) activities'.[28] This differs significantly from the interwar period, when the State acted solely an arbiter of the command and control economy and was highly protective of domestic industries through tariffs, industry assistance[29] and State ownership of corporations.

Since this period, Western liberal democracies, such as Australia, the UK, Canada and the USA, have steadily moved to a market-led economy creating a 'deregulated', 'modern regulatory state'[30] where the State acts as a 'light touch' regulator to facilitate and monitor the activities of a market-led economy.[31] From the post-war period onwards, regulatory reform encouraged the lowering of tariff barriers and abolition of State-controlled trade boards, and reinterpreted the role of the State in the public and private sphere, which 'efficiently delivered and exposed (market) competition'.[32] Guided by the emergence of neoliberal deregulation philosophies, regulators have ascribed to regulatory norms, proposing that governments must 'steer' rather than 'row' economies in regulatory policymaking.[33] This policy impetus runs the risk of holding an export-first focus for the State, thus increasing demand and prices for domestic consumers, and illustrating a high level of policy influence by private stakeholders and market forces.[34]

'Steering' the flow of events in regulation necessarily involves decisions about outcomes as they relate to the public and private sphere – that is, what is 'good' and, consequently, 'effective' regulation. A response to the question of 'good regulation' is offered by Breyer.[35] Breyer's methodology is essentially linear and functional in defining a problem, method and objective in regulation. However, this is a broad framework focusing on economic regulation and does not offer a sophisticated analysis of the impact of the regulatory objective. Breyer's methodology has been criticised as insufficient to manage contemporary regulatory issues that are presented to legislators.[36]

Kolieb introduces the concept of enforcement as an aspect of regulatory intention, to compel and hold accountable actors within the regulatory system.[37] However, while compliance is an element of the regulatory process, it is not, in and of itself, the definition of 'good regulation' regulation.[38] The 'better regulation' movement has emerged over the past two decades as an attempt to make a terminological and philosophical switch from deregulation to 'better regulation'.[39] This argues that 'better regulation' is necessary and that only bad regulation, as opposed to regulation in general, is burdensome. Freiberg identifies the development of 'better' regulatory strategies, including responsive regulation and risk-based regulation which has given recognition to the importance of 'evidence and consultation in the regulatory cycle and a more systematic and principled approach to compliance and enforcement'.[40] In contrast, Levi-Faur confines the definition of regulation to 'prescriptive rules and the monitoring and enforcement of these rules by social, business and political sectors on other social, business and political actors'.[41]

The elements to encourage 'better' and more effective regulation include:

1   clarifying regulatory objectives and definitions of problems;
2   ensuring that regulatory objectives are achieved effectively; and
3   identifying alternative options for achieving desired objectives.[42]

Regulatory failures

Regulatory failure, considered broadly, occurs when 'the law imposes unnecessary complexity costs on diffuse, poorly organized groups to the advantage of politically influential groups that benefit from that complexity ... by re-evaluating existing regulations agencies can simplify their rules and ease unnecessary regulatory burdens'.[43] Black classifies regulatory failures as a subset of policy failures resulting from the 'unintended and unforeseen consequences of the design and/or operation of a regulatory system and its interactions with other systems'.[44] Regulatory failures typically arise from poor regulatory design, poor regulatory tool choice and failure to achieve a broadly defined goal or set of goals.[45]

Regulatory failure then broadly encompasses the inability to satisfy underlying policy objectives, creating unnecessary costs and eroding the general confidence in the regulatory system and the rule of law itself.[46] Freiberg identifies the key attributes of regulatory failure as:

1   regulatory bad design;
2   inadequate consultation;
3   poor or ineffective implementation;
4   conflict and confusion;
5   failure to clearly identify appropriate targets for regulation;
6   poor tool choice;
7   poor or ambiguous rules or laws;
8   ambiguity in forms of regulation;
9   procedural injustice; and
10 regulatory capture.[47]

These 10 attributes provide a comprehensive framework to identify and codify regulatory failure.

Thus, regulatory failures inherently stem from poor regulatory design, poor tool choice, and result in ambiguous rules, poor or ineffective implementation, failure of clear targets as well as conflict, confusion and procedural injustice.

First, regulatory conflict emerges when the same type of conduct is regulated by numerous different regulators at the local, state and federal level. This can result in differing objectives, techniques and tools which may create duplication, complexities and contradictions that work against the achievement of the regulatory outcome.[48] Second, poor tool choice emerges where the regulatory tool chosen does not align with or deliver a successful and effective regulatory framework to achieve the specific policy objective of the State.[49]

Learning from regulatory failure due to poor tool choice and bad regulatory design creates a basis for functional exchange between jurisdictions with similar institutional structures, socio-historic foundations and sector-specific regulatory imperatives, forming readily transferrable, alternative, regulatory designs.[50] By undertaking an analysis of differing regulatory frameworks in the jurisdictions analysed within this book, it is possible to learn from any regulatory failures to ensure improvement in future regulation of land use and land access.

Regulatory gaps appear when regulatory responsibilities between multiple legislative instruments and governmental agencies are not clear, and regulatory failure due to bad regulatory design and poor tool choice is evident.[51] For example, one of the outcomes of regulatory gaps in the petroleum sector is highlighted by the Australian Petroleum Production and Exploration Association (APPEA):

There has been considerable public confusion around the respective respon-sibilities of the Federal and State Governments, and of State Departments, for overseeing regulatory inspections, assessments and approvals. It is essential for effective and efficient regulation of critical supply industries, and for effective and efficient governance, that the public and other stakeholders are able to identify and have confidence in the responsible regulator.[52]

By identifying regulatory gaps and reviewing regulations, regulatory burdens can be identified to ascertain the basis of regulatory needs which may not be met by stakeholders.

Unnecessary regulatory burdens are defined as 'burdens that can be removed without compromising desirable outcomes such as relating to resource manage-ment, the environment, heritage, development, land access and occupational health and safety'.[53] The unconventional gas sector has regulatory gaps and burdens as identified by the Australian Productivity Commission (APC), which proposes criteria to reduce unnecessary regulatory burden through regulatory reforms:

- streamline regulation by reducing the need for multiple agency approvals and remove duplication of assessment and reporting requirements;
- avoid, where possible, arrangements that involve multiple agencies and, where multiple agencies have to be involved, have in place clear administrative arrangements to avoid or minimise unnecessary overlap in regulatory functions;
- avoid unnecessary inconsistencies in regulatory requirements or decision-making within and across jurisdictions;
- provide regulators with clear regulatory objectives and minimise unneces-sary conflicts of interest; and
- consolidate specialist expertise and promote efficient use of resources.[54]

Although these reforms were proposed and outlined by the APC in relation to the upstream offshore petroleum sector in Australia, the concepts of regulatory burden and gaps are equally applicable to Queensland's upstream onshore unconventional gas due to the similarities in licensing procedures, departmental oversight and reg-ulatory overlap in both offshore and onshore unconventional gas sectors in Australia.

The upstream petroleum sector licensing regime involves a number of pro-cesses as described by Bunter:

The identification by government of potential (upstream) petroleum investment opportunities in the national territory, their subdivision into discrete contract areas of prospective size, their offering to the international oil companies by a suitable tendering process and the establishment and negotiation of technical, financial and contractual terms and conditions (for award) consistent with their petroleum prospectively and with the national interest.[55]

Wiseman categorises 'regulatory islands' within the USA, particularly in relation to oil and development, as regulatory states with inherent regulatory burden and gaps due to poor regulatory design. The classification of regulatory islands recognises the complexity of overlapping regulatory tools as 'each state [in the USA] has hundreds of regulations and statutory provisions, housed within hundreds of different portions of codes and statutes. Many local governments, too, have detailed oil and gas codes'.[56]

Unconventional gas regulation

Unconventional gas is highly political and strategic. At stake are export earnings, economic stability and the potential failure of the State if it cannot guarantee an energy source. The jurisdictions analysed comparatively within this book fall within three socio-legal regulatory approaches diverging between prioritising domestic energy security and seeking a balance between both agricultural land use and unconventional gas development, and banning unconventional gas exploitation. Each State holds national sovereignty over its natural resources, including unconventional gas, as conferred in international law instruments such as the United Nations General Assembly Resolution on Permanent Sovereignty over Natural Resources.[57] Therefore, each State has at its disposal a suite of regulatory tools to implement its petroleum policy objectives. Petroleum titles, approvals and contracts are examples of regulatory tools for oil and gas activities involving overlapping and interjecting regulations.[58]

The regulatory tool often applicable to natural resource sectors, including that of petroleum (incorporating unconventional gas), is authorisation. Authorisation generally refers to the process of permitting a certain activity to take place, the absence of which would be a contravention of regulation.[59] The petroleum industry authorises activities through a permissioning framework to allow certain activities to be undertaken and permitted, while some authorisations are exempt from certain activities. Exploration licences are a licence to 'explore for oil and gas in a particular area issued to a company by the governing jurisdiction'[60] and a production licence is a licence to 'produce oil or gas in a particular area issued to a company by the governing state authority'.[61] Depending on a State's regulatory framework, a number of State government departments, administrative authorities and judicial bodies may enforce and issue petroleum licences and permits. Authorisations recognise and legitimise State power over many forms of activities and ownership and control over natural resources. Authorisation is also a primary means of addressing information asymmetries, creating an environment of trust and is an essential element in preventative risk management.

Freiberg identifies multiple objectives of an effective licensing system:

- protecting the community;
- addressing information asymmetries;
- enhancing probity;
- promoting market stability;

- minimising or preventing harm;
- promoting order and facilitating enforcement;
- providing redress; and
- recovering costs.[62]

Petroleum licensing systems, therefore, require regulators to assess the risk posed by a petroleum activity and differing levels of licenses and exclusions applicable based on the level of risk identified. The consequences of discrete risk levels relate to different regulatory oversight measures, differential fees and publication of petroleum licences for public scrutiny. This is compared with negative licensing as a form of regulation, where no licence or permit is required to enter a market, but a serious breach will be reprimanded with exclusive sanctions.[63] Finally, the common elements and procedures in licensing schemes are as follows:

- a regulatory authority;
- application procedures, setting out the methods of granting, amendment, transfer, renewal restoration and replacement of licences;
- determination of applications;
- specification of minimum standards;
- awarding of licences;
- specification of conditions attached to a licence vary according to the requirements of the activity or industry;
- enforcement and sanctioning provisions in licensing schemes, containing specific enforcement and sanctioning mechanisms. Internal disciplinary procedures may be provided for either by the regulatory department of by semi-judicial tribunals and provide for the cancellation, suspension and variation of licences and disqualification form obtaining licence in the future; and
- appellate or review systems or bodies.[64]

Petroleum legislation is a particular and specific form of natural resource regulation and consists of primary and subordinate legislation, policy decisions and guidelines. Petroleum legislation must not only regulate petroleum activities, but also optimise the extraction of unconventional gas in a manner that is 'transparent, predictable and consistent with the overarching petroleum objectives of that State'.[65]

The dilemmas in creating effective petroleum regulation are the estimation of risks and benefits, which are often based on limited scientific knowledge, particularly in relation to hydraulic fracturing technology. Thus, the regulatory conundrum presents itself as follows, 'regulating too soon may stifle innovation, but regulating too late may result in significant or irreversible damage'.[66] Such an approach is very much reactive, as the State responds to regulatory issues as they occur, rather than trying to anticipate and legislate for problems prior to the activity taking place. For example in Queensland, regulating unconventional gas activities 'too soon' has led to the search for a regulatory mechanism to manage coexistence.

The numerous unconventional gas regulatory reviews, legislative amendments and land access regime assessments commencing in the early 2000s all demonstrate the adaptive management approach of 'learning by doing'[67] in Queensland. This has led to the assessment of Queensland's unconventional gas regulatory regime as arguably one of the most complex, prescriptive and rule-based natural resource states in Australia. The rule-based unconventional gas framework in Queensland requires the enactment of new legislation each time a new regulatory issue arises as 'the PGPSA outlines, in minutiae, the "rules" for the extraction of CSG. Such rule-based regulation relies on legislatively entrenched rules to regulate petroleum activities'.[68] As at 2017, the *Petroleum and Gas (Production and Safety) Act 2004* (Qld) (PGPSA) consists of over 808 pages and 992 sections and is just one of the seven resource Acts applicable to Queensland's oil and gas regulatory regime.[69] The *Petroleum Act 1923* (Qld), as the first instance of petroleum legislation in Queensland, was seen as inadequate for the regulation of the development of unconventional gas, leading to the introduction of the much longer and more detailed amendments in the PGPSA.[70] As a result of the changes to petroleum licensing and activities, the detailed PGPSA has required over 1,000 amendments with more than 40 consolidated versions released.[71]

Adaptive management

One of the most critical elements of unconventional gas regulation is recurrent decisions – decisions that need to be made on a regular basis in response to changing conditions and priorities with the aim of reducing ecological uncertainty.[72] Adaptive management holds its origins as a technical, scientific environmental management methodology for the management of complex ecosystems by 'monitoring the results of a suite of management initiatives'.[73] Adaptive management was first described by Beverton and Holt in the environmental sector as an alternative decision-making process for fisheries.[74] Holling[75] and Walters and Hilborn[76] further refined and theorised the framework for adaptive management, which continues to be applied frequently to the monitoring and mitigation of risks in the fisheries, forestry and engaged species sectors.[77]

Adaptive management is traditionally an environmental management theoretical approach which 'seeks insights into the behavior [sic] of ecosystems … and incorporates and integrates concepts such as social learning, operations research, economic values, and political differences with ecosystem monitoring, models, and science'.[78] The concept is designed to support action in environmental management issues facing limitations of scientific knowledge and complexities of large ecosystems. Therefore, adaptive management is typically utilised as an overarching management goal without the context of creating environmental management policies to assist in learnings of 'complex ecological systems by monitoring the results of a suite of management initiatives'.[79]

Adaptive management theory was expanded significantly by Lee, who positioned adaptive management beyond the environmental sector in his application of the theory to the political and social sciences.[80] However, according to Williams, 'many in natural resources conservation now claim, sometimes inappropriately, that adaptive management is the approach they commonly use in meeting their resource management responsibilities'.[81] Williams' statement alludes to the fact that adaptive management is not correctly categorised as an approach to guide regulation – rather, it is an environmental management protocol.[82]

Two alternating approaches in the application of adaptive management framework by States are passive or active adaptive management.[83] Passive adaptive management outlines a single preferred course of action based on existing information and understanding. Outcomes of management actions are then monitored and subsequent decisions are adjusted based on the outcomes.[84] This approach contributes to environmental management, but it is limited in its ability to enhance scientific and management capabilities for conditions that exceed the course of action selected.[85] By contrast, an active adaptive management approach reviews information before management actions are taken. A range of competing alternative system models of ecosystem and related responses (e.g., demographic changes and recreational uses), rather than a single model, are then developed. Utilising an active adaptive management approach, options are then chosen based upon evaluations of these alternative models.

Adaptive management has been identified as a concept of environmental management that is 'widely promoted and widely misunderstood'.[86] This is due to adaptive management being adopted by States without clear identification of appropriate measures and targets for regulation, creating ineffective implementation and ambiguity in the application of the concept. To succeed, adaptive management must have clear measures and objectives, rather than serving as a measure to create excessive regulation leading to burdens and gaps. As stated by Jones, adaptive management:

> [it is] an approach that ensures management not only plans and carries out actions to achieve objectives, but also measures the results so that everyone can see what's working and what's not, and consequently make informed decisions and adjustments to enhance the achievement of objectives and the delivery of desired outcomes.[87]

Therefore, when applying an adaptive management approach to regulation, it is important to ensure sufficient flexibility and responsiveness within the broader regulatory framework. Active adaptive management as an approach must 'embrace complexity'[88] by presenting broad objective principles that allow for responsiveness to a range of regulatory conditions without the State needing to amend its regulatory approach. The absence of an appropriately flexible adaptive management regulatory scheme can create 'costly implementation failures',[89] for example, through legislative overhauls and the administration of multiple regulatory agencies.

Although a key motivation of adaptive management is to improve regulation by reducing structural uncertainty, its success can be impeded by a failure to adapt to social and institutional changes that inevitably occur over time. A well-designed active adaptive management regulatory approach provides the opportunity for learning at social and institutional levels, recognising that learning often occurs on different time scales. Thus, technical learning occurs in a context of relatively short-term objectives, alternatives and predictive models. However, learning about the decision process itself occurs through periodic, but less frequent assessment of these factors as they evolve in response to management actions and environmental conditions.

Consequently, effective active adaptive management regulation has been typically confined to small-scale projects limited to environmental management issues. Regulation with an inflexible, prescriptive and rule-based approach is unlikely to provide effective petroleum regulation, as adaptive environmental management is not a 'one size fits all solution'.[90] Swayne also identifies that adaptive management is not automatically classified 'an active decision-making framework nor does it make the decision making process easier'[91] as an active adaptive management approach 'evaluates alternative options on the assumption that decisions will be made and enacted, rather than avoided'.[92] Swayne identifies the hallmarks of evaluation for successful and active adaptive managing including clearly defining:

> What are the management objectives and the key desired outcomes for the regulatory system?
> What are the appropriate strategies and actions to be taken to achieve the objectives and key desired outcomes?
> What range of potential performance indicators can be used to monitor or measure the effectiveness of the management approach?
> How will what is learnt be used in deciding what to do? And critically, who will be responsible for adjustments in the management approach in response to the results of the evaluation?[93]

As a regulatory technique, adaptive management was adopted by Queensland regulatory bodies as an 'off the shelf' solution to the technical nature of natural resource legislation. This regulatory approach was probably selected since adaptive management theory had its origins in scientifically challenging policy and regulatory issues operating in a regulatory environment of uncertainty surrounding the cumulative impacts and multiple stakeholder interests the State is required to manage. Issues of scientific uncertainty have continually arisen in unconventional gas activities, as evidenced by the establishment of the *Independent Expert Scientific Committee on Coal Seam Gas and Large Coal Mining Development* in 2012, under the *Environment Protection and Biodiversity Conservation Act 1999* (Cth) (EPBCA), providing scientific advice to the State on the impact that CSG activities may have on Australia's water resources.

Queensland states that its adaptive management approach recognises the uncertain impacts of unconventional gas activities and puts in place a system 'to monitor the industry and instigate change where required'.[94] The object of this approach is to 'ensure the government is able to respond to what happens on the ground and protect the environment'.[95] Queensland's current unconventional gas regulatory framework operates a myriad of amendments, superimposed onto the existing legal duties. Queensland's adoption of adaptive management as the overarching approach to its unconventional gas regulation may be characterised as passive and inflexible, as evident in its enactment of numerous retrospective and overlapping regulations after specific unconventional gas inquiries, such as the *Mineral and Energy Resources (Common Provisions) Act 2014* (Qld) (MERCPA) in response to the Land Access Committee Report of 2012.[96]

Adaptive management, as the regulatory approach adopted by Queensland, Australia and British Columbia, Canada for the regulation of unconventional gas will be analysed within Chapters 5 and 6 of this book. The contrasting approaches of both jurisdictions provide examples of an active and a passive adaptive management approach to highlight any potential regulatory 'failures' in the context of the current unconventional gas regulatory regimes of both countries.

The precautionary principle

The precautionary principle was originally coined within environmental policy literature in Germany during the 1970s and is founded upon the principle of sustainable development, an essential element of natural resource governance and management. Sustainable development is concerned with each State's right to development in order 'to equitably meet developmental and environmental needs of present and future generations'.[97] According to Paterson:

> In the face of uncertainty, this (precautionary) principle urges caution and provides some comfort to those who are fearful that the potential short-termism of political and economic actors will lead to unwise risk-taking. Its proponents would, therefore, insist that it serves precisely to prevent recklessness in the face of uncertainty.[98]

The precautionary principle has been increasingly incorporated as a 'common sense' approach to decision-making[99] in national and international legislation, to embolden environmental protection from anthropogenic impacts in relation to agriculture, waste, climate change and marine pollution.[100] One of the goals of the precautionary principle is to 'guide action in situations of scientific uncertainty regarding environmental impacts'.[101] The most often cited formulation of the precautionary principle is that of Principle 15 of the *Rio Declaration on Environment and Development* which states:

In order to protect the environment, the precautionary approach shall be widely applied by States according to their capabilities. Where there are threats of serious or irreversible damage, lack of full scientific certainty shall not be used as a reason for postponing cost effective measures to prevent environmental degradation.[102]

However, Godden and Peel critique the Rio Declaration definition of the precautionary principle as 'weak', by simply discouraging regulators to rely on 'lack of scientific certainty' as grounds for failing to enact regulation.[103] Despite ongoing debates about the validity of the precautionary principle, it has been adopted as a legal approach, demonstrating a principle-based approach to regulatory decision making.

A more rigid and stringent definition of the precautionary principle is found in the *United Nations World Charter for Nature* which suggests where 'potential adverse effects are not fully understood, the activities should not proceed'.[104] The *United Nations Framework Convention on Climate Change*,[105] *Protocol to the 1979 Convention on Long Range Trans Boundary Air Pollution*[106] and the *Montreal Protocol on Substances that Deplete the Ozone Layer*[107] all apply and reference the precautionary principle. Therefore, Hohman[108] and Cameron and Abouchar[109] argue the prevalence of the precautionary principle in international forums has led to the adoption of the precautionary principle as part of customary international law.

To ensure that the precautionary legislative measures adopted are reasonable, practical considerations should be considered, including 'the effectiveness of the selected precautionary measure, proportionality, cost-effectiveness (economic and social costs) and consistency'.[110] During consideration of the precautionary legislative instrument, the input of the broader community and stakeholders should be accommodated as part of the democratic decision-making feature of the principle.

As van der Meulen points out, a precautionary approach applies in the presence of scientific uncertainty: 'That is to say, when risk assessment is inconclusive but gives scientific reasons to suspect a food safety risk, public authorities are entitled to base protective measures on a worst case scenario'.[111] This principle is of particular relevance to issues related to farm rehabilitation and productive land capacity, where there is an ongoing scientific debate about the safety of extracting unconventional gas on farmland.

According to Gullet, advancing precaution would best be achieved by mandatory consideration for decision-makers.[112] Thus, this progression highlights the increasing focus of environmental protection in decision making. For example, in France, pursuant to Law No. 2011–835 of July 2011, the exploration and exploitation of hydrocarbons by hydraulic fracturing, all exclusive licenses to prospect unconventional gas were revoked within two months of the passage of the law in an act of advancing precaution. The underlying basis of Law No. 2011–835 is Article 5 of the French Charter for the Environment which states:

When the occurrence of any damage, albeit unpredictable in the current state of scientific knowledge, may seriously and irreversibly harm the environment, public authorities shall, with due respect for the *principle of precaution* and the areas within their jurisdiction, ensure the implementation of procedures for risk assessment and the adoption of temporary measures commensurate with the risk involved in order to preclude the occurrence of such damage.[113]

Harding argues in situations where there is a greater level of scientific uncertainty or more significant threat of environmental harm, stronger and proportional precautionary measures are needed.[114] Sedeleer attempts to clarify the decision-making process by arguing that the principle of proportionality is particularly important in how decision-makers will weigh competing interests.[115] That is, when the goal of environmental protection is pitted against developmental or social agendas, decision-makers will need to ensure that the precautionary measures that are imposed on a project are 'appropriate, even in the absence of conclusive scientific evidence that damage will occur'.[116] The question of what is appropriate involves determining if a measure can 'demonstrate a causal link to the purpose being pursued'.[117] Practically speaking, this is a consideration of whether an action is suitable to achieve the goal being pursued, in this instance unconventional gas and agricultural land coexistence. Deville and Harding note that monitoring is an 'essential precautionary measure and should be built into the development and ongoing operation of all projects'.[118]

In the USA, the precautionary approach is evident in the enactment of a permanent ban on hydraulic fracturing activities in New York State due to the threat of serious or irreversible environmental harms, including in the deterioration of rural communities in the state. This is coupled with the confirmation that hydraulic fracturing activities would adversely impact the sentiment found within the New York State Constitution article XIV, section 4, to 'encourage the development and improvement of its agricultural lands for the production of food and other agricultural products [which] … shall include the protection of agricultural lands'.[119] Consequently the New York State Department of Environmental Conservation (DEC) found 'the only means of completely eliminating the risks of impacts to farmlands and livestock is to employ the No-Action alternative'.[120]

Chapters 7, 8 and 9 of this book examine the UK, New York State and France in their precautionary approach to unconventional gas regulation. Their respective moratoriums or ban of hydraulic fracturing activities are examined and serve as comparators in their respective approaches to prevent adverse and irreversible effects of unconventional gas activities on agricultural land.

Statist regulation

The State, as regulator of unconventional gas resources, holds two key governance responsibilities in the development of its natural resources. First, the State should govern its unconventional gas in a manner that creates and maintains benefits for the State itself and its citizens. Second, the State must create an access regime, such as a licensing regime where proprietary rights of unconventional gas extraction are assigned to private companies in exchange for economic benefits to the State. Furthermore, the State may choose to participate in the extraction of unconventional gas within the regulatory framework. In all natural resource-producing jurisdictions, the State intervenes in the extraction of the resource, seeking to control the activity to achieve its policy objectives. Such State intervention is categorised by Nelsen as *minimal intervention, regulatory intervention* and *participatory intervention.*[121] Therefore, a State selects the form of regulation according to the degree of control it wishes to exert over the development of its unconventional gas resources. This categorisation is based on levels of State intervention in the contestation between differing stakeholders, including landholders[122] and licenseholders.

An example of *minimal intervention* is within Western Australia, where no specific legislative provision exists to govern negotiations between petroleum titleholders and private landholders, unless the activity occurs on Native Title land.[123] *Regulatory intervention* is evident in Queensland, within *The Land Access Code 2016* (Qld), containing both mandatory and voluntary provisions to encourage 'good relations'[124] between titleholders and landholders.[125] Finally, *participatory intervention* is found in China where State oil companies, as the major petroleum titleholders, participate in the exploration for and production of unconventional gas on construction land in China, with State oil companies assuming the operatorship, guiding the development of shale gas.[126]

The broad definition of statist regulation is 'dominated, financed or coordinated by the state'.[127] While the definition of statist regulation has evolved over time, the concept of the State as the central actor acting and directly participating in economic activities remains the linchpin of the concept. The statist approach is thus complex, as the State has the additional role of dealing with contestation between private parties. As detailed by Vivoda, a statist-based approach echoes a securitisation approach to energy whereby, 'an energy policy problem is an energy security issue if it is presented and perceived as affecting the stability (and in critical situations, the survival) of a nation, the functioning and continuity of the economy, or the realization of major national values and objectives'.[128] Taking Nelson's three-part classification, *minimal intervention* is based on the neoliberal market model which can be classified as non-statist. According to this approach, the State minimally participates in regulating the free market, utilising bidding for licences, allowing exploration and production of unconventional gas to be based on corporate profit imperatives within a regulatory framework.

On the other side of the spectrum, *participatory intervention* is founded on principles of socialism, where the State seeks to insert itself in the activities and operations that are conducted by private oil companies in addition to regulating the activity itself. This is grounded in a statist-based regulatory approach, whereby the State has sanctioned the activities of unconventional gas operations, and is directly participating in the exploration and production of the resource. Therefore, the State takes a more socialist and participatory approach to unconventional gas as an active participant. The primary difference between a non-statist and statist approach is the high level of State intervention and the State's level of involvement in creating, enforcing, and monitoring regulations as an active shareholder in its natural resources. In seeking to balance the private shareholder-driven interests of private companies with those of the State, a statist approach provides a multifaceted role of the State. It creates the conditions for State ownership, the development of appropriate regulatory frameworks for the appropriate return from the production of petroleum, with benefits of the exploitation of unconventional gas going to the State as the owner and regulator of the resource.

At an international level, the statist approach is enshrined within the United Nations General Assembly Resolution on the Permanent Sovereignty over National Resources, which establishes 'The right of peoples and nations to permanent sovereignty over their natural wealth and resources must be exercised in the interest of their national development and of the well-being of the people of the State concerned'[129] and in relation to natural resource development by States, 'Nationalization, expropriation or requisitioning shall be based on grounds or reasons of public utility, security or the national interest which are recognized as overriding purely individual or private interests, both domestic and foreign'.[130] These articles enshrine, and indeed promote, a State's right to exercise natural resource development from a socio-economic perspective of ensuring national security as well as a State's energy security.

The International Energy Agency (IEA) defines energy security as 'the uninterrupted availability of energy sources at an affordable price'.[131] Energy security concerns have been heightened amongst States with an increasing imbalance between demand and supply of unconventional gas. Thus, as Vogler and Stephan have noted, 'secure access may be at risk because increasing scarcity implied greater international competition and encourages a move away from market allocation towards "statist" forms of energy security'.[132] Therefore, countries taking a statist approach to unconventional gas exploitation may place security of energy supply above all other state policies, including those that relate to agriculture.

A statist approach views the path to natural gas extraction on agricultural land as directly tied to a State's unfettered right to exploit domestic natural resources. Therefore, the concept is based on the belief that the State's primary task is to ensure energy for its citizens, utilising State-owned corporations where necessary. Chapters 10 and 11 of this book examine the statist approach to unconventional gas exploration and extraction in both China and Poland, motivated by the underpinnings of energy security and independence.

Conclusion

The foundational theoretical constructs of this book have been detailed within this chapter. The hallmarks of an adaptive management, precautionary and statist approach are highlighted in examining the relevant benefits and burdens to each approach, prior to applying each socio-legal theory incrementally in a comparative analysis that follows in Part II of the book.

Finally, this chapter observes the notion of public and private ownership applicable to petroleum under the guidance of State to exploit and develop its unconventional gas resources 'in the public interest'. In doing so, this chapter addresses the many roles of the State as arbiter, legislator, advocate, monitor, contractor and service provider in the context of unconventional gas development and the shifting role of the State in responding to changing political, social, economic and environmental factors.

Notes

1   Anthony Ogus, 'Regulation' in Bronwen Morgan and Karen Yeung (eds), *An Introduction to Law and Regulation: Text and Materials* (Cambridge University Press, 2007) 1, 18.
2   *Permanent Sovereignty over Natural Resources*, GA Res 1803 (XVII), UN GAOR, 17th sess, 1194th plen mtg, UN Doc A/RES/1803(XVII) (14 December 1962). The principles for sovereignty over national resources are found within eight articles concerning the exploration, development and disposition of natural resources and other related issues.
3   Stephan Schott and Graham Campbell, 'National Energy Strategies of Major Industrialized Countries' in Hugh Dyer and Maria Julia Trombetta (eds), *International Handbook of Energy Security* (Edward Elgar, 2013) 174.
4   Samantha Hepburn, 'Public Resource Ownership and Community Engagement in a Modern Energy Landscape' (2017) 34 *Pace Environmental Law Review* 379, 379.
5   Rudolf Klein and Theodore Marmor, 'Reflections on Policy Analysis: Putting it Together Again' in Robert E. Goodin (ed.), *The Oxford Handbook of Political Science* (OUP, Oxford, 2011) 7.
6   Catherine Althaus, Peter Bridgman and Glyn Davis, *The Australian Policy Handbook* (Allen & Unwin, 2012) 6.
7   Paulo de Sa, 'Mineral Policy: A World Bank Perspective' in E. Bastida, T. Walde and J. Warden-Fernandes (eds), *International Comparative Mineral Law and Policy* (Wolters Kluwer, 2005) 492, 494.
8   Richard Dworkin, *Taking Rights Seriously* (Harvard University Press, 1997) 77.
9   Ibid.
10 Ibid.
11 Tina Hunter, 'Law and Policy Frameworks for Local Content in the Development of Petroleum Resources: Norwegian and Australian Perspectives on Cross-Sectoral Linkages and Economic Diversification' (2014) 14(2–3) *Mineral Economics* 115, 120.
12 Tina Hunter and John Chandler, *Petroleum Law in Australia* (LexisNexis, 2013) 41.
13 Ibid.
14 Robert Baldwin, Martin Cave and Martin Lodge, *Understanding Regulation: Theory, Strategy, and Practice* (OUP Oxford, 2012) 68.
15 Richard Gordon, *Regulation and Economic Analysis: A Critique Over Two Centuries* (Springer, 1994).
16 Robert Baldwin, Martin Cave and Martin Lodge, *Understanding Regulation: Theory, Strategy, and Practice* (OUP Oxford, 2012) 27.
17 Peter Drahos, *Regulatory Theory: Foundations and Applications* (ANU Press, 2017).
18 Jacint Jordana and David Levi-Faur, *The Politics of Regulation: Institutions and Regulatory Reforms for the Age of Governance* (Edward Elgar, 2004).
19 Council of Australian Governments, *Best Practice Regulation: A Guide for Ministerial Councils and National Standard Setting Bodies* (2007) 3.
20 Phillip Selznick, 'Focusing Organisational Research on Regulation' in Roger Noll (ed.), *Regulatory Policy and the Social Sciences* (University of California Press, 1985) 363, 363.

21 Arie Freiberg, *Regulation in Australia* (The Federation Press, 2017) 4.
22 Julia Black, 'Regulatory Conversations' (2002) 29(1) *Journal of Law and Society* 163, 170.
23 Arie Freiberg, *Regulation in Australia* (The Federation Press, 2017).
24 Robert Baldwin, Martin Cave and Martin Lodge, *Understanding Regulation: Theory, Strategy and Practice* (Oxford University Press, 2nd edn, 2011) 243.
25 Sarah Maddison and Richard Dennis, *An Introduction to Australian Public Policy: Theory and Practice* (Cambridge University Press, 2nd edn, 2013).
26 John Braithwaite, Cary Coglianese and David Levi-Faur, 'Can Regulation and Governance Make a Difference?' (2007) 1(1) *Regulation and Governance* 1, 3.
27 Phillip Selznick, 'Focusing Organisational Research on Regulation' in Roger Noll (ed.), *Regulatory Policy and the Social Sciences* (University of California Press, 1985) 363–364.
28 Reut Snir, 'Trends in Global Nanotechnology Regulation: The Public–Private Interplay' (2014) 17 *Vanderbilt Journal of Entertainment & Technology Law* 1, 19.
29 *Industries Assistance Omission Act 1973* (Cth).
30 John Braithwaite, *Neoliberalism or Regulatory Capitalism* (RegNet Occasional Paper No. 5, 2005). <https://www.anu.edu.au/fellows/jbraithwaite/_documents/Articles/Neoliberalism_Regulatory_2005.pdf>. For further theoretical discussions of the Modern Regulatory State of deregulation, see James E. Anderson, *Emergence of the Modern Regulatory State* (Public Affairs Press, 1963).
31 David Osborne and Ted Gaebler, *Reinvesting Government* (Addison-Wesley, 1992).
32 Paul Kelly, *The End of Certainty: Power, Politics and Business in Australia* (Allen & Unwin, 2008).
33 Brian Head and Elaine McCoy, *Deregulation or Better Regulation? Issues for the Public Sector* (Macmillan Education, 1991).
34 For example, in Australia, the high level of long-term export contracts for gas has created a domestic gas price and gas security crisis in its East Coast States as typified by the ACCC report on gas pricing whereby a shortfall in domestic gas supply is evident of between 55–108 petajoules in 2018. See Australian Competition and Consumer Commission, *Gas Inquiry Interim Report 2017–2010* (2017) <https://www.accc.gov.au/system/files/Gas%20inquiry%20December%202017%20interim%20report.pdf>.
35 Stephen G. Breyer, *Regulation and its Reform* (Harvard University Press, 2009); Stephen G. Breyer, *Breaking the Vicious Circle: Toward Effective Risk Regulation* (Harvard University Press, 1993).
36 Eric J. Gouvin, 'A Square Peg in a Vicious Circle: Stephen Breyer's Optimistic Prescription for the Regulatory Mess' (1995) 32 *Harvard Journal on Legislation* 473.
37 Johnathon Kolieb, 'When to Punish, When to Persuade and When to Reward: Strengthening Responsive Regulation with the Regulatory Diamond' (2015) 41 (1) *Monash Law Review* 136.
38 Ibid, 145.
39 Stephen Weatherill, *Better Regulation* (Bloomsbury Publishing, 2007).
40 Arie Freiberg, *Regulation in Australia* (The Federation Press, 2017), 157.
41 David Levi-Faur, *Handbook on the Politics of Regulation* (Edward Elgar Publishing, 2011) 122.
42 Robert Baldwin, 'Better Regulation: Tensions Abroad the Enterprise' in Stephen Weatherill (ed.), *Better Regulation* (Bloomsbury Publishing, 2007) 27, 34.
43 Peter Schuck, *Why Government Fails So Often: And How It Can Do Better* (Princeton University Press, 2014) 405.
44 Julia Black, *Learning from Regulatory Disasters, Society and Economics Working Papers* 24/2014 (London School of Economics, Department of Law, 2014) 11.
45 Ibid, 11.
46 Organisation for Economic Co-Operation and Development (OECD), *Regulatory Performance: Ex Post Evaluation of Regulatory Tools and Institutions* (OECD, 2004).

47   Arie Freiberg, *Regulation in Australia* (The Federation Press, 2017) 489.
48   Peter Grabosky, 'Counterproductive Regulation' (1995) 17(3) *Law and Policy* 257.
49   Arie Freiberg, *Regulation in Australia* (The Federation Press, 2017).
50   Clifford Winston, *Government Failure Versus Market Failure: Microeconomics Policy Research and Government Performance* (AEI-Brookings Joint Center for Regulatory Studies, 2006).
51   Australian Productivity Commission, *Review of Regulatory Burden on the Upstream Petroleum (Oil and Gas) Sector: April 2009* (Productivity Commission, 2009) 209 <https://www.pc.gov.au/inquiries/completed/upstream-petroleum/report/upstream-petroleum.pdf>.
52   APPEA (Australian Petroleum Production and Exploration Association), *Australia's Upstream Oil and Gas Industry: A Platform for Prosperity, Issues Paper* (2006) 18.
53   Australian Productivity Commission, *Review of Regulatory Burden on the Upstream Petroleum (Oil and Gas) Sector: April 2009* (Productivity Commission, 2009) 209 <https://www.pc.gov.au/inquiries/completed/upstream-petroleum/report/upstream-petroleum.pdf>.
54   Ibid.
55   Michael Bunter, *The Promotion and Licensing of Petroleum Prospective Acreage* (Kluwer Law International, 2002) xxii.
56   Hannah Wiseman, 'Regulatory Islands' (2014) 89 *New York University Law Review* 1697.
57   *Permanent Sovereignty over Natural Resources*, GA Res 1803 (XVII), UN GAOR, 17th sess, 1194th plen mtg, UN Doc A/RES/1803(XVII) (14 December 1962).
58   Hannah Wiseman, 'Risk and Response in Fracturing Policy' (2013) 84 *University of Colorado Law Review* 730.
59   Tina Hunter and John Chandler, *Petroleum Law in Australia* (LexisNexis, 2013) 77.
60   Australian Productivity Commission, *Review of Regulatory Burden on the Upstream Petroleum (Oil and Gas) Sector: April 2009* (Productivity Commission, 2009) XVII <https://www.pc.gov.au/inquiries/completed/upstream-petroleum/report/upstream-petroleum.pdf>.
61   Ibid.
62   Arie Freiberg, *Regulation in Australia* (The Federation Press, 2017) 309.
63   Stephen Rimmer, 'Best Practice Regulations and Licensing as a form of Regulation' (2005) 65(2) *Public Administration* 3, 13.
64   Arie Freiberg, *Regulation in Australia* (The Federation Press, 2017).
65   Tina Hunter, 'The Development of Shale Gas and Coal Bed Methane in Australia: Best Practice for International Jurisdictions?' (2016) 38(2) *Houston Journal of International Law* 367, 370.
66   Arie Freiberg, *Regulation in Australia* (The Federation Press, 2017) 142.
67   Nicola Swayne, 'Regulating Coal Seam Gas in Queensland: Lessons in an Adaptive Environmental Management Approach?' (2012) 29(2) *Environmental and Planning Law Journal* 163, 165.
68   Tina Hunter, 'The Development of Shale Gas and Coal Bed Methane in Australia: Best Practice for International Jurisdictions?' (2016) 38(2) *Houston Journal of International Law* 367, 394.
69   The others being the *Environment Protection and Biodiversity Conversation Act 1999* (Cth), *Petroleum Act 1923* (Qld), *Environmental Protection Act 1994* (Qld) (EPA), *Water Act 2000* (Cth), *Water Supply (Safety and Reliability) Act 2008* (Qld) and *Gasfields Commission Act 2013* (Qld) (GCA).
70   CEDA, *Australia's Unconventional Energy Options* (Report, September 2012) <https://www.ceda.com.au/Research-and-policy/All-CEDA-research/Research-catalogue/Australia-s-Unconventional-Energy-Options>.
71   Tina Hunter, 'The Development of Shale Gas and Coal Bed Methane in Australia: Best Practice for International Jurisdictions?' (2016) 38(2) *Houston Journal of International Law* 367, 372.

72 Byron Williams and Eleanor Brown, 'Adaptive Management: From More Talk to Real Action' (2014) 53(2) *Environmental Management* 465; Byron Williams and Fred Johnson, 'Confronting Dynamics and Uncertainty in Optimal Decision Making for Conservation' (2013) 8(2) *Environmental Research Letters* 1.

73 Robin Gregory, Dan Ohlson and Joe Arvai, 'Deconstructing Adaptive Management: Criteria for Applications to Environmental Management' (2006) 16(6) *Ecological Applications* 2411, 2412.

74 Raymond Beverton and Sidney Holt, *On the Dynamics of Exploited Fish Populations* (Chapman and Hall, 1957).

75 Crawford Stanley Holling, *Adaptive Environmental Assessment and Management* (Wiley, 1978).

76 Carl Walters and Ray Hilborn, 'Ecological Optimization and Adaptive Management' (1987) 9 *Annual Review of Ecology and Systematics* 157.

77 Carl Walters, *Adaptive Management of Renewable Resources* (The Blackburn Press, 2002).

78 National Research Council, *Adaptive Management for Water Resources Project Planning* (The National Academies Press, 2004) 19.

79 Robin Gregory, Dan Ohlson and Joe Arvai, 'Deconstructing Adaptive Management: Criteria for Applications to Environmental Management' (2006) 16(6) *Ecological Applications* 2411, 2412. Carl Walters, *Adaptive Management of Renewable Resources* (The Blackburn Press, 2002).

80 Kai N. Lee, *Compass and Gyroscope: Integrating Science and Politics for the Environment* (Island Press, 1993).

81 Byron Williams, 'Adaptive Management of Natural Resources – Framework and Issues' (2011) 92 *Journal of Environmental Management* 1346, 1356.

82 Ibid.

83 Carl J. Walters and Crawford Stanley Holling, 'Large-scale Management Experiments and Learning by Doing' (1990) 71(6) *Ecology* 2060; Carl Walters, *Adaptive Management of Renewable Resources* (The Blackburn Press, 2002); United States Department of Agriculture (USDA), *Adaptive Management of Natural Resources: Theory, Concepts, and Management Institutions* (USDA, 2005) <https://www.wrrb.ca/sites/default/files/18.%20Stankey%20Adaptive%20Management%20PNW.pdf>.

84 Ibid.

85 Ibid.

86 Nicola Swayne, 'Regulating Coal Seam Gas in Queensland: Lessons in an Adaptive Environmental Management Approach?' (2012) 29(2) *Environmental and Planning Law Journal* 163, 165.

87 Glenys Jones, 'The Adaptive Management System for the Tasmanian Wilderness World Heritage Area – Linking Management Planning with Effectiveness Evaluation' in Catherine Allan and George Stankey (eds), *Adaptive Environmental Management* (Springer, Netherlands, 2009) 227, 380.

88 Robert Argent, 'Components of Adaptive Management' in Catherine Allan and George Stankey (eds), *Adaptive Environmental Management* (Springer, 2009) 26.

89 Robin Gregory, Dan Ohlson and Joe Arvai, 'Deconstructing Adaptive Management: Criteria for Applications to Environmental Management' (2006) 16(6) *Ecological Applications* 2411, 2433.

90 Catherine Allan and George Stankey (eds), *Adaptive Environmental Management* (Springer, Netherlands, 2009) 20.

91 Nicola Swayne, 'Regulating Coal Seam Gas in Queensland: Lessons in an Adaptive Environmental Management Approach?' (2012) 29(2) *Environmental and Planning Law Journal* 163, 166.

92 Robert Argent, 'Components of Adaptive Management' in Catherine Allan and George Stankey (eds), *Adaptive Environmental Management* (Springer, 2009) 26.

93 Nicola Swayne, 'Regulating Coal Seam Gas in Queensland: Lessons in an Adaptive Environmental Management Approach?' (2012) 29(2) *Environmental and Planning Law Journal* 163, 166.

94 Queensland Government, *Adaptive Management* (Queensland Department of Heritage Protection, 2017) <https://www.ehp.qld.gov.au/management/non-mining/adaptive-management.html>.

95 Ibid.

96 Tina Hunter, 'The Development of Shale Gas and Coal Bed Methane in Australia: Best Practice for International Jurisdictions?' (2016) 38(2) *Houston Journal of International Law* 367, 370.

97 Report of the United Nations Conference on the Human Environment, *Rio Declaration on Environment and Development*, UN Doc A/CONF,151/26, (vol. I) / 31 ILM 874 (1992) <https://cil.nus.edu.sg/rp/il/pdf/1992%20Rio%20Declaration%20on%20Environment%20and%20Development-pdf.pdf>.

98 John Paterson, 'Sustainable Development, Sustainable Decisions and the Precautionary Principle' (2007) 42(3) *Natural Hazards* 515, 516.

99 Arie Freiberg, *Regulation in Australia* (The Federation Press, 2017) 142.

100 Lee Godden and Jacqueline Peel, *Environmental Law: Scientific, Policy and Regulatory Dimensions* (Oxford University Press, 2010) 61.

101 Ronnie Harding and Elizabeth Fisher, 'Introducing the Precautionary Principle' in R. Harding and E. Fisher (eds), *Perspectives on the Precautionary Principle* (The Federation Press, 1999) 3.

102 Report of the United Nations Conference on the Human Environment, *Rio Declaration on Environment and Development*, UN Doc A/CONF,151/26, (vol. I) / 31 ILM 874 (1992) <https://cil.nus.edu.sg/rp/il/pdf/1992%20Rio%20Declaration%20on%20Environment%20and%20Development-pdf.pdf>.

103 Lee Godden and Jacqueline Peel, *Environmental Law: Scientific, Policy and Regulatory Dimensions* (Oxford University Press, 2010) 61.

104 United Nations General Assembly, *World Charter for Nature*, GA Res 37/7, 48th sess.

105 United Nations General Assembly, *United Nations Framework Convention on Climate Change 1992*, UN Doc FCCC/INFORMAL/84.

106 *Protocol to the 1979 Convention on Long Range Trans Boundary Air Pollution on Further Reduction of Sulphur Emissions*, opened for signature 14 June 1994, UN Doc GE.94.31969 (entered into force 2 September 1987).

107 *Montreal Protocol on Substances that Deplete the Ozone Layer*, opened for signature 19 September 1987, 1522 UNTS 29 (entered into force 1 January 1989).

108 Harald Hohmann, *Precautionary Legal Duties and Principles of Modern International Environmental Law: The Precautionary Principle: International Environmental Law Between Exploitation and Protection* (Springer, 1994).

109 James Cameron and Juli Abouchar, 'The Precautionary Principle: A Fundamental Principle of Law and Policy for the Protection of the Global Environment' (1991) 14 *Boston College International and Comparative Law Review* 1.

110 D. Resnik, 'Is the Precautionary Principle Unscientific?' (2003) 34 *Studies in History and Philosophy of Science* 329.

111 Bernd M.J. van der Meulen and Menno van der Velde, *European Food Law Handbook* (Wageningen Academic Publishers, 2008) 322.

112 Warwick Gullet, *The Precautionary Principle in Australia: Policy, Law and Potential Precautionary EIAs* (University of Wollongong Research Online, 2000) <http://ro.uow.edu.au/cgi/viewcontent.cgi?article=1138&context=lawpapers>.

113 The French Constitutional Act of 1 March 2005, relating to the 2004 Environmental Charter Art 5.

114 Ronnie Harding, *Environmental Decision-Making* (The Federation Press, 1998).

115 Nicholas de Sadeleer, *Environmental Principles – From Political Slogans to Legal Rules* (Oxford University Press, 2002).

116 UK Biodiversity Action Plan, Department of the Environment (1994) (para. 6.8) <http://jncc.defra.gov.uk/PDF/UKBAP_Action-Plan-1994.pdf>.

117 Nicholas de Sadeleer, *Environmental Principles – From Political Slogans to Legal Rules* (Oxford University Press, 2002) 293.

118 Adrian Deville and Ronnie Harding, *Applying the Precautionary Principle* (Federation Press, 1997) 52.

119 New York State Department of Environmental Conservation (NYSDEC), *Final Supplemental Generic Environmental Impact Statement on the Oil, Gas and Solution Mining Regulatory Program: Findings Statement* (NYSDEC, 2015) 23.

120 Ibid, 175.

121 Brent F. Nelsen, *The State Offshore: Petroleum, Politics, and State Intervention on the British and Norwegian Conventional Shelves* (Praeger Frederick, 1991) 8.

122 Landholders being a generic term encompassing and ranging from private land-holders or rural collectives.

123 See s 17 *Petroleum and Geothermal Energy Resources Act 1967* (WA).

124 *Land Access Code 2016* (Qld) pt 1.

125 As explored within Chapter 1 of this book, which examines the *Land Access Code 2016* (Qld) in detail.

126 Refer to Box 11, section 5 of the 13th Five Year Plan for goals relating to shale gas exploration and development <http://en.ndrc.gov.cn/newsrelease/201612/P020161207645765233498.pdf.>

127 *The Australian Macquarie Dictionary* (2018, online).

128 Further, according to Vivoda, a statist approach 'emphasizes state control of resources and favours a major role by the government in sponsoring energy-related activities, such as support for specific energy sources, direct participation in domestic energy production, and equity participation in the upstream sectors of producer states' and is typically exemplified by States that have suffered an energy crisis, such as the oil crises in 1973 severely effecting Japanese policymakers, to secure access to energy imports at reasonable prices as a net energy-importing country. See Vlado Vivoda, 'State–Market Interaction in Hydrocarbon Sector: The Cases of Australia and Japan' in Andrei Belyi and Kim Talus (eds), *States and Markets in Hydrocarbon Sectors* (Palgrave Macmillan, 2015) 240, 241–242.

129 *Permanent Sovereignty over Natural Resources*, GA Res 1803 (XVII), UN GAOR, 17th sess, 1194th plen mtg, UN Doc A/RES/1803(XVII) (14 December 1962) Art I, 1.

130 Ibid, Art I, 4.

131 International Energy Agency (IEA), *Energy Policies of IEA Countries – Australia 2018 Review* (IEA, 2018).

132 John Vogler and Hannes R. Stephan, 'Governance Dimensions of Climate and Energy Security' in Julia Maria Trombetta and Julia Dyer (eds), *International Handbook of Energy Security* (Edward Edgar, 2013) 297, 299.

Part II
Socio-regulatory approaches in a comparative context

5 Queensland, Australia

Introduction

Australia is a top global supplier of liquefied natural gas (LNG), experiencing a 50 per cent rise in the volume of LNG produced since 2015. To meet the seemingly insatiable energy appetite of Japan, South Korea and China, it has exported 45 million tonnes of LNG to Asia in 2016 alone.[1] Indeed, the size of Australia's onshore unconventional gas resources, in combination with its conventional offshore gas reserves, has seen it move from being the world's sixth-largest LNG exporter in 2013 to the second-largest LNG exporter as of June 2017.[2]

Whilst Australia's offshore conventional reserves have been available for development for decades, the potential of onshore unconventional gas reserves have only been realised in the last decade. There is a clear division in the location of unconventional gas resources in Australia, with CSG dominating the east coast (particularly Queensland and New South Wales (NSW)), and shale gas dominating the central and western areas of Australia. To date only the CSG reserves[3] of Queensland have been commercially exploited, and will therefore be the focus of this chapter. More than 90 per cent of Australia's CSG reserves (approximately 42,000 petajoules) are located in Queensland's agriculturally productive Bowen and Surat Basins, while smaller reserves are located in the Clarence–Moreton, Gunnedah, Gloucester and Sydney Basins in NSW.[4] All of Queensland's unconventional gas resources are located on privately held agricultural land in the Darling Downs agricultural region.[5]

The development of the CSG industry in Queensland has created an intense public debate, centred on the coexistence of agricultural land and unconventional gas extraction. It has thrown into sharp relief the role of the State in creating effective natural resource regulation, particularly in relation to the interconnection between economic development, land use and conflicting interests with unconventional gas exploitation. It has also raised the question of what the appropriate role a State should take in the development of its resources. Should it be minimal, regulatory or participatory intervention?[6]

To answer this question, this chapter analyses the current agricultural land protection and petroleum regulatory frameworks of Queensland, which represents an example of adaptive management with its incumbent less protective measures for agricultural land and landholders during CSG activities. Further, it addresses attempts at natural resource governance in Queensland through the Gasfields Commission and Land Access Code, analysing the socio-political context of farmer land access and compensation agreements in Queensland.

The Australian legal framework regulating coal seam gas

Historically a colony of the UK, the Australian legal system is derived from English common law, drawing heavily on English constitutional law, statute and case-based precedent. Until 1985, the Australian states continued to be influenced by the UK, which retained legal sovereignty over the Australian colonies and extended legislation enacted over the colonies so that the colonial parliaments could not legislate inconsistently.[7] Prior to 1986,[8] the legislation of the Australian colonies was subject to the scrutiny of the UK Government and the Judicial Committee of the Privy Council functioned as the final court of appeal from the Supreme Court of each Australian colony until 1988.[9]

As a consequence of the Federation of six Australian colonies, as well as the creation of two territories, Australia currently comprises nine jurisdictions, consisting of six states – New South Wales, Queensland, South Australia, Tasmania, Victoria and Western Australia – two territories – the Northern Territory and the Australian Capital Territory – and the overarching federal legal system. Each of the states has its own Constitution concerned with the powers, structure and process of the state's institutions.[10] The reception of English law into the Australian legal context means that the concept of Royal Prerogative and Crown Reservation of minerals is also imbued within the legal framework relating to petroleum resources in Queensland.[11]

In Queensland, there is Crown Reservation over mineral resources and the fee simple titleholder is required to grant access to petroleum titleholders. Unconventional gas development has a dispersed geospatial footprint, as a result of the distribution and geology of shale reservoirs, and a high environmental impact due to the need to undertake hydraulic fracturing to extract the gas. In addition, the commercialisation of LNG for export and the concurrent development of pipelines and ports in the region have had a significant impact on local communities, including economic impacts and changes to demographics and social structures. This has had flow-on effects, evident in measures of community wellbeing.[12] Land access and land use in the exploitation of unconventional gas in Queensland's agricultural regions are significant regulatory issues. However, in contrast to coal mining, where the land to be mined is purchased and all other activities halted due to the large-scale strip mining, existing land uses continue to coexist with unconventional gas extraction — that is, unconventional gas wells are being drilled on active farms and grazing properties. The overlap of activities exposes a larger number of people to the social and economic impacts of the resource development.

The regulation of unconventional gas activities and its impact on agricultural land use has been one of the most pressing concerns for lawmakers, including those in Australia. Queensland's use of adaptive management as a regulatory approach has led to significant legislative reform since the early 2000s, as a consequence of greater understanding of the consequences of CSG extraction and its impact on agricultural land and its landholders.[13] The selection of adaptive management as a regulatory approach in Queensland's unconventional gas industry has created a regulatory environment that requires constant, additional regulations, quasi-regulations and guidance notes. This has resulted in a regulatory landscape where the complexity creates 'unintended consequences and perverse incentives as the original outcomes are buried under sedimentary layers of fresh red tape'.[14]

Because of its traditional application as an environmental management system, scholars do not classify adaptive management as a theoretically accepted regulatory approach.[15] Therefore, the misinterpretation of adaptive management in guiding regulation can create adverse, irreversible effects on agricultural activities. This is due to the substantial lag time that exists between the undertaking of CSG extraction and the observing/measuring of the consequential adverse effects of the rule-based regulation utilised. Such an approach is very much reactive, as the State responds to regulatory issues as they occur, rather than trying to anticipate and legislate for problems prior to the activity taking place. For example, unconventional gas exploration and production licensing processes do not have a single overarching regulatory oversight body or agency in Queensland to provide support for individual proponents to navigate the major approvals required from 12 different state and federal government agencies.[16] This lack of an overarching coordinating mechanism results in 'uncertainty, complexity, time delays and costs for proponents'.[17]

As illustrated in Figure 5.1, the unconventional gas regulatory framework of Queensland primarily consists of the *Petroleum and Gas (Production and Safety) Act 2004* (Qld) (PGPSA) and *Petroleum Act 1923* (Qld) (PA Act) and its subordinate regulation – the Petroleum and Gas (Production and Safety) Regulation 2007 (Qld) and its associated guidelines and codes of practice regulating petroleum activities. Environmental protection obligations relating to unconventional gas resources (UGR) are imposed through the *Environmental Protection Act 1994* (Qld) (EPA) stipulating that petroleum tenures may only be granted once the petroleum license holder has received an EA categorised as 'environmentally relevant' activities under the EPA. Agricultural land access regulation is governed primarily through the *Regional Planning Interests Act 2014* (Qld) (RPIA) and Regional Interests Development Approvals (RIDA) and the Land Access Code (LAC) established by the PGPSA. Finally, water management activities are regulated by the PGPSA in relation to hydraulic fracturing and *Water Act 2000* (Qld) (WA) for UGR activities affecting underground and aquifer systems.

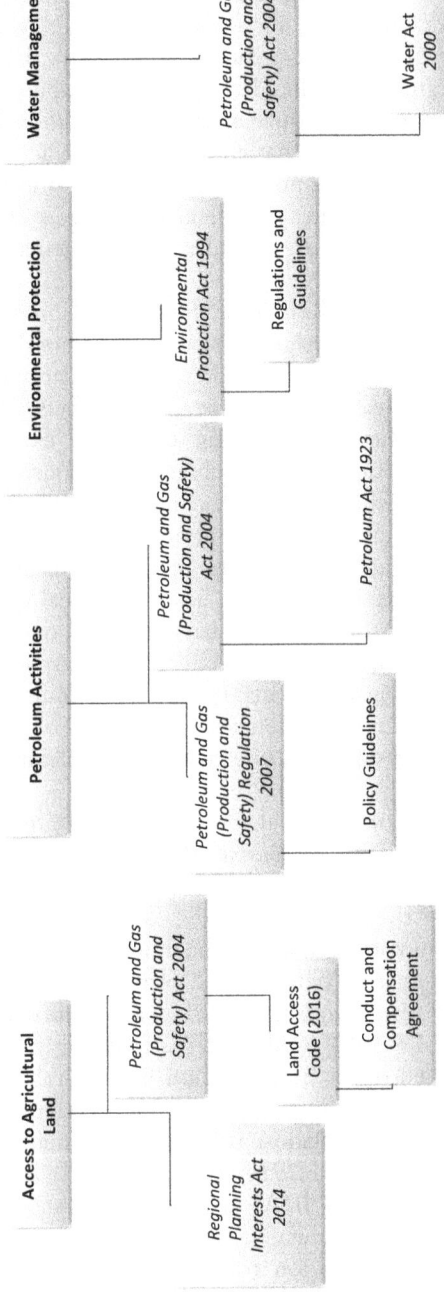

Figure 5.1 **Overarching unconventional gas regulatory framework of Queensland**
Source: Compiled by author.

In Queensland, petroleum titleholders must negotiate land access agreements with landholders, including Indigenous Native Titleholders,[18] as regulated by the LAC and the PGPSA. In Queensland, an authority to prospect is a type of personal property right[19] rather an interest *in rem*[20] while it remains in force. This authorises the titleholder to explore exclusively for petroleum; to carry out such operations and execute such works as are necessary for that purpose in the title area.[21] Unlike other Australian states and territories, Queensland does not limit the area over which the title is renewed or the prescribed portion of the area over which the exploration title was originally granted. An annual expenditure requirement for drilling and operations must be conducted 'in a good and skilful manner in accordance with recognised and approved methods and practice to the satisfaction of the Minister'.[22] The holder of a lease must take all reasonable precautions to prevent 'waste of petroleum; and must carry out all reasonable directions of the Minister specifically regarding prevention of waste; and generally, regarding methods of operation'.[23]

The *Queensland Gas Supply and Demand Action Plan Discussion Paper* states that the current unconventional gas policy is to maximise Queensland's unconventional gas resource potential and create a strong export market. The policy's intent is to address and balance the needs of landholders, local communities and traditional owners while ensuring environmental safeguards are maintained.[24] The aspirational policy goals to be met by 2025 include Queensland becoming a 'best-practice leader in environmental and social performance, and an important contributor to local community wellbeing; highly attractive to domestic and foreign direct investment; typified by a high level of innovation and collaboration and actively exploring frontier/greenfield basis'.[25] Further, the vision for Queensland's unconventional gas sector is to 'maximise its potential and be internationally competitive, balancing the needs of landholders, local communities and traditional owners while ensuring environmental safeguards are maintained'.[26] It is significant that the policy framework includes aspects of a principles-based regulatory framework mentioning 'collaboration', 'community' and 'best practice leadership'.[27] Both the petroleum and agricultural sectors are of critical importance in Queensland and are often managed in conjunction, frequently with conflicting policy objectives to either preserve and productively farm land or exploit it.

An adaptive approach to regulating land use conflict: Protecting priority agricultural land

In Queensland, following the review of the repealed *Strategic Cropping Land Act 2011* (Qld) (SCL Act) and the *Land Access Review Implementation Report* (2013),[28] the RPIA regime was enacted.[29] The RPIA represents the regulatory response aiming to promote coexistence between multiple land zoning regulations by creating 'areas of regional interest'.[30] These areas of interest create a framework for the zoning of multiple land uses including agriculture, communities and the environment. As stated in the RPIA, its purpose is to:

(a)   identify areas of Queensland that are of regional interest because they contribute, or are likely to contribute, to Queensland's economic, social and environmental prosperity;

(b)   give effect to the policies about matters of State interest stated in regional plans; and

(c)   manage, including in ways identified in regional plans—

   (i)   the impact of resource activities and other regulated activities on areas of regional interest, and

   (ii)   the coexistence, in areas of regional interest, of resource activities and other regulated activities with other activities, including, for example, highly productive agricultural activities.[31]

To achieve this purpose, the RPIA aims to provide a transparent and accountable process for the impact of proposed resource activities on areas of regional interest.[32] The RPIA and the Regional Planning Interests Regulation 2014 (Qld) (RPI Reg) aims to deliver a 'responsive adaptive management'[33] regulatory framework for the protection of agricultural land and the cumulative impact of unconventional gas extraction. It prescribes a new approvals process in the creation of RIDAs for 'resource activities' and other 'regulated activities' that are carried out in 'areas of regional interest', unless the person holds or is acting under RIDA. Carrying out a 'regulated activity' is defined as likely to 'have a widespread and irreversible impact on the area of regional interest and prescribed under a regulation[34] for the area'.[35] In contrast, a 'resource activity' is defined as an 'activity for which a resource authority is required to lawfully carry out … or an authorised activity for the authority or proposed authority … under the relevant resources Act'.[36] Two of the mentioned 'Resource Acts' are the PGPSA and the PA Act, both regulating CSG activities.[37] Further, a resource authority is stated as including a petroleum prospecting licence, a petroleum lease, a pipeline licence and a petroleum facility licence.[38]

   The RPIA acknowledges the use of regional plans to promote coexistence of resource activities in areas of regional interest. The RPI Reg framework identifies and protects areas of regional interests to ensure a balance between protecting 'priority land uses' (such as farming on highly fertile land) and supporting diverse economic development.[39] The four areas of regional interests are Priority Agricultural Areas (PAAs), Priority Living Areas (PLAs), Strategic Cropping Areas (SCAs) and strategic environmental areas.[40] PAAs are defined as areas used for a priority agricultural land and either shown in a regional zoning plan as a priority agricultural land area or prescribed under a regulation.[41] A priority agricultural land use is 'highly productive agriculture of a type identified in a regional plan for an area of regional interest or of a type prescribed under a regulation for an area of regional interest'.[42]

   A petroleum activity on a PAA is exempt from needing a RIDA where the resource activity applicant:

a) enters agreement of the land owner and the activity is not likely to have a significant impact on the priority agricultural area or area that is in the strategic cropping area and the activity is not likely to have an impact on land owned by a person other than the land owner;
b) the activity is carried out for less than 1 year; or
c) the activity is pre-existing before the RPIA was introduced in 2014.[43]

A petroleum survey licence, a data acquisition authority or a water monitoring authority under the PGPSA is exempt from the RPIA and the RIDA approvals regime as it is not defined as a 'resource activity'.[44] A resource activity is defined as having an impact on the PAA if the resource activity has an impact on the suitability of the land to be used for a PAA or, in the case of SCL, if a resource activity has an impact on those factors (such as the land's soil, climate and landscape features) that make the area highly suitable, or likely to be highly suitable, for cropping.[45] A significant impact is broadly defined as being an impact 'that is important, notable or of consequence, having regard to its context or intensity'.[46] For an activity to be 'likely' to have a significant impact, 'a lack of scientific certainty about the potential impacts of an activity will not, in and of itself, justify declaring the activity unlikely to have a significant impact on the area of regional interest'.[47]

This broad definition of a likely 'significant impact' requires each RIDA application to be subjectively assessed on a case-by-case basis. A subjective standard is arguably a lower threshold to prove than an objective-based definition or a land class system, such as in British Columbia where types of resource activities that do constitute a significant impact are specified. For example, Appendix I of British Columbia's ALC–OGC Delegation Agreement[48] outlines specific resource activities and facilities that are objectively permitted on Agricultural Land Reserve (ALR) lands according to objective criteria. This scheme is analysed in Chapter 6.

The RPIA also adopts and integrates the previous SCL Act policy framework of protecting SCL areas defined as 'land that is, or is likely to be, highly suitable for cropping because of a combination of the land's soil, climate and landscape features',[49] or identified by the SCL trigger map as being areas of regional interest. The Darling Downs, for example, is located within the Eastern Darling Downs SCL Zone and the Western SCL Zone extends to Roma.[50] Both of these zones are located over the Surat Basin. If an SCL area is situated within a PAA, then the PAA criteria and RIDA application procedure apply. That is, whether the SCA criteria are met or not is irrelevant in deciding that part of the application where the overlap occurs. However, the SCA criteria must be met for all areas where no overlap occurs.[51]

Schedule 2 of the RPI Reg provides criteria for assessment or decisions of a proposed resource activity (such as CSG extraction) in the SCL areas.[52] The criteria require an RIDA applicant to identify whether an activity will have a 'permanent impact' on SCL. An activity has a 'permanent impact' on SCL if, when the activity is carried out, the land cannot be restored to its pre-activity condition. Pre-activity condition is defined as 'the condition of the land's soil as identified and analysed within one year before the making of a RIDA application to be carried out on the land'.[53] Therefore, the SCL requirement for restoration of lands to their pre-activity condition is higher than that of a PAA likely significant impact as 'restoring the land means that the land is not only returned to its pre-activity use but that it is also returned to its pre-activity productive capacity or potential productive capacity'.[54]

As unconventional gas resource facilities and wells are classified as temporary infrastructure,[55] it is not likely that unconventional gas activities will have an evident permanent impact on SCL or significant impact on PAA land immediately. This is because the well itself may be remediated and the soil rehabilitated, but the flow back of produced water could arguably create permanent impacts to surface and groundwater systems if not adequately treated. In a study by Al-Ibrahim, Strezov, Davies and Wright, water found downstream of unconventional gas activities pointed to high metal content including aluminium, iron, manganese, nickel and zinc.[56] This would likely impact future farm soil and cropping viability of agricultural lands reliant on surface and underground water aquifers. Therefore, even though unconventional gas wells are temporary in nature, the impact on underground aquifers is long term and permanent.

The other significant exemption of the RPIA is that of pre-existing activities. Resource activities are exempt from requiring a RIDA where the resource authority for the petroleum activity was issued or granted before 30 January 2012.[57] Prior to 2012, the majority of unconventional gas activities and LNG export contracts had already commenced in some agricultural areas of Queensland, for example, in the Darling Downs.[58] The Darling Downs is recognised as the 'food bowl' of Queensland, accounting for an estimated quarter of the state's agricultural production. This fertile black volcanic soil region lies above some substantial unconventional gas basins – the Surat, Bowen and Clarence–Moreton. The Surat Basin is a geological basin that extends across an area of 270,000 km^2, with two thirds of the Basin occupying a large part of southeast Queensland and the remainder in northern NSW. The communities in this region are situated above the Great Artesian Basin (GAB), the world's largest and deepest artesian basin. These communities rely on the GAB for access to subsurface water for agricultural activities due to the arid nature of the land and low rainfall. The RPIA includes regional plans that address a broad range of land use regulation and zoning, of which agriculture is one use among an array of other equally important planning land uses. For example, PLAs are identified through regional planning governed by the *Planning Act 2016* (Qld)[59] and include the existing settlement area of a city, town or other community that is to be protected for the future growth of the existing settled area.[60]

The RIDA process seems to provide a lack of detail and enforceability regarding Gasfields Commission (GC) advice to the chief executive, the transparency of RIDA applications, and reasoning for development approvals.[61] Enforceability of GC advice may improve the strength of internal processes of the RIDA system to establish an independent and transparent regulatory process in a legally enforceable framework that may also provide agricultural community assurance.[62] Transparency regarding the RIDA application process by creating mandatory public disclosure processes for proposed resource activity approvals on PAAs may also assist with providing the unconventional gas resource sector with certainty and predictability. Further, regulatory complexity and duplication, as evident in the three differing governmental agencies[63] with oversight over the RPIA regime, can create an environment for 'creative compliance'[64] as unconventional gas operators operate in an uncertain regulatory environment. A lack of objective standards and appropriate measurement criteria to determine whether objectives and standards have been met encourages regulatory gaps and the operation of ineffective regulatory tools.

An example of the regulatory burden found in the RPIA is the relationship between the chief executive and the GC. Pursuant to s 46 of the RPIA, the chief executive is required to seek advice from the GC, Queensland's administrative unconventional gas resource oversight body, about an assessment application if the application relates

i)   to an activity for which a resource authority is required and where the activity is proposed,

ii)   in a PAA, a SCA or a PLA and either the application is modifiable or,

iii) in the chief executive's opinion, the expected surface impacts of the resource activity are significant.[65]

As Queensland's RPI regulatory framework is based on a passive adaptive management regulatory approach, the RPIA is a highly unusual system of land use regulation, illustrated by its broad RIDA approval scheme for 'areas of regional interests' governed as a purely planning instrument – yet managing competing and differing land use zoning types. It provides broad, uninterrupted rules and definitions without interpretation, such as 'coexistence'[66] and 'significant impact'.[67] This leads to uncertainty in the interpretation of its legislative objectives. This prescriptive regulatory framework for resource activities on PAAs and SCLs has led to unnecessary regulatory gaps and duplication. As a result, there have been multiple regulatory reviews, confusion and uncertainty for agricultural landholders in Queensland and unnecessarily complex regulatory processes for the petroleum industry. It is important to note the RPIA is just one regulatory framework relating to land use coexistence in Queensland. While the RPIA is administered by the Department of Planning and Infrastructure, the GC is enacted via the *Gasfields Commission Act 2013* (Qld) (GCA),[68] and the GC Review was carried out under the portfolio of the Queensland Department of State Development.[69] The land access laws operating between agricultural landholders and oil and gas companies are administrated by the Queensland Department of National Resources and Mines based on the LAC, established under the PGPSA, and the CSG Compliance Unit.[70]

Water use and access

The PGPSA permits petroleum titleholders to take or interfere with underground water in the area of the tenure, if the taking or interference happens during the course of, or results from, the carrying out of another authorised activity for the tenure. There is no limit to the volume of water that may be taken by the petroleum titleholder[71] and underground water taken by a petroleum titleholder is deemed 'associated water', which may be used for any purpose and within or outside the area of tenure.[72] The allowance of groundwater access and use by a titleholder is because 'water is a by-product and is not used directly in the resource extraction process'.[73]

When using or accessing groundwater, petroleum titleholders have an obligation to comply with the underground water management framework under the *Water Act 2000* (Qld) (WA). The Office of Groundwater Impact Assessment is established by the WA and regulates unconventional gas water management issues. These issues include those relating to hydraulic fracturing fluids and cumulative groundwater impacts via 'make good' arrangements.[74] A make good measure for a water bore includes:

a)   ensuring the bore owner has access to a reasonable quantity and quality of water for the bore's authorised use or purpose, or

b)   carrying out a plan to monitor the bore, including, for example, by undertaking periodic bore assessments, or

c)   giving the bore owner monetary or non-monetary compensation for the bore's impaired capacity.[75]

The CSG Compliance Unit is administered via the Queensland Department of Natural Resources and Mines, with the aim of responding to landholders' complaints and to coordinate landholder groundwater monitoring by landholders themselves. This monitoring occurs through the online tool 'CSG Net'[76] – which is a reporting unit rather than a unit dedicated to overseeing complaints. The role of the GC, in comparison with its counterpart in British Columbia (the OGC) warrants further comparative analysis to determine whether amendments to the GC operation could deliver its desired objective of coexistence between landholders and the onshore unconventional gas industry in Queensland.[77]

Administrative authorities and regulatory oversight

The Gasfields Commission and the management of resource use conflict

The GC was enacted by the GCA as a response to conflict between agricultural land uses, landholders and unconventional gas activities in Queensland.[78] It was intended that the GC would 'manage and improve the sustainable coexistence of landholders, regional communities and the onshore gas industry in Queensland'.[79]

The GC operates as an independent oversight administrative body to facilitate complaints of landholders and advise relevant ministers in certain circumstances:

a)   facilitating better relationships between landholders, regional communities and the onshore gas industry;
b)   reviewing the effectiveness of government entities in implementing regulatory frameworks that relate to the onshore gas industry;
c)   advising Ministers and government entities about the ability of landholders, regional communities and the onshore gas industry to coexist within an identified area.[80]

Figure 5.2 provides the framework of the relationship between the RIDA and GC approvals process for unconventional gas on PAAs.

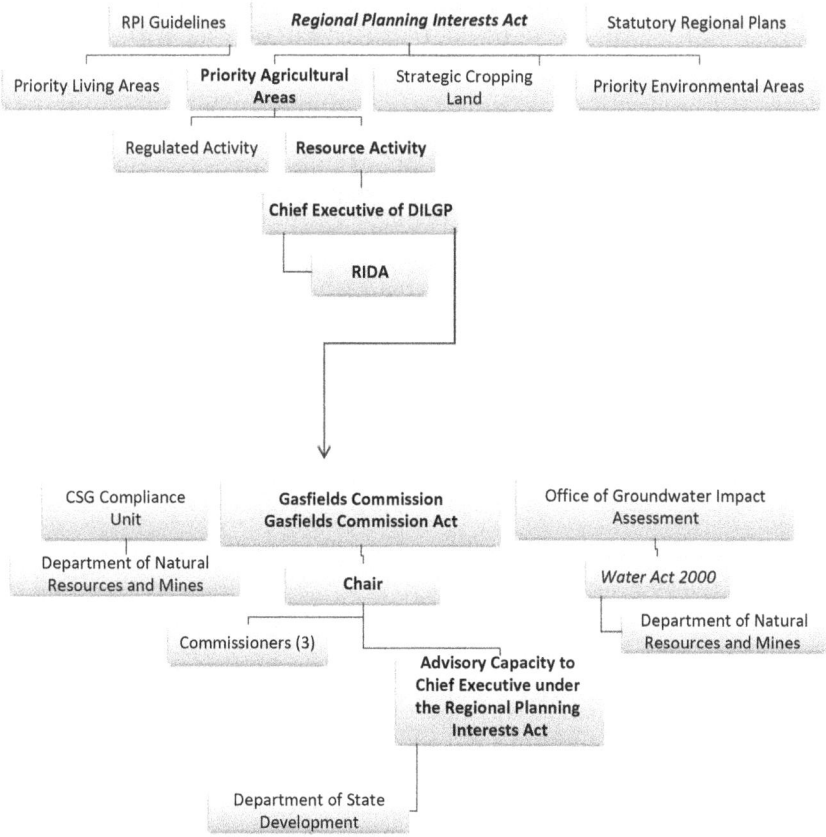

Figure 5.2 **Relationship of RIDA process and GC in Queensland**
Source: compiled by author.

The GC and Office of Groundwater Impact Assessment (OGIA) bodies hold advisory and facilitative powers only. Both bodies hold the capacity to 'advise the chief executive' on matters relating to UGR water impacts[81] or RIDA assessments.[82] Similarly, the CSG Compliance Unit aims to investigate and facilitate landholder complaints relating to CSG land access. Over two separate government departments – the Department of Natural Resources and Mines and the Department of Infrastructure, Local Government and Planning (DILGP) – the three UGR oversight bodies in Queensland do not hold regulatory powers over UGR activities. The DILGP Chief Executive holds the only law-making power in relation to approving UGR activities in PAAs. The regulatory capacity of UGR decisions remains with the Department of Natural Resources and Mines Minister in granting authorities to prospect[83] and petroleum leases by competitive tender who must then work in conjunction with the minister for the DILGP for an approval for non-excluded UGR activities in PAAs.[84]

The primary role of the GCA lies in its facilitation of relationships between landholders and unconventional gas activities. The GC consists of one Commissioner, who acts as chair, and up to six part-time Commissioners with differing portfolios including communications, policy and engagement and corporate services. The delivery of the *Independent Review of the Gasfields Commission Queensland and Associated Matters* (Independent Review) has led to a reorganisation of the GC, with Commissioners formerly having positions in the Queensland Farmers Federation[85] and Cotton Australia. However, one Commissioner remains the chief technical officer to the Australian Petroleum Production and Exploration Association (APPEA).[86]

According to the GC, it has facilitative powers and certain specific powers, such as the power to compel onshore gas companies (and their contractors) and landholders to provide the GC with information or documents required for the effective and efficient carrying out of the Commission's functions.[87] However, it is clear the GC does not hold any regulatory power nor does it adopt a role 'of being an advocate for landholders, nor of addressing individual cases, neither of which (is) required by statute, both of which were, and still are, expected by many landholders'.[88] Criticisms of the GC and its independence raised by landholders are:

- the Commission does not represent landholders but represents the CSG industry and government;
- commissioners are not members of landholder peak bodies yet one commissioner is a member of APPEA;
- some commissioners are conflicted in the discharge of their duties; and
- when issues are raised with the Commission the response is often in a noncommittal 'form' letter and advice is not forthcoming.[89]

There is a lack of understanding between the GC, its 'powers' and its tangible ability to assist landholders with conflict relating to unconventional gas activities on their agricultural land. The GC does not hold regulatory powers to investigate, mediate, arbitrate or make binding decisions concerning individual disputes between agricultural landholders and unconventional gas activities. Rather, the GCA[90] establishes an avenue for review of the policies of government entities and quasi-regulation of the unconventional gas sector. However, the other stated purposes of the GCA provide evidence of the potential advisory role of the GC, providing advice and aid to the relevant ministers and government. It does not explicitly state a built-in policy adjustment regulatory framework.

The lack of powers and advisory role of the GC creates a lack of 'regulatory teeth', and has given rise to stakeholders identifying issues of confusion about the roles and responsibilities of the GC. There is a lack of awareness about the work of the Commission behind-the-scenes, a lack of an independent and accessible source of information for landholders dealing with CSG companies and a need for an alternative way of reaching agreement when dealing with disputes about conduct and compensation, than having to resort to court.[91] Such confusion led to the creation of the Terms of Reference for an independent review into the GC and its effectiveness in relation to managing multiple interests between the unconventional gas industry, regional communities and landholders.[92]

On 18 December 2015, the Queensland Government commissioned the Independent Review. This review was managed by the Queensland Department of State Development with Robert Scott appointed as the independent reviewer. The purpose of the review, as outlined in the terms of reference, was to:

a)   evaluate whether the GC is achieving its purpose
b)   evaluate whether the functions given to the GC are sufficient to allow it to effectively manage disputes about land access and other disputes between resource companies and landholders
c)   evaluate whether the functions given to the GC should include a role in managing or facilitating responses to public health and community concerns arising from onshore gas activities
d)   investigate whether an alternative model, such as an independent Resources Ombudsman, is needed to provide a mechanism for dispute resolution between resource companies and landholders
e)   investigate whether harmonisation between the CSG Compliance Unit and the GC would provide efficiencies and improve dispute resolution between resource companies and landholders
f)   any other relevant matters the reviewer considers appropriate.[93]

On 1 December 2016, the Minister for State Development and Minister for Natural Resources and Mines announced the release of the independent report and the government response. The government's response detailed a range of measures to be adopted based on the Independent Review report, including:

•   a renewed focus by the Commission on extension and communication activities to improve the availability of information on the CSG industry particularly for landholders;

- establishing a Land Access Ombudsman to deal with disputes between landholders and resource companies in relation to conduct and compensation agreements;
- structural and operational changes to the GC that will enable it to work more effectively; and
- developing in consultation with stakeholders improved approaches to negotiation and alternative ways to resolve land access disputes.[94]

Among the Independent Review recommendations was the refocussing of the GC to 'be the trusted advisor to government and stakeholder representative bodies on strategic issues including the status of the coexistence model'.[95] The Independent Review also recommended an overhaul of the dispute resolution framework to establish arbitration as an alternative to Land Court litigation. This would provide a simpler and legally binding resolution for both parties and established an independent dispute relation body to assist with disputes about Conduct and Compensation Agreements (CCAs).[96]

Further, the harmonisation of communication between the GC and CSG Compliance Unit was recommended, seeking to ensure more effective communication and greater strategic functions. In addition, there was a recommendation that the CSG Compliance Unit be equipped with legal powers to issue penalty infringement notices for breaches of the mandatory provisions of the LAC pursuant to the PGPSA.[97] The final report noted that there were

> significant opportunities for improvement in the overall operation of the GC and in the perception of it by a large number of stakeholders particularly landholders … (including) the failure to adequately address some of its statutory functions; and the sub-optimal strategic and operational planning and reporting.[98]

There has been no substantive legislative amendment to the GCA since the Review nearly a year after its report was released.[99] It is evident that Queensland has been slow to react to the mounting criticism and concern relating to the operation and purview of the GC. The scope of the GC is to manage and facilitate complaints by landowners in relation to CSG activities. The Review noted the Commission had reported over 90 enquiries from 44 landlords between 2014 and 2015, but had failed to demonstrate how these enquiries were managed and whether they were resolved.[100]

The Land Access Code 2016 (Qld)

According to s 804 of the PGPSA, a person who carries out an authorised activity for a petroleum authority must not unreasonably interfere with anyone else carrying out a lawful activity. To ensure this duty is carried out, the LAC was introduced in 2010, to provide a 'best practice' guide for CCA negotiations and mandatory provisions concerning the conduct of authorised activities on private land. The stated intention of the LAC is to balance the interests of the agricultural and resources sectors, including through best practice guidelines for effective regulation and 'good faith' between operators and the owners/occupiers of private land.[101]

The requirement for a LAC and Land Access Agreement is set out in s 36 of the *Minerals and Energy Resources (Common Provisions) Act 2014* (Qld) (MERCPA):

> A regulation may make 1 or more codes for all Resource Acts (each a land access code) that—
>
> (a) states best practice guidelines for communication between the holders of resource authorities and owners and occupiers of land, public land authorities and public road authorities; and
> (b) imposes on resource authorities mandatory conditions concerning the conduct of authorised activities on land.[102]

A review of the previous *Land Access Code 2010* (Qld), in accordance with the Land Access Framework (LAF) review process, included comments from land-holders and companies who believed that limited outcomes were achieved through dispute resolution:

> There is little incentive for good faith negotiations or timely resolution of a dispute. Stakeholders noted that, where negotiations broke down, the process tended to drag on indefinitely, frustrating all involved and costing time and money ... landholders indicated they were generally more concerned about the conduct of resource companies on their property as it relates to their business rather than just the issue, *per se*, of compensation. In summary, stakeholders did not see the Land Court as a viable, timely and cost-effective way to resolve disputes.[103]

The LAF review also revealed that stakeholders requested further guidance as to what constitutes negotiating in 'good faith'.[104] Government-authorised officers undertaking conferences in dispute resolution were criticised as inadequately trained for mediation or other dispute resolution methodologies, limiting the value of this part of the process. A number of recommendations were received by the LAF review to 'require that only trained and accredited mediators undertake conferences and implementing an arbitration process that is binding rather than a guiding mediation that may not result in an outcome'.[105] The key criticism of alternative dispute resolution (ADR) in the land access process is that 'there is no decision making power vested in the Government officer or person holding the ADR, therefore there is no certainty of an outcome'.[106]

The Association of Mining and Exploration Companies states:

> Landholder rights relate to the use of the surface of the land. However access to those mineral rights often means infringing on the rights of the

landholder. Therefore, negotiation between the owner of the mineral rights and the landholder rights takes place such that the infringement on the right is appropriately compensated.[107]

The Queensland Resource Council noted that the previous *Land Access Code 2010* (Qld) focused on maximising compensation, rather than building effective working relationships in good faith:

> Unfortunately, a perverse outcome of Queensland's land access laws is that the land access process has become focused on maximising compensation with little priority on building effective working relationships to ensure there is a minimal impact on the landholder business or enjoyment of the land.[108]

The imbalance in bargaining position under the current state laws was also noted by the National Farmers' Federation (NFF), which stated:

> The NFF's view is that a forced negotiation, where the landholder does not have the option to refuse an agreement, is not an equal or fair negotiation. Fixed outcome negotiation provides an unfair advantage to well-resourced mining and gas companies, which employ skilled professionals to negotiate these types of agreements on a regular basis.[109]

Rarely is there agreement or accord between bodies involved in unconventional gas exploration, especially those representing as diverging interests as the Association of Mining and Exploration Companies and the NFF. However, both these bodies agree that the current regime is inadequate in representing the interests of landholders. The purpose of the updated version of the LAC is to 'balance the interests of the agricultural and resource sectors to address issues related to land access for resource exploration and development'.[110]

The LAC provides generalised guidelines for resource acts in a broad 'one size fits all' regime for tenement holders and prospective tenement holders seeking to obtain access to landowner properties for the purpose of exploration and extraction.[111] Therefore, the LAC is not formulated to manage the unique needs of agricultural landholders and CSG petroleum tenement holders. The LAC contains two key sections. The first, entitled 'Good Relations', is based on general voluntary principles for communications between parties, negotiation agreements, communication before and during the carrying out of activities and after completion of activities.[112] The second is on 'Mandatory Conditions' on activities conducted under the resource authority in Part 3 of the LAC. Mandatory Conditions of the LAC include induction training for staff and contractors; using existing access points, roads and tracks if possible on a property; minimising disturbance to people, livestock and property; taking reasonable steps to ensure there is no spread of weeds and pests; prior agreement of camp locations; collecting rubbish or waste produced in carrying out authorised activities; and closing gates, grids and fences.[113]

The LAC is limited in specific detail and guidelines to assist resource companies and landowners in resolving conflicts in a situation when land access is contested. Rather, it provides a general framework of principles for land access agreement based on general principles of 'good faith', 'adequate consultation and negotiation', 'transparency' and 'cooperation':

> Good relationships between these groups, assisted by adequate consultation and negotiation, will improve transparency, equity and cooperation across the sectors involved and creates a more level playing field for all.[114]

The requirement to negotiate in good faith encompasses 'the notion of fidelity (or faithfulness) to the bargain'.[115] As stated by Allsop P in the NSW case of *United Group Rail Services v Rail Corporation of New South Wales*:

> A promise to negotiate … genuinely and in good faith with a view to resolving claims to entitlement by reference to a known body of rights and obligations, in a manner that respects the respective contractual rights of the parties, giving due allowance for honest and genuinely held views about those pre-existing rights is not vague, illusory or uncertain.[116]

There are certain inherent difficulties when discussing the concept of 'good faith', which relate primarily to the role of the State. On one hand, the State is owner of the public asset, in this case the unconventional gas resources. It is also the grantor of the interest in land to a petroleum titleholder and recipient of royalties that this activity will produce. On the other hand, the State also has an obligation to represent the interest of private landholders under whose land the unconventional gas asset is situated. This potential conflict of interest has been observed in numerous qualitative studies[117] and the National Office of the Chief Economist's *Review of the Socioeconomic Impacts of Coal Seam Gas in Queensland*.[118] This has resulted in the recommendation that 'Regulation of the [unconventional gas] sector should support coexistence, including ensuring that the landholder's agreement is sought for access to their property, that landholders are fairly compensated, and that prime agricultural land and water resources are not compromised by development activity'.[119]

The aim of providing fairness for landholders may provide the impetus to establish an independent regulatory tool, to provide assistance and appellant applications for landholders to navigate the 'good faith' requirement in this difficult and complex regulatory regime.[120] This is proposed to manage the generality and scarcity of obligations within the regulatory framework, as well as the lack of significant conduct standards for the parties. One option is to adopt a principles-based, independent regulatory tool to address any disadvantages facing landholders that currently exist under the MERCPA, which may otherwise be construed as disproportionately favouring petroleum titleholders.

The framework to manage negotiation of CCAs is limited to a paragraph in the LAC:

> Agreements between the landholder and holders should clearly articulate what has been agreed to between the parties and comply with the relevant resource Acts. In the course of negotiations, the parties should endeavour to stay in regular contact and work together to reach a mutually acceptable and practical agreement.[121]

Resource companies must 'minimise disturbance' to people and a landholder's livestock and property, although no further guidance is given as to what constitutes 'minimal disturbance'. The LAC also states:

1) If, in carrying out authorised activities, a relevant person becomes aware of any potential adverse impact, caused by the activities, on a landholder's livestock or property, the relevant person must immediately notify the landholder of the potential impact;
2) If a relevant person injures or kills a landholder's livestock, the relevant person must immediately notify the landholder of the injury or death of the livestock; and
3) If a relevant person damages a landholder's property, the relevant person must—

> a) Immediately notify the landholder of the damage; and
> b) Repair the damage as soon as practicable.[122]

Therefore, a resource holder must 'immediately notify' a landholder where the activity holds a potential adverse impact, rather than prohibiting the authorised activities to take place, where adverse impacts are likely to affect people, livestock or property. A resource holder must only immediately notify the landholder of the injury or death of livestock, rather than providing adequate compensation for damage caused. This is of significant concern for agricultural landholders, particularly dairy and other husbandry farmers. Further, if damage is inflicted to landholder's property, the resource holder must immediately notify and repair the damage. Guidance is neither given on the extent of repair required, nor on whether rehabilitation must take place to ensure the landholder's property is repaired to its original state prior to resource activities. Moreover, there is also no definition of 'as soon as practicable'.[123]

The LAC is a generalised approach to land access as it applies to all types of resource companies and, therefore, does not consider the specific environment and contested nature of land access agreements with CSG petroleum tenement holders. Land access has historically been the subject of dispute between resource companies and agricultural landowners. As stated by Nader QC:

If the law is to proceed on the basis that it does now, namely no agreement then arbitration, this thing is not going to make any difference to it ....The hard, cold bottom line is still what it always has been. As long as the act contains these arbitration clauses, farmers are virtually at the mercy of the miners.[124]

Although the LAC provides a unique and aiding approach for land access agreements in providing 'best practice' guidelines for petroleum titleholders, it does not place any statutory requirements on landholders.[125] During the *Land Access Review Report of 2012*, it was recommended that more stringent requiring on obligations on landholders and petroleum titleholders is introduced in relation to timely responses to notification requests and defining 'good faith' negotiations.[126]

The general voluntary principles found in Part 2 of the LAC concerning good relations between parties, including advising the landholder of any significant changes to operations or timing, promptly paying compensation agreed with the landholder and being responsible for all authorised activities and actions being undertaken by employee. This is intended to ensure mandatory regulations concerning negotiation procedure will be enforced.[127] Although the LAC is based on achieving transparent and effective land access agreements, the lack of enforcement and compliance via mandatory provisions create a lack of enforceability by government agencies. Further, petroleum titleholders are not adequately informed of their obligations in accessing land and how to ensure compliance with LAC mandatory provisions as there is no enforcement of its provisions.

Conduct and Compensation Agreements for affected landholders

Where CSG exploration and extraction activities are to be undertaken, negotiation between the petroleum titleholder and the landholder is required in a process designed to balance the interests of the stakeholders prior to a titleholder entering the landholder's land to undertake 'advanced activities' that have a significant impact on the landholder's use of the land.[128] The content of a CCA is partly structured through the LAC, incorporating guidelines that establish standard requirements. The negotiation of a CCA under the MERCPA is regulated under ss 84–91 of the MERCPA. It must be noted that a CCA is unenforceable to the extent of inconsistency with the PGPSA, MERCPA, or the LAC.[129] For example, the holder of an authority to prospect does not need to enter into a CCA if the authority holder already has a right to enter the land, if entry is to preserve life or property, if an emergency that exists or may exist occurs, or if the landholder is an applicant or respondent to a Land Court application which seeks to determine compensation.[130]

Where a CCA is agreed upon by both parties, the RPIA exempts this circumstance from requiring a RIDA approval. In effect, this removes CCAs from the regulatory scope of the RPIA in circumstances that cover PAAs. If parties cannot negotiate a land access agreement, either party may apply to the Land Court for review of agreed compensation where the material circumstances have changed since that date of agreement or determination.[131] The PGPSA does not restrict the terms or conditions of the CCA. However, a CCA must address the compensation liability that is owed by the holder of the petroleum authority to the landholder for any 'compensatable effect', including:

- deprivation of the possession of the land's surface;
- diminution of the land's value;
- diminution of the use made of the land or any improvement on it;
- severance of any part of the land from other parts of the land or from other land of the landowner; and
- any cost, damage or loss arising from the carrying out of authorised activities on the land.[132]

A CCA may cover both monetary and non-monetary forms of compensation,[133] as well as accounting, legal and valuation costs necessarily and reasonably incurred by the landholder in negotiating the agreement.[134]

Shortcomings in the CCA framework have also been identified in *Australia Pacific LNG Pty Ltd v Golden & Ors*.[135] The case concerned Australia Pacific LNG (APLNG) seeking of access to two properties west of the township of Wandoan for the purpose of drilling, constructing, and operating a number of petroleum wells, associated infrastructure and flow lines on each property. After unsuccessful negotiations, APLNG issued formal negotiation notices under the negotiated access and compensation regime of the PGPSA. The parties were unable to reach agreement within the set negotiation period (20 business days following the issue of the negotiation notices). Accordingly, after expiry of that period, the landholders issued an election notice, nominating for the negotiations to be referred to arbitration for resolution.[136] Muir JA granted the injunction to withhold the landholders from forcing APLNG to attend arbitration to settle on terms of the land access. However, as the parties signed a CCA after the granting of the injunction, the Court of Appeal was not required to make a determination due to the interpretation and inconsistency of provisions in the MERCPA concerning ADR.[137]

This confusion is created because the notice given under s 88 of the MERCPA apparently contemplates the parties agreeing to an ADR process, whereas the requirement under s 90 of the MERCPA is for the parties to 'use reasonable endeavours' to finish the ADR process. Parties also have and the right to apply to the Land Court under s 96 of the MERCPA and both provisions operate by reference to the expiry of a period after delivery of the election notice. Neither of the latter provisions expressly requires agreement on the ADR process to have been reached or the other party to attend. Further confusion is created by inconsistency between the intent of the legislation and the provisions seeking to give effect to that intent. As Muir JA recognised, the MERCPA appears to be directed towards requiring parties to reach a negotiated agreement (through ADR, if required) and, failing that, for the Land Court to determine compensation. However, the legislation provides an example of a form of ADR (namely arbitration) that is not directed towards facilitating negotiations, but which can instead result in a quasi-judicial determination of rights between parties, potentially in their absence, with very limited rights of appeal.[138]

In relation to compensation and land access amounts, in *C.M. Fitzgerald & Anor v Struber & Anor*,[139] Member PA Smith determined compensation in respect of mining areas at an annual rate of $10/ha per year and access areas at $5/ha per year for the current landowners. In the later case of *Eacham Abrasive Blasting Pty Ltd v Gundersen & Anor*,[140] Member Smith granted compensation sums of $10/ha per year for the area covered by mining and $5/ha per year for access in respect of the renewal of a mining lease in the Mareeba area. *Wallace & Ors v Bottomer & Ors*[141] and *Pryce v Stuber & Anor*[142] determined compensation for the mining area in question be payable at $10/ha per year in respect of the mining area and $5/ha per year in respect of the access area.

It is important to note that, with the exception of *Nothdurft*,[143] current case law concerning compensation and land access concern mineral tenements rather than petroleum tenements. To date, there is limited case law of unconventional gas CCA compensation in Queensland. However, the similarity between compensation provisions in the *Mineral Resources Act 1989* (Qld) (MRA) and PGPSA reflect the low compensation figures provided in these cases. This is instructive and demonstrates the judicial interpretation applicable to the LAF, which does not recognise the cumulative impacts to landholders and the 'aversion (to unconventional gas activities) in the rural property market'.[144]

Conclusion

In Queensland, land use associated with petroleum activities is subject to development consent requirements under the *Environmental Protection Act 1994* (Qld).[145] An authority to prospect,[146] and subsequent petroleum lease,[147] is authorised under the PGPSA. At the same time, an application for development consent under the *Planning Act 2016* (Qld) may be also be required.[148] If an unconventional gas activity falls outside the area of a petroleum lease (e.g., ancillary facilities), then it will require both an assessment under the PGPSA, RPIA and relevant local planning scheme for development approvals.[149]

The broad range of overlapping land use and planning legislation applicable to unconventional gas on agricultural land in Queensland has created a complex array of regulatory instruments and oversight bodies. This has given rise to 'substantial scope for unnecessary regulatory burden to be imposed'.[150] Such regulatory burden affects the State's ability to effectively manage, protect and rehabilitate priority agricultural lands where CSG extraction occurs. Such a regulatory framework may, as stated by the Queensland Competition Authority, 'create unnecessary compliance costs, delays or uncertainties. There are tangible deadweight economic losses associated with the poorly designed regulations'.[151]

Notes

1 Commonwealth of Australia, Department of Industry, Innovation and Science, *Gas Resources and Energy Quarterly June 2017* (2017).
2 Ibid, 20.

3   There is a difference between the exploitation of CSG reserves and that of shale gas. CSG arises from the dewatering of coal seams to release the gas contained within the coal cleats. It often does not require the use of hydraulic fracturing to stimulate the well. Conversely, shale gas arises from the hydraulic fracturing of shales to create porosity and permeability, which enables the gas to be recovered. For a more comprehensive discussion on hydraulic fracturing, prefer to Chapter 1.

4   Commonwealth of Australia, Department of Industry, Innovation and Science, *Gas Resources and Energy Quarterly June 2017* (2017), 11.

5   This land is a combination of land owned in fee simple, or leasehold land held by the landholder under a pastoral lease, typically either a perpetual pastoral lease, or a long-term lease (often up to 99 years). For more information on land ownership and Tenure in Australia refer to Brendan Edgeworth, *Butt's Land Law* (Thompson Reuters, 2017). The collective term 'landholder' will refer to those with an interest in land under which the unconventional gas is located, and may be either a landowner, or a holder of a leasehold interest.

6   A discussion on these roles is undertaken in Chapter 4.

7   *Colonial Laws Validity Act 1865* (Imp), 28 & 29 Vict, c 63, s 2.

8   See *Australia Act 1986* (Cth) and *Australia Act 1986* (UK). The passage of both Acts through the respective parliaments terminated the power of the Parliament of the United Kingdom to legislate for Australia.

9   *Judicial Committee Act 1833* (IMP).

10 *The Constitution of Queensland 2001* (Qld) is the governing constitutional act of Queensland.

11 The vesting of ownership of petroleum in the state is set out in s26 of the *Petroleum and Gas (Production and Safety) Act 2004* (Qld).

12 Tina Hunter and John Chandler, *Petroleum Law in Australia* (LexisNexis, 2013) 1.

13 The conflict arising from the impact of CSG activities is well documented in the literature. Jonathan Fulcher and Martin Klapper, 'Coal Seam Gas Exploration and Production in NSW: The New Access Argument' (2011) *The APPEA Journal* 51 (2) 688, 688; Linda Connor and Phil McManus, 'What's Mine is Mine(d): Contests over Marginalisation of Rural Life in the Upper Hunter, NSW' (2013) 22(2) *Rural Society* 166.

14 Catherine Allan, *Adaptive Management of Natural Resources*, Proceedings of the 5th Australian Stream Management Conference. Australian Rivers: Making a Difference. Charles Sturt University, Thurgoona, New South Wales 26.

15 Nicola Swayne, 'Regulating Coal Seam Gas in Queensland: Lessons in an Adaptive Management Approach?' (2012) 29(2) *Environment and Planning Law Journal* 163–185, 163. Refer to Chapter 4 for types of regulation and a discussion of adaptive management and its failings as a regulatory instrument.

16 Government departments responsible for the regulation of unconventional gas at the State level in Queensland include: The Department of Environment and Heritage Protection, Department of Natural Resources and Mines, the Department of Energy and Water Supply and the Department of Infrastructure, Local Government and Planning.

17 Queensland Department of Natural Resources and Mines, *Queensland Gas Supply and Demand Action Plan Discussion Paper* (2016) 20.

18 In the form of Indigenous Land Use Agreements (ILUAs). Indigenous and native title rights in relation to unconventional gas contestation is beyond the scope of this book.

19 *Petroleum and Gas (Production and Safety) Act 2004* (Qld) s 30.

20 Meaning in the land. *TEC Desert Pty Ltd v. Commissioner of State Revenue* [2010] HCA 49 [28–36].

21 *Petroleum and Gas (Production and Safety) Act 2004* (Qld) s 32.

22 Ibid, s 47.

23 Ibid.
24 Queensland Government, *Department of Natural Resources and Mines, Queensland Gas Supply and Demand Action Plan Discussion Paper* (November 2016) i.
25 Ibid, 10.
26 Ibid.
27 Ibid.
28 Land Access Implementation Committee, Parliament of Queensland, *Land Access Implementation Committee Report* (30 August 2013) <http://www.parliament.qld. gov.au/Documents/TableOffice/TabledPapers/2014/5414T5893.pdf>.
29 The RPI regime consists of the *Regional Planning Interests Act 2014* (Qld), Regional Planning Interests Regulation 2014 and creates and regulates 12 statutory regional zoning plans in total, namely: the ShapingSEQ: South East Queensland Regional Plan (2017); Cape York Regional Plan (2014); Central West Regional Plan (2009); Darling Downs Regional Plan (2013); Far North Queensland Regional Plan (2009); Gulf Regional Development Plan (2000) (non-statutory); Mackay, Isaac and Whitsunday Regional Plan (2012); Maranoa-Balonne Regional Plan (2009); North West Regional Plan (2010); South West Regional Plan (2009) and Wide Bay Burnett Regional Plan (2011). The RPI regime has also released 11 policy guidelines in total.
30 *Regional Planning Interests Act 2014* (Qld) s 3.
31 Ibid.
32 Ibid, s 3(2).
33 Poh-Ling Tan, David George and Maria Comino, 'Cumulative Risk Management, Coal Seam Gas, Sustainable Water, and Agriculture in Australia' (2015) 31 *International Journal of Water Resources Development* 682.
34 *Regional Planning Interests Act 2014* (Qld) s 12.
35 Ibid, s 17.
36 Ibid, s 12.
37 The *Petroleum Act 1934* (Qld) has been retained after the introduction of the PGPSA due to legal requirements pertaining to Native Title.
38 *Regional Planning Interests Act 2014* (Qld) s 13(e).
39 Ibid, sub-div 2.
40 Ibid, s 7.
41 Ibid, s 8(1).
42 Ibid, s 8(2).
43 Ibid, div 2.
44 Queensland Government, Department of Infrastructure, Local Government and Planning, *RPIA Statutory Guideline 04/14* (2017) 2.
45 *Regional Planning Interests Act 2014* (Qld) s 22.
46 Queensland Government, Department of Infrastructure, Local Government and Planning, *RPIA Statutory Guideline 02/14* (2017) 2.
47 Ibid.
48 British Columbia, Oil and Gas Commission, *ALC–OGC Delegation Agreement* (2013) <https://www.bcogc.ca/node/5759/download>.
49 *Regional Planning Interests Act 2014* (Qld) s 10(2).
50 Queensland Government, Department of Natural Resources and Mines, Strategic Cropping Land Zone Map (2017) <https://www.dnrm.qld.gov.au/__data/assets/ pdf_file/0006/171564/scl-zone-map.pdf>.
51 Queensland Government, Department of Infrastructure, Local Government and Planning, *RPIA Statuary Guideline 03/14* (2017).
52 Regional Planning Interests Regulation 2014 (Qld) Schedule 2.
53 Queensland Government, Department of Infrastructure, Local Government and Planning, *RPIA Statuary Guideline 09/14* (2017) 3.
54 Ibid, 3.

55   *Petroleum and Gas (Production and Safety) Act 2004* (Qld) s 442.
56   Ali Al-Ibrahim, Vladimir Strezov, Peter Davies and Ian Wright, 'Environmental Impact of Coal Mining and Coal Seam Gas Production on Surface Water Quality in the Sydney Basin, Australia' (2017) 189(9) *Environmental Monitoring* Assessment 408.
57   *Regional Planning Interests Act 2014* (Qld) s 99.
58   John Williams, *An Analysis of Coal Seam Gas Production and Natural Resources Management in Australia Issues and Ways Forward* (2012) <http://www.aie.org.au/AIE/Documents/Oil_Gas_121114.pdf>.
59   *Planning Act 2016* (Qld), Ch 2.
60   Queensland Department of Infrastructure, Local Government and Planning, *RPIA Statutory Guideline 04/14* <https://www.dilgp.qld.gov.au/resources/planning/planning/statutory-guideline-04-14.pdf> 3.
61   *Gasfields Commission Act 2013* (Qld) s 7(d); *Regional Planning Interests Act 2014* (Qld) s 46.
62   As established by the *Gasfields Commission Act 2013* (Qld).
63   The Department of State Development, Infrastructure and Planning; the Department of Natural Resources and Mines; the Department of Environment and Heritage Protection.
64   Robert Baldwin, *Rules and Government* (Clarendon Press, 1995); Arie Freiberg, *Regulation in Australia* (Federation Press, 2017).
65   *Regional Planning Interests Act 2014* (Qld) s 46.
66   Ibid, s 3(1)(ii).
67   Ibid, s 22(b).
68   *Gasfields Commission Act 2013* (Qld); Gasfields Commission, *About Us* (2017) <http://www.gasfieldscommissionqld.org.au/about-us/>.
69   Department of State Development, *Gasfields Commission Review* (2016) <https://www.statedevelopment.qld.gov.au/industry-development/gasfields-commission-review.html>.
70   Queensland Government, *Department of Natural Resources and Mines, Compliance and Enforcement* (2017) <https://www.ehp.qld.gov.au/management/non-mining/enforcement-compliance.html>.
71   *Petroleum and Gas (Production and Safety) Act 2004* (Qld) s 185(3).
72   Ibid, s 185.
73   Queensland Government, Department of Environment and Heritage Protection, *Underground Water* (2017) <https://www.ehp.qld.gov.au/management/non-mining/groundwater.html>.
74   *Water Act 2000* (Qld) pt 5.
75   Ibid, s 421.
76   Information relating to CSG Net see Queensland Government, Department of Natural Resources and Mines, *CSG Net and CSG Online: Monitoring CSG Impacts on Groundwater* (2017) <https://agforceprojects.org.au/file.php?id=319&open=yes>.
77   *Gasfields Commission Act 2013* (Qld) s 3.
78   Robert P. Scott, *Independent Review of the Gasfields Commission Queensland and Associated Matters* (Department of State Development (Qld), July 2016) <https://www.statedevelopment.qld.gov.au/resources/report/gasfields-commission-review-report.pdf>.
79   Ibid, 5.
80   *Gasfields Commission Act 2013* (Qld) s 7.
81   *Water Act 2000* (Qld) s 456(1)(a).
82   *Regional Planning Interests Act 2014* (Qld) s 7(d).
83   *Petroleum and Gas (Production and Safety) Act 2004* (Qld) div 2.
84   Ibid, div 3.
85   The new Commission Chair, Ruth Wade, was appointed in 2017.
86   Rick Wilkinson is a Senior Associate with EnergyQuest, an energy consultancy, provides technical advice as the Chief Technical Officer for the Australian

Petroleum Production and Exploration Association (APPEA) and has been a GC commissioner since 2013.

87 *Gasfields Commission Act 2013* (Qld) s 26.
88 Robert P. Scott, *Independent Review of the Gasfields Commission Queensland and Associated Matters* (Department of State Development (Qld), July 2016) <https://www.statedevelopment.qld.gov.au/resources/report/gasfields-commission-review-report.pdf> 28.
89 Ibid.
90 *Gasfields Commission Act 2013* (Qld) s 7.
91 Robert P. Scott, *Independent Review of the Gasfields Commission Queensland and Associated Matters* (Department of State Development (Qld), July 2016) <https://www.statedevelopment.qld.gov.au/resources/report/gasfields-commission-review-report.pdf> 5.
92 Department of State Development, Gasfields Commission Independent Review, *Terms of Reference* (2016) <https://www.statedevelopment.qld.gov.au/resources/terms-of-reference/terms-of-reference-gfcq-review.pdf>.
93 Ibid.
94 Gasfields Commission Queensland, *Annual Report 2016–2017* (2017) <http://www.gasfieldscommissionqld.org.au/resources/documents/Annual%20Report%202016-17%20%20FINAL%20-%20ONLINE.pdf>.
95 Robert P. Scott, *Independent Review of the Gasfields Commission Queensland and Associated Matters* (Department of State Development (Qld), July 2016) <https://www.statedevelopment.qld.gov.au/resources/report/gasfields-commission-review-report.pdf> 22.
96 Ibid, 46.
97 Ibid, 6.
98 Queensland Department of State Development, *Government Response to the Independent Review of the Gasfields Commission* (2016) <https://www.statedevelopment.qld.gov.au/resources/report/government-response-to-the-independent-review.pdf>.
99 Ibid.
100 Robert P. Scott, *Independent Review of the Gasfields Commission Queensland and Associated Matters* (Department of State Development (Qld), July 2016) <https://www.statedevelopment.qld.gov.au/resources/report/gasfields-commission-review-report.pdf>.
101 *Land Access Code* (2016) pt 2.
102 'Resource Acts' means the *Mineral Resources Act 1989* (Qld), *Petroleum and Gas (Production and Safety) Act 2004* (Qld), *Petroleum Act 1923* (Qld), *Geothermal Energy Act 2010* (Qld) and the *Greenhouse Gas Storage Act 2009* (Qld).
103 Land Access Review Panel, *Land Access Framework 12 Month Review Report* (2012) <https://www.dnrm.qld.gov.au/__data/assets/pdf_file/0004/193090/land-access-review-panel-report.pdf> 12.
104 Ibid.
105 Ibid, 18.
106 Ibid, 11.
107 AMEC, Submission No 34 to Productivity Commission, *Inquiry into the Non-Financial Barriers to Mineral and Energy Resource Exploration*, March 2013, 8.
108 Queensland Resources Council, Submission No 13 to Productivity Commission, *Inquiry into the Non-financial Barriers to Mineral and Energy Resource Exploration*, 2013, 3.
109 National Farmers' Federation, Submission 171 to Select Committee on Unconventional Gas Mining, *Inquiry into Unconventional Gas Mining*, 14 March 2016, 3.
110 *Land Access Code* (2016) pt 1.
111 Ibid, pt 2.
112 Ibid, pt 1.
113 Ibid, pt 2.

114   Ibid, pt 1.
115   Jack O'Connor, 'The Enforceability of Agreements to Negotiate in Good Faith' (2010) 29(2) *University of Tasmania Law Review* 177, 202.
116   *United Group Rail Services v Rail Corporation of New South Wales* [2009] NSWCA 177, 639.
117   Cindy Chen and Alan Randall, 'The Economic Contest Between Coal Seam Gas Mining and Agriculture on Prime Farmland: It May Be Closer than We Thought' (2013) 15(3) *Journal of Economic and Social Policy* 1; Phil McManus and Linda H. Connor, 'What's Mine is Mine(d): Contests over Marginalisation of Rural Life in the Upper Hunter, NSW' (2013) 22(2) *Rural Society* 166.
118   Department of Industry, Innovation and Science (Cth), Office of the Chief Economist, *Review of the Socioeconomic Impacts of Coal Seam Gas in Queensland* (Commonwealth of Australia, 2015) <https://www.industry.gov.au/sites/g/files/net3906/f/June%202018/document/pdf/review_of_the_socioeconomi-c_impacts_of_coal_seam_gas_in_queensland.pdf>.
119   Ibid, 2.
120   Laurence Boulle, Tina Hunter, Michael Weir and Kath Kurnow, 'Negotiating Conduct and Compensation Agreements for Coal Seam Gas Operations: Developing the Queensland Regulatory Framework' (2014) 17(1) *Australasian Journal of Natural Resources Law and Policy* 75.
121   *Land Access Code* (2016) pt 2 s 6.
122   See for example, s 14(2) of the *Land Access Code* (2016). Note that this term of as soon as practicable differs from the term used in safety in environment and defined in the *Edwards v National Coal Board* [1949] 1 All E. R. 743 'as soon as reasonably practicable'.
123   *Land Access Code* (2016) pt 2 s 14.
124   ABC News, 'New Land Access Code Described as a "Con"' (15 January 2014) *ABC News* (online) <http://www.abc.net.au/news/2014-01-16/new-land-access-agreement-code-described-as-a-27con27/5202442>.
125   *Land Access Code* (2016).
126   Land Access Review Panel, *Land Access Framework – 12-Month Review: Report of the Land Access Review Panel, February 2012* (2012). <http://www.mellorolsson.com.au/Media/Default/News/News%20Documents/Land_-Access_Review_Panel_report.pdf> 18, 46.
127   Queensland Government, Department of Natural Resources and Mines, *Land Access Code* (2016) 41.
128   *Mineral and Energy Resources (Common Provisions) Act 2014* (Qld) s83.
129   *Petroleum and Gas (Production and Safety) Act 2004* (Qld) s 533(2).
130   Ibid, s 500A(a),(e),(f).
131   *Mineral and Energy Resources (Common Provisions) Act 2014* (Qld) s 96.
132   Samantha Hepburn, *Mining and Energy Law* (Cambridge University Press, 2015) 199.
133   *Petroleum and Gas (Production and Safety) Act 2004* (Qld) s 543(2)(b)(i).
134   Ibid, s 532(4)(b).
135   *Australia Pacific LNG Pty Ltd v Golden & Ors* [2013] QCA 366.
136   *Petroleum and Gas (Production and Safety) Act 2004* (Qld) s 537A.
137   *Australia Pacific LNG Pty Ltd v Golden & Ors* [2013] QCA 366. Queensland Parliamentary Committee, *Mineral and Energy Resources (Common Provisions) Bill 2014* (Report No. 46 Agriculture, Resources and Environment Committee) 42.
138   James Plumb and Andrew Shute, *Negotiated Access to Land in Queensland – Is This the End of ADR?* (2014) <http://www.carternewell.com/page/Publications/Archive/Negotiated_access_to_land_in_Queensland_is_this_the_end_of_ADR/>.
139   *Fitzgerald & Anor v Struber & Anor* [2009] QLC 0076.
140   *Eacham Abrasive Blasting Pty Ltd v Gundersen & Anor* [2014] QLC 38.
141   *Wallace & Ors v Bottomer & Ors* [2015] QLC 23.

142 [2016] QLC 1.

143 *Nothdurft v QGC Pty Ltd* [2017] QLC 41.

144 *Peabody West Burton Pty Ltd v Mason* [2012] QLC 0023.

145 *Petroleum and Gas (Production and Safety) Act 2004* (Qld) s 316; See generally, *Environmental Protection Act 1994* (Qld) ch 3 pt 2.

146 *Petroleum and Gas (Production and Safety) Act 2004* (Qld) s 32.

147 Ibid, div 7.

148 Ibid, s 33; See generally the *Planning Act 2016* (Qld) ch 3.

149 Note that, *prima facie*, an activity authorised under the PGPSA and subject to a petroleum lease is exempt from local government planning schemes.

150 Queensland Competition Authority (Qld), *Final Report: Coal Seam Gas Review* (January 2014) <http://www.qca.org.au/getattachment/aaaeab4b-519f-4a95-8a65-911bc46cc1d3/CSG-investigation.aspx> 26.

151 Ibid.

6 British Columbia, Canada

Introduction

Under the *Natural Gas Strategy*, British Columbia aims to become a 'global leader in secure and sustainable gas investment, development and export',[1] vigorously pursuing the creation of an LNG export industry to supply the Asian market. The province also has a profitable and well-developed agricultural industry, worth Canadian dollar (CAD) $14 billion, which has historically been protected through provincial legislation to ensure its viability.[2] Protection of arable land is implemented through the Agricultural Land Reserve (ALR); a system of land use 'clustering' and zoning which prohibits non-farm activities on agricultural land to encourage farming and safeguarding farmland.[3] The agricultural industry has public support to maintain the ALR, stemming from a desire to secure local food production, maintain the local agricultural economy and, by extension, enable food security and food sovereignty within the state.

While British Columbia has a strong regulatory system to protect agricultural land, it faces similar challenges to those faced by other liberal democracies with growing unconventional gas industries. For example, Queensland, Australia, which is further down the path of gas exploitation with a more developed CSG industry, demonstrates the challenges that lie ahead for Canada. This is particularly pertinent for British Columbia which, it is anticipated, will need to perform a policy 'balancing act' to manage the interests of an entrenched agricultural sector and the ambitions of natural gas titleholders to exploit the resources under arable land.

The challenges are evident when viewed from a socioeconomic standpoint. The rights of farmers and landowners are recognised and protected at the provincial level in British Columbia through the *Right to Farm Act 1996* [4] and *the Canadian Agricultural Partnership*, [5] providing legislative protection for farming activities, agricultural pricing and farmer rights. This creates a unique challenge for the provincial government which has not yet faced challenges to land tenure and multiple-land use. Added to this is the potential for environmental degradation as a result of a more aggressive unconventional gas extraction industry. Elsewhere, the legacy of developed countries with a similarly strong agricultural base is to advocate protection strategies for food security and protection of rural communities. For example, in New York State, USA, shale gas activities were deemed incompatible with sustainable rural communities and agricultural activities due to the high likelihood of significant adverse environmental and social impacts. This resulted in the permanent ban of hydraulic fracturing in 2015, as examined in Chapter 8.

British Columbia has adopted an adaptive management approach that seeks to safeguard arable land while developing an unconventional gas industry. This chapter examines the broad legal framework and property rights of Canada as a basis for gas activities on protected ALR lands. In further analysis, this chapter reviews the adaptive management approach of British Columbia in the creation and maintenance of the ALR and its collaborative Delegation Agreement[6] with the Oil and Gas Commission (OGC) as the key regulatory instrument to manage land use conflicts.

The Canadian legal framework regulating unconventional gas

Canada consists of 10 provinces – Alberta, British Columbia, Manitoba, New Brunswick, Newfoundland and Labrador, Nova Scotia, Ontario, Prince Edward Island, Quebec and Saskatchewan – and three territories – the Northwest Territories, Nunavut and Yukon. With the exception of Quebec as a former French colony,[7] all provinces were all former British colonies and thus are common law-based legal systems.[8] Each Canadian province has its own constitution and holds legislative power over the province or territory.[9] Provincial courts hold a wider ambit of powers in the establishment and amendment of both federal and provincial law-making. This is in contrast to other Commonwealth jurisdictions, where State courts are limited to the regulatory powers of State-based legislation and cases.[10] All of Canada's provinces have constitutionally enumerated powers, with the Federal Government having plenary powers. Consequently, in Canada, the *Constitution Act, 1867* [11] provides additional rights to the provinces with respect to lands, mines, minerals and royalties. Parliament does not have authority pursuant to s 91(24) of the *Constitution Act, 1867* [12] to acquire provincial land for mining, since mining activities fall within provincial jurisdiction.

The key natural resource states – namely, the provinces of Alberta, Manitoba, Saskatchewan and British Columbia – acquired sovereignty over mineral and petroleum rights from the Federal Government in 1930 by virtue of the *Natural Resources Transfer Acts, 1930.* [13] The Canadian Constitution provides for the allocation of exclusive 'Heads of Power' between the federal government and provincial governments.[14] Section 92(5) of *The Constitution Act, 1867* provides provincial governments with the power to regulate the management and sale of provincial public lands, including natural resources lands.[15] Other sources of provincial constitutional authority to regulate energy projects include municipal institutions[16]; local, municipal and provincial revenue[17]; and the enumerated enforcement powers.[18] Additionally, s 92A of the Constitution confers on each provincial legislature the exclusive authority to make laws regulating non-renewable natural resources.

However, s 92A does not derogate power from the authority of the Canadian Parliament to enact laws in relation to natural resources, and where such a law of the federal parliament and a law of a province conflict, the federal jurisdiction will prevail.[19] In addition to its general power over the sea, coast and fisheries, the Canadian Parliament may also obtain jurisdiction over certain provincial works by virtue of its declaratory power. Certain natural resources, which would typically fall within provincial jurisdiction, can be federally regulated through Parliament's declaratory power. Parliament can declare works situated in the province 'to be for the general advantage of Canada or for the advantage of two or more of the Provinces'.[20] The regulatory authority of the Federal Government relates to interprovincial and international trade, falling within the regulation of trade and commerce.[21]

Under the *Canadian Environmental Assessment Act 2012* (Can),[22] unconventional gas projects on federal lands may be subject to federal environmental assessments, particularly in relation to projects with potential for 'significant', adverse, environmental effects.[23] Projects include offshore natural gas, oil exploration and production, gas processing plants, LNG facilities, and natural gas pipelines regulated by the National Energy Board (NEB).[24] The *National Energy Board Act 1985*[25] establishes the NEB as the overarching responsible authority, with respect to federal environmental assessment reports that require certificates for an energy project to proceed. Therefore, the NEB provides an advisory and authoritative power over the exploration for and the production of energy.

The *Constitution Act, 1867*[26] gives the provinces jurisdiction over works and undertakings within their boundaries. In situations where oil and gas facilities are within a province but part of a pipeline system constructed between two provinces, the regulation will fall under federal jurisdiction. The Supreme Court of Canada has indicated how oil and gas facilities such as pipelines, gathering and tie-in facilities all within one province may come under federal jurisdiction. According to *Westcoast Energy Inc v Canada (National Energy Board):*[27]

> It is well settled that the proposed facilities may come within federal jurisdiction under s 92(10)(a) of the *Constitution Act 1867* in one of two ways.[28] First, they are subject to federal jurisdiction if the Westcoast mainline transmission pipeline, gathering pipelines and processing plants, including the proposed facilities, together constitute a single federal work or undertaking. Second, if the proposed facilities do not form part of a single federal work or undertaking, they come within federal jurisdiction if they are integral to the mainline transmission pipeline.

Petroleum rights granted in respect of Crown-owned petroleum rights are, by virtue of Crown leases, granted at the provincial level. In British Columbia, the Ministry of Energy and Mines (Title Division) dispenses Crown petroleum and natural gas rights, and dispositions (permits, drilling licences and leases) monthly by public tenure.[29] The *Land Act*[30] also expressly states that no granting of Crown land conveys the right to any petroleum, gas, coal, mineral or geothermal resource found in the land.[31]

Petroleum property rights in British Columbia

The doctrine of tenure and the subsequent legislative implementation of these principles in Canada bestows ownership of all land to the Crown,[32] while private landholders may hold estates in fee simple. As a consequence, unconventional gas licenses are typically issued over fee simple estates of private landholders or Crown Land.[33] Title to petroleum is generally transferred from the Crown to an oil and gas company with the issue of a licence or lease and 'ownership' of that resource transfers from the Crown to the titleholder at the wellhead. This is also the point in time at which royalties are calculated and paid to the State. The State's petroleum legislation then requires a petroleum licenceholder to pay a petroleum royalty and the annual rent prescribed under the relevant regulation; *Petroleum and Natural Gas Act*, 1996 (BC) (PNGA).[34]

In British Columbia, unconventional gas development is primarily governed by the OGC. It is a three-phase approval process, established in accordance with the PNGA,[35] together with the *Oil and Gas Activities Act 2008* (OGAA) and the Code of Practice for the Discharge of Produced Water from Coalbed Gas Operations promulgated under the *Environmental Management Act*.[36] The OGAA enables regulation of surface land use, primarily through the Environmental Protection and Management Regulation (EPMR). The EPMR regulates the actions a permit holder and a person carrying out an oil and gas activity must take, or refrain from taking, to protect and/or effectively manage the environment.

Under the PNGA, property over petroleum and natural gas[37] in British Columbia is vested in the Crown. The rights granted to petroleum titleholders depend on the nature of the Crown disposition, however, a lease granted to a lessee establishes an exclusive right to produce both petroleum and natural gas[38] from the lease area. The freehold petroleum and natural gas lease is the document that governs the relationship between a freehold owner of mineral rights and a party contracting to exploit and develop the petroleum and natural gas as well as related resources owned by the freehold owner. The standard freehold, petroleum and natural gas lease in use in the Canadian oil and gas industry is the Canadian Association of Petroleum Landmen (CAPL) lease, 'which enjoys almost universal acceptance in the industry'.[39]

The fundamental purpose of a freehold petroleum and natural gas lease is to establish a contractual arrangement between the registered owner, or the party entitled to become the registered owner, and the lessee. The lease allows the lessee to explore for petroleum and natural gas and to produce the petroleum and natural gas if those resources are found.[40] The freehold lease must provide security to the lessee that the lease will continue if a successful well is drilled and must ensure from the lessor's perspective that the lands are developed or become available for re-leasing. Therefore, the main provisions of the lease are concerned with balancing the interests of the lessor and lessee and the competing land uses of a private fee simple titleholder and a petroleum lessee.

Legal framework regulating unconventional gas activities

The Agricultural Land Commission

Agricultural land protection in British Columbia operates as a form of provincial-level zoning that takes priority over local land use regulations by creating comprehensive land use regulations to protect the agricultural land base. The provincial ALR comprises land that was zoned for agricultural purposes by the relevant local government authority, as at the establishment of the reserve in 1973, plus additions and removals to the zone by the Agricultural Land Commission (ALC) since. The ALC is an independent administrative authority commissioned to 'preserve agricultural land; encourage farming on agricultural land in collaboration with other communities of interest;[41] and to encourage local governments, first nations, the government and its agents to enable and accommodate farm use of agricultural land'.[42] In general, land in the ALR may not be subdivided or used for a non-farm use without the approval of the ALC and permitted under the under the *Agricultural Land Commission Act*, SBC 2002, c 36 (ALCA).[43]

British Columbia's pioneering province-wide implementation of the ALR is one of the earliest examples of a legislated agricultural land use protection framework in North America. According to the Census of Agriculture for British Columbia, of the 19,759 farms comprising the total farm area in British Columbia in 2011, 61.7 per cent was pasture land (tame or seeded pasture and natural land for pasture), while cropland accounted for an additional 23 per cent.[44] As at June 2015, there was 4,620,858 ha (11,418,388.79 ac) included in the ALR in British Columbia, representing 5 per cent of the total provincial area, only half of which is currently engaged in agricultural production.[45]

Twenty-seven per cent of British Columbia's total ALR land is located in the Peace River Regional District (PRRD), where most shale gas development in the province takes place.[46] The largest amount of ALR land in the Regional District is in the communities of Fort St John and Dawson Creek. According to the 2011 Census of Agriculture, 823,498 ha (2,034,907.87 ac) are being farmed in the PRRD, accounting for 64 per cent of the region's ALR.[47] Fort St John and Dawson Creek are situated above the Montney Shale Gas Basin. Consequently, resource development of shale gas on ALR lands has been at the forefront of regulation for the province. Geologically, the province holds an estimated 400 trillion cubic feet (Tcf) of shale gas, primarily situated in the northeast region of the Horn River Basin, the Montney, the Liard Basin and the Cordova Embayment.[48] As of 2012, 1,400 shale gas wells produce over 2 billion cubic feet of gas per day in British Columbia – equating to 24,325 cumulative wells drilled since 1952.[49]

In its early years, the ALC's main function was fine-tuning of land parcels included in the ALR and applying a precautionary approach, rarely granting permission for non-farm use activities on ALR protected land.[50] However, since the commercial viability of unconventional gas has been proven in the province, the administration of the ALR has changed to adapt regulation and thus challenge its historically precautionary approach. In 2014, the passing of Bill 24 signalled the province's desire to apply an adaptive management approach in permitting unconventional gas activities on ALR land. Bill 24 divides the ALR into two zones, requiring the ALC to provide more 'flexibility in land use in Zone 2 to allow activities such as food processing and potential oil and gas development'.[51] The use of regional ALR plans and sub-regional and issue-specific plans are designed with a view to outline the various land and resource management goals for a particular area based on an assessment of either Zone 1 or Zone 2 areas.[52]

Zone 1 includes the Island Panel Region, the Okanagan Panel Region and the South Coast Panel Region. For the purpose of s 4.1(d), Zone 2 consists of:

a) The geographic area of British Columbia within the boundaries of the following regional districts and regional municipalities, as those boundaries existed on January 1, 2014:

  i) Regional District of Bulkley-Nechako
  ii) Regional District of Fraser-Fort George
  iii) Regional District of Kitimat-Stikine
  iv) Northern Rockies Regional Municipality
  v) Peace River Regional District (PRRD)
  vi) Skeena-Queen Charlotte Regional District

b) All the land in British Columbia that is not within the boundaries of a regional district or a regional municipality, as those boundaries existed on January 1, 2014.[53]

Zone 2 consequently includes the North Panel Region, the Interior and Kootenay, which produces 15 per cent of British Columbia's agricultural output and contains 90 per cent of provincial ALR lands (4,132,308 ha or 10,211,155.45 ac).[54] Zone 2 contains 85 per cent of the best soils (class 1 to 4), of which 72 per cent are located in the Peace River Region. Land located in Zone 1 represents 10 per cent of the ALR lands (489,391 ha or 1,209,311.5 ac).[55]

In making recommendations, the ALC must give weight to a differing mandate in Zone 1 and Zone 2.[56] In Zone 1, weight must be given to the following values in descending order of priority:

a agricultural values, including the preservation of agricultural land and the promotion of agricultural purposes;
b environmental and heritage values, but only if:

  i those values cannot be replaced or relocated to land other than agricultural land, or
  ii giving weight to those values results in no net loss to the agricultural capabilities of the area;

c economic, cultural and social values.[57]

In making recommendations to land located in Zone 2, the board must give weight to the considerations set out in ss 4.3(a)–(d), in descending order of priority:[58]

a   the purposes of the commission set out in section 6;
b   economic, cultural and social values;
c   regional and community planning objectives;
d   other prescribed considerations.[59]

Encouraging farming and protecting agricultural land is no longer the first priority of the ALC when making decisions concerning ALR lands in Zone 2. Criterion (d) has particularly raised concerns of threats to food security and food sovereignty due to its broad scope and wording relied upon to prioritise oil and gas activities taking place on ALR lands to align with economic values.

Six panels are established, representing the six panel regions – Interior, Island, Kootenay, North, Okanagan and South Coast.[60] Each panel has at least two members, including the vice chair from the panel region plus all other members of the Commission who reside in the panel region.[61] These panel representatives have full authority to make final regulatory decisions in their own panel region. The North Panel Region is the hub of shale gas activity in British Columbia and holds a significant proportion of ALR lands regulated by the ALC. Subject to s 11.2, whereby the Chair of the ALC may refer an application to the executive committee, the Chair of the ALC must refer an application under ss 17(1)(b) or (c) or 17 (3), 20(3), 21(2), 29(1) or 30(1), in relation to land located in a panel region, to the panel established for the panel region.[62] Notably, there has been both criticism and support for the effects of Bill 24 in creating a two-tier zoning system for the ALR.[63] However, the creation of regional panels as an act of flexibility and transparency in regulatory decision-making is arguably evident within the ALR system.

The Agricultural Land Tribunal

Any non-farm use of ALR land that is not designated in the ALCA as farm use or identified as a use permitted in an ALR is prohibited, unless that use is otherwise allowed under the ALCA.[64] All oil and gas developments and activities are classified as non-farm uses.[65] Since 1976, the ALC has facilitated oil and gas activities on ALR land by working 'collaboratively with the industry to develop a process of allowing the non-farm use of land in the ALR for oil and gas activities'.[66] The passing of General Order #4473/76 in 1976, facilitated 'accommodation' of the oil and gas industry on ALR lands less than 2 ac, and stated:

> General application to all land within the designated Agricultural Land Reserve Plan of the Peace River-Liard Regional District to the effect that oil and gas sites and ancillary buildings and equipment occupying an area less than 2 acres, exploratory sites and ancillary buildings and sump pumps, and required road and gathering and flow line rights-of-way be allowed, provided that the well site or exploratory site is rehabilitated to its original or better

topographical and soil conditions when abandoned and any pipeline that is constructed for gathering purposes does not unduly restrict the agricultural use of the land and that during construction of the pipeline the topsoil is conserved and replaced on the surface of the trench when the pipeline is backfilled.[67]

General Order #4473/76 required sites of oil and gas activities on ALR between 1976 and 1995 to broadly 'be restored to a condition as good or better than existed prior to the development.'[68] As this is a broad reclamation requirement, flexibility is encouraged in the review of Schedule B of a surface lease requirements used to assess the reclamation of existing developments and all Schedule B reports are to be submitted to the OGC for review.[69] Provisions of the ALCA in relation to these lands include issuing a stop work order, prescribing additional remedies to restore the land, seeking a court order or levying a penalty if the ALCA is contravened and if the soil is not adequately reclaimed or protected.[70]

Since 1976, the ALC has viewed oil and gas activities in the ALR as being 'temporary' in nature and vitally important to the economic wellbeing of British Columbia.[71] Despite this, the ALC acknowledges an adaptive approach to mitigating harm to arable land quality:

> The ALC has, and continues to view the land use as temporary, albeit likely long term, and its accommodation was predicated on the commitment from industry to restore the land back to an agricultural standard equal to, or better than, that which existed prior to development.[72]

As agricultural planning and zoning regulation plays a significant role in how unconventional gas affects the land access and land use of agricultural lands in British Columbia, the ALC has the function of a quasi-judicial tribunal. This judicial function of the ALC Tribunal seeks to provide 'the cornerstone of planning for agriculture and heightening certainty for persons engaged in farm businesses and support industries'.[73] Important elements in the legislation include a clear mandate for the ALC that is focused specifically on protecting farmland.

The policy language in the legislation includes provision for consistency between local government plans and the ALCA, providing a necessary link to extend the provincial legislation into the domain of local land use planning and decisions. The ALCA provides a mechanism for landowners, including governments, to apply to the ALC to exclude or include land in the ALR, to approve subdivisions and to permit non-farm uses.

Official Community Plans (OCPs), as enforceable regulatory plans, are the foundation of stability for local frameworks in British Columbia. Typically, OCPs include vision or goal statements, agricultural objectives and specific policies for agricultural lands. The OCP is supported by the zoning by-laws which provide regulations for designated agricultural land uses, contributing to the stability of the framework. An example of an integrated and comprehensive legislative framework is in the South Peace River Regional Area, where a local development plan, regional agricultural plan, and the ALR and Community Planning Guidelines operate to maintain and secure a productive agricultural resource base. The South Peace River Area is situated on the Montney unconventional gas tenement.

The Oil and Gas Commission

In 1998, the British Columbia OGC was created to substantially increase production of UGR and to encourage exploration in the Western Canadian Sedimentary Basin.[74] Therefore, the OGC was enacted to simplify approvals required for oil and gas exploration and development, avoiding overlapping legislation, inconsistent legislative application and an overly complex approval processes. The OGC is staffed with a group of officers dedicated to the consideration of applications necessary for upstream activity with single-window authority over all of the principal approvals required for oil and gas development as the means to achieve a streamlined approval process.

The result of this policy objective was a Memorandum of Understanding (MOU) between government and the oil and gas industry, signed in February 1998.[75] The MOU provided for an Oil and Gas Initiative, the goal of which was to 'make British Columbia one of the most attractive places in North America for oil and gas investment'.[76] Once the MOU was signed, it was quickly determined that the best means of streamlining the regulatory approval process would be to allow all essential approvals be available from a 'single window'. According to Rankin et al., this 'allowed greater control by government in an industry which has a significant impact on government revenue and public policy'.[77]

The OGC, a Crown Corporation and agent of the Crown, collaborates and regulates diverse stakeholders, including First Nations and the petroleum industry, 'to provide efficient and effective oversight of oil and gas activity'.[78] The OGAA mandates the OGC to regulate the oil and gas industry to ensure sound development of British Columbia's oil and gas resources. Accordingly, the OGC is responsible for developing processes to accept and review industry applications related to oil and gas activities and/or pipeline activities falling within provincial jurisdiction. To approve such applications, the OGC must ensure that the application is in the 'public interest', having regard to environmental, economic and social effects of the activities.[79]

The OGC acts as a review board and manages conflicts between agricultural landholders and petroleum titleholders to resolve enquiries and complaints. The OGC also holds powers to refer complaints to the Oil and Gas Appeal Tribunal, an independent legal body that reviews decisions, certain orders, administrative penalties and findings made by the OGC within 30 days.[80] A myriad of powers are granted to the OGC, acting as a 'single-window regulatory agency, with responsibilities for overseeing oil and gas operations in British Columbia, including exploration, development, pipeline transportation and reclamation'.[81] Further, the core role of the OGC includes:

> Reviewing and assessing applications for industry activity, consulting with First Nations, ensuring industry complies with provincial legislation and cooperating with partner agencies. The public interest is protected through the

objectives of ensuring public safety, protecting the environment, conserving petroleum resources and ensuring equitable participation in production.[82]

Consequently, the OGC is involved in each step of the UGR activity cycle, ranging from consultation with industry, applications, permits review and assessment, compliance. Ultimately, to minimise impact on agricultural land, any shale gas activities on ALR land must take into account 'the optimal combination of total area disturbed and location of the activity in relation to current and planned agricultural operations and agricultural capability of the land'[83] in British Columbia. The OGC is also responsible for reviewing and approving land tenure, water use, forest harvesting, waste disposal and potential heritage impacts. Figure 6.1 outlines the relationship and process of non-farm use approvals for ALR lands for oil and gas activities as approved by the OGC.

As illustrated in Figure 6.1, the ALC and OGC both hold regulatory, policy and judicial powers relating to ALR lands and UGR activities in British Columbia.

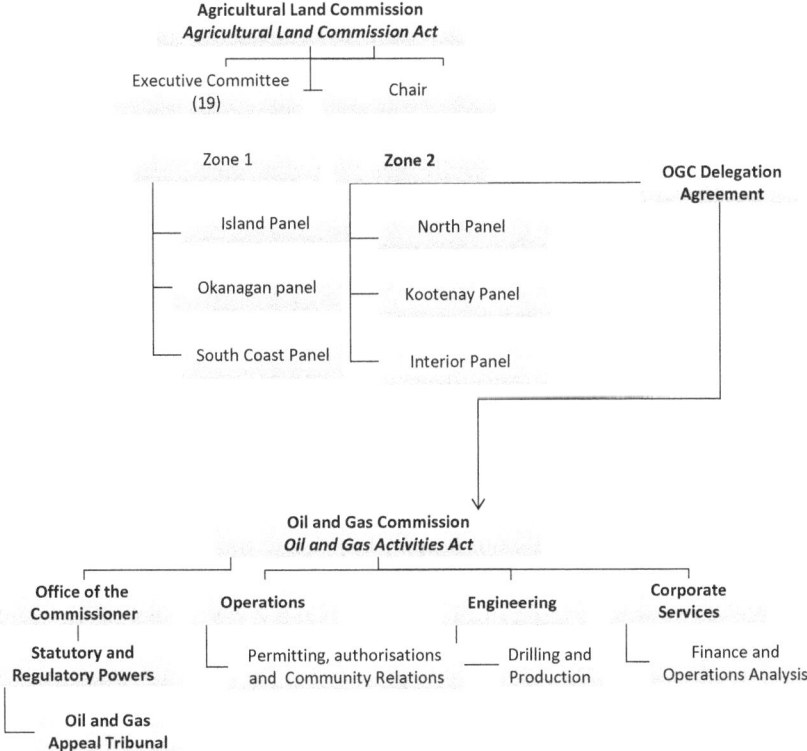

Figure 6.1 **Relationship of non-farm use approvals ALR process and OGC in British Columbia**
Source: compiled by author.

The facilitation of the ALC – OGC Delegation Agreement provides for a 'streamlined' single-window approvals and regulatory process for UGR activities on ALR lands in the Peace River Region (Zone 2). Both regulatory oversight and tribunal bodies act as independent bodies, with the OGC acting on behalf of the ALC and making decisions guided by the ALCA and regulations.

Petroleum permits are granted by the OGC in conjunction with approval for UGR activities on ALR lands. Oil and gas non-farm use applications to the OGC are then referred to local government and the Ministry of Agriculture independently for any comment.

The Oil and Gas Appeal Tribunal was established by the OGAA.[84] Eligible persons have the ability to request a review of specified administrative and permitted decisions.[85] Requests for review must be made within 30 days of the determination in question and reviews are carried out by designated officials in the Commission. The Oil and Gas Appeal Tribunal provides appeal process systems independent of the OGC. The OGC has a clear and transparent mandate to facilitate, evaluate and report on the outcomes of complaints and enquiries. The operation of the OGC is reviewed regularly by the Canadian Standards Association and this formal review process is publicly available for review and comment.

The stated purposes of the OGC are to regulate oil and gas activities in British Columbia in a manner that:

i   Provides for the sound development of the oil and gas sector, by fostering a healthy environment, a sound economy and social well-being;
    Conserves petroleum and natural gas resources;
ii Ensures safe and efficient practices; and
iii Assists owners of petroleum and natural gas resources to participate equitably in the production of shared pools of petroleum and natural gas.[86]

The OGC holds the power to issue key approvals in relation to oil and gas activities and pipelines.[87] Further, the OGC holds plenary powers relating to heritage, environmental and water management from the *Forest Act*,[88] *Forest Practices Code of British Columbia Act*[89] (the 'Forest Practices Code'), *Heritage Conservation Act*,[90] *Land Act*,[91] *Environmental Management Act*[92] and *Water Sustainability Act*.[93]

According to s 11 of the OGAA, the OGC may establish and appoint an advisory committee of the Commission. In general, the advisory committee has a mandate to assist the Commission in discharging its responsibility to consider or inquire into any matter and to report its findings and advice to the board.[94] Division 2 of the OGAA formed the Oil and Gas Appeal Tribunal to create an appellate tribunal system for landholders or any eligible persons to review OGC decisions.[95] Eligible persons have the ability to request a review of specified administrative and permitted decisions.[96]

Further, the Surface Rights Board of British Columbia (SRB) assists in resolving disputes between landowners and companies that require access to private land in order for exploration, development or production of Crown-owned subsurface resources such as oil, gas, coal, minerals and geothermal. The SRB provides an additional avenue of independent appeals and decision-making relating to compensation, compliance, terms of entry onto land and rent renegotiation.

The Board of the OGC has the power to make regulations under the OGAA, primarily of a technical nature, and has exercised its regulatory power to make regulations related to pre-application consultation and notification requirements, geophysical exploration (seismic activities), drilling and production activities, pipelines and LNG facilities, levies and security. On finding a contravention, the OGC may exercise a further new power to impose monetary penalties, referred to as 'administrative penalties' in the OGAA. The monetary penalty provisions are backed up with substantial civil liability sanctions. For example, the OGC may enforce a penalty by registering it with the Supreme Court of British Columbia and such registration is deemed to be a judgment of the court for the payment of a debt. The OGC may also make any orders it views as necessary to mitigate public safety risks. As a consequence of these regulatory powers, the OGC has greater regulatory reach, as well as adequate 'teeth' to address concerns pertaining to UGR development on agricultural lands in Canada.

An adaptive approach to managing conflicting interests: The ALC–OGC Delegation Agreement

The overall unconventional gas regulatory framework of British Columbia is illustrated in Figure 6.2 and consists primarily of the *Petroleum and Natural Gas Act* and the *Oil and Gas Activities Act* and the associated policy guidelines and regulatory direction notices by the OGC. Environmental protection at the provincial level is regulated by the *Environmental Management Act 2003* and the EPMR, which applies only to Crown land and does not apply to private subsurface oil and gas activities associated with an operating area. The EPMR provides the statutory authority to the OGC for the management and protection of environmental values and water management. Access to agricultural land is governed by the ALC–OGC Delegation Agreement 2013 and the ALCA. The Delegation Agreement grants regulatory powers to the OGC to decide on applications for non-farm use of identified ALR lands for oil and gas activities and ancillary activities.

The 2002 Amendment to the ALCA[97] provided the ALC with the ability to delegate decision-making powers to an 'authority'.[98] This provided the regulatory platform for the ALC to adjust its powers and collaborate with another administrative authority and to implement decisions more adaptively over specified non-farm uses on ALR lands. The ALC has exercised its delegation powers to enter into an agreement with the OGC relating to certain oil and gas non-farm uses in the ALR.[99] Therefore, British Columbia asserts provincial oversight and control through regulatory checks and balances beginning from agricultural land zoning in the ALR through to the ALC–OGC Delegation Agreement to regulate oil and gas activities on agricultural land.[100]

However, it is noted that the ALC–OGC Delegation Agreement applies only to ALR land within the Northern Rockies and Peace River Regional Districts. This is different from the Zone 1 and Zone 2 designation granted by the

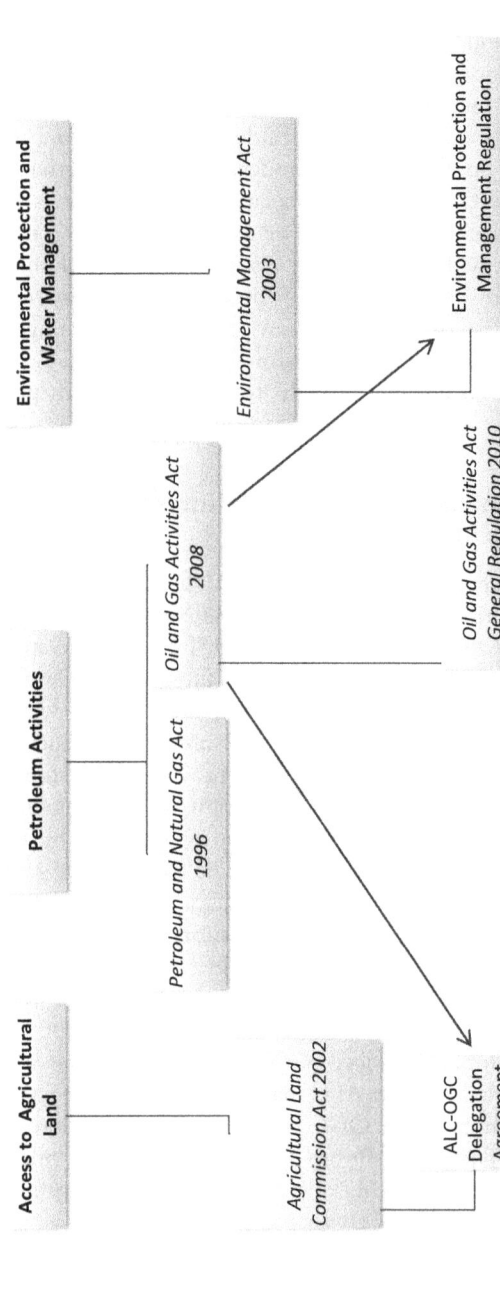

Figure 6.2 Overarching UG regulatory framework of British Columbia
Source: Compiled by author.

Agricultural Land Commission Amendment Act 2014, [101] as the ALC–OGC Delegation Agreement applies only to one of the panel regions in Zone 2, the PRRD and the Northern Rockies Regional Municipality.[102] Therefore, applicants submitting applications outside of these areas that impact ALR lands must acquire ALC approval prior to the OGC adjudicating on the application.[103] The OGC has assumed these powers, through the Delegation Agreement, to make decisions relating to UGR activities on ALR lands according to the purpose of the ALC – which is to preserve agricultural land and encourage and enable farming.[104] The aim of the ALC–OGC Delegation Agreement is to encourage, enable and accommodate farming on agricultural land while sustainably developing onshore shale gas activities on ALR land.[105] Ultimately, to minimise impact on agricultural land, any ALR land on which shale gas activities take place must take into account 'the optimal combination of total area disturbed and location of the activity in relation to current and planned agricultural operations and agricultural capability of the land'[106] in British Columbia.

The OGC is charged with balancing a broad range of environmental, economic and social considerations. In achieving its aim to provide 'oil and gas regulatory excellence for British Columbia'[107] the OGC therefore oversees all regulatory aspects of UGR operations. This includes exploration, development, pipeline transportation and reclamation. Regulatory responsibility of the Commission extends from the exploration and development phases, through to facilities operation and, ultimately, decommissioning, while landholder appeals are heard and addressed by the Oil and Gas Appeal Tribunal. Petroleum permits are granted by the OGC in conjunction with approval for UGR activities on ALR lands. Oil and gas non-farm use applications to the OGC are then referred to local government and the Ministry of Agriculture independently for any comment.

In British Columbia, the state's role is to arbitrate between the two land uses and the administrative regulatory bodies evaluate the value of both sectors and make a determination at the agency level rather than at the individual level. The legislative paradigm that underpins the OGC's role acts as either an advisor at 'arms-length' or as a regulator of approved UGR activities on ALR lands by assisting those who are affected by those activities.

The Delegation Agreement provides a comprehensive set of regulations that define permitted oil and gas uses on agricultural land as well as conditions and procedures for when the ALC must be involved in the application processes in the Northern Rockies and Peace River Regions. The key purpose of the Delegation Agreement is to create a regulatory framework which facilitates the adaption of ALR lands to permit unconventional gas development in order to 'streamline and improve the review and approval processes for oil and gas activities and ancillary activities on agricultural reserve lands while preserving agricultural lands and encouraging the farming of agricultural lands'.[108] The ALCA process requires public consultation, disclosure and comment from local governments. All applications for non-farm use in the ALR are submitted to the local government before being submitted the OGC. Local governments then review the application and determine if they will forward the application to the ALC for decision with or without a recommendation or comment.[109] Local government zoning, therefore, plays a role in determining land use and a proposed use may require re-zoning at the local level in meeting one of the key purposes of the ALCA, to 'encourage local governments ... to enable and accommodate farm use of agricultural land and uses compatible with agriculture in their plans, bylaws and policies'.[110]

Conclusion

For more than 40 years, British Columbia has focused on implementing its policy of developing unconventional gas resources, while aiming to provide strong regulatory oversight and protection mechanisms. The ALR system is likely to face further challenges as unconventional gas exploration ramps up to meet global demand. This will place the current regulatory framework to manage coexistence under pressure. British Columbia awards unconventional gas activity approvals on ALR lands under a discretionary system while stipulating approval conditions within the ALC–OGC Delegation Agreement. This discretionary system has, to date, facilitated a collaborative relationship between agricultural and unconventional gas land uses.

The ALC–OGC Delegation Agreement provides a unique and innovative approach to achieve its objective in protecting ALR lands, while coexisting with unconventional gas activities.[111] This involves a continuing regulatory dialogue between the two regulatory administrative bodies in enacting regulation and decision-making, while maintaining transparency to ensure collaboration between unconventional gas activities in agricultural zoned areas. This system has served and safeguarded the interests of agricultural landowners and gas titleholders since the 1970s.

Ostensibly, British Columbia has the institutional capacity and bodies in place to manage land use contestation. Yet, the ALR and the OGC have not been fully tested against the sheer scale and size of the potential pressures that may be exerted by the unconventional gas extraction industry to acquire greater access to agricultural lands. So far, public opposition has restricted the development of large scale LNG ports in coastal regions, including the cancelled Pacific NorthWest project. LNG ports are seen by stakeholders as a representation of likely dominance by the shale gas industry, in direct competition to agricultural land viability over ALR zoned lands in the Peace River Regional District. This backlash against the growth of the gas sector demonstrates the challenge to harmonisation of both sectors.

Despite public sentiment, the Federal Government and NEB has approved all 17 proposed port export licences and 9 federal environmental assessments are underway to facilitate the gas export industry. In an effort to provide balance, these approvals are limited by restrictive parameters, including project assessments at the federal and provincial level before LNG port approval. However, this balance may be at risk, as British Columbia continues to support the unconventional gas industry in the Peace River Region, where most of its ALR land is located. It is likely British Columbia will continue to rely upon the ALR and OGC and an adaptive management approach, in amending its Delegation Agreement to facilitate development of its ALR in Zone 2 and regulate the on-going contestation between the two land uses.

Notes

1 Government of British Columbia, *British Columbia's Natural Gas Strategy* (GBC, 2012) <http://www.gov.bc.ca/ener/popt/down/natural_gas_strategy.pdf> 1.
2 Government of British Columbia, *Sector Snapshot 2016: B.C. Agrifood & Seafood* (GBC, 2017).
3 Robert Androkovich, Ivan Desjardins, Gordon Tarzwell and Peter Tsigaris, 'Land Preservation in British Columbia: An Empirical Analysis of the Factors Underlying Public Support and Willingness to Pay' (2008) 40(3) *Journal of Agricultural and Applied Economics* 999.
4 *Farm Practices Protection (Right to Farm) Act*, RSBC 1996, c 131.
5 The *Canadian Agricultural Partnership* is a 5-year, $3 billion investment by federal, provincial and territorial governments to strengthen the agriculture and agri-food sector. Created in 2018, the Partnership aims to continue to help the sector grow trade, advance innovation while maintaining and strengthening public confidence in the food system, and increase its diversity.
6 British Columbia Oil and Gas Commission, ALC–OGC Delegation Agreement (2013) <https://www.bcogc.ca/node/5759/download>. The delegation agreement is limited in that it only applies to the oil and gas sector.
7 Quebec holds a juridical legal system under which civil matters are regulated by French civil law. However, public law, criminal law and other federal law operates according to Canadian common law.
8 Although Canada was under French dominion from 1534–1763, commencing with the exploration of Newfoundland it came under British dominion from 1763–1931. See, eg, Bob Bothwell, *Penguin History of Canada* (Penguin Canada, 2007), for an in depth analysis of Canadian history.
9 *Constitution Act 1867* (IMP), 30 & 31 Vict, c 3, s 96 Ch V.
10 Ibid, s 92(14).
11 Ibid, Vict, c 3. A fundamental principle of law in Canada is the supremacy of the Constitution which is enshrined in the *Constitution Act, 1982*. All laws, whether common or legislative, must comply with the Constitution. *The Constitution Act, 1982*, being Schedule B to the *Canada Act 1982* (UK), 1982, c 11.
12 *Constitution Act 1867* (IMP), 30 & 31 Vict, c 3.
13 Which by scheduled and individual Memorandums of Agreement, transferred natural resources regulation to the various Western Canadian provinces.
14 Allan Ingelson, 'Strategic Planning for Energy Development in Canada' (2015) 6 *Journal of Energy and Environmental Law* 35, 38; *Constitution Act 1867* (IMP), 30 & 31 Vict, c 3, s 92A(b-c).
15 *Constitution Act 1867* (IMP), 30 & 31 Vict, c 3, s 92(5).
16 Ibid, s 92(8).
17 Ibid, s 92(9).
18 Ibid, s 92(15).
19 Ibid, s 92A(3).
20 Ibid, s 92(10)(c).
21 Ibid, s 91(2).
22 *Canadian Environmental Assessment Act*, SC 2012, c 19, s 52.
23 Ibid, c 19, ss 52 and 67.
24 *National Energy Board Act*, RSC 1985, c 7.
25 Ibid.
26 *Constitution Act 1867* (IMP), 30 & 31 Vict, c 3.

27 *Westcoast Energy Inc v Canada (National Energy Board)* [1998] 27 SCJ.

28 *Constitution Act 1867* (IMP), 30 & 31 Vict, c 3.

29 John Bishop Ballem, *The Oil and Gas Lease in Canada* (University of Toronto Press, 2008).

30 *Land Act*, RSBC 1996, c 245.

31 Ibid, c 245, s 50.

32 This reservation is found in s 26 of the *Petroleum and Gas (Production and Safety) Act 2004* (Qld) and s 49 of the *Land Act*, RSBC 1996, c 245 in British Columbia.

33 Mark Thompson and Martin George, *Thompson's Modern Land Law* (Oxford University Press, 2017).

34 *Petroleum and Natural Gas Act*, RSBC 1996, c 361, pt 10.

35 Ibid, c 361, c 361.

36 *Environmental Management Act*, SBC 2003, c 53.

37 The PNGA defines petroleum as meaning 'crude petroleum and all other hydrocarbons, regardless of gravity, that are or can be recovered in liquid form from a pool through a well by ordinary production methods or that are or can be recovered from oil sand or oil shale'. Natural gas is defined as 'all fluid hydrocarbons, before and after processing, that are not defined as petroleum, and includes hydrogen sulphide, carbon dioxide and helium produced from a well'. *Coalbed Gas Act*, SBC 2003, c 18 s 4(1) states that 'a natural gas tenure, whether made before or after the coming into force of this Act, includes any coalbed gas rights'.

38 *Petroleum and Natural Gas Act*, RSBC 1996, c 361, s 50(1) provides that '[a] lease shall be a petroleum and natural gas lease'.

39 John Bishop Ballem, *The Oil and Gas Lease in Canada* (University of Toronto Press, 2008), 4.

40 Alastair R. Lucas and Constance Hunt, *Oil and Gas Law in Canada* (Carswell, 1990).

41 Communities of Interest are established within the following zones outlined in *An Act to Amend the Agricultural Land Commission Amendment Act*, SBC 2002 (2nd Sess), c 36, s 4.2, '(a) Zone 1, consisting of the Island Panel Region, the Okanagan Panel Region and the South Coast Panel Region; (b) Zone 2, consisting of all geographic areas of British Columbia not in Zone 1'.

42 *An Act to Amend the Agricultural Land Commission Amendment Act*, SBC 2002 (2nd Sess), c 36, s 6.

43 Ibid, s 20.

44 Statistics Canada, *2011 Census of Agriculture* (SC, 2011) <https://www.statcan.gc.ca/eng/ca2011/index>.

45 Agricultural Land Commission, *Provincial Land Commission Annual Report 2014/2015* (ALC, 2015) <http://blogs.ubc.ca/alrmap/files/2016/02/annual_report_2014-2015.pdf>.

46 Ryan Green, *Case Studies of Agricultural Land Commission Decisions: The Need for Inquiry and Reform* (University of Victoria Environmental Law Centre, 2006) <http://www.elc.uvic.ca/documents/ALR%20Final%20Report%20(FINAL-2).pdf>.

47 Peace River Regional District, *Regional Agricultural Plan Background Report* (Don Cameron Associates, 2014) <http://prrd.bc.ca/wp-content/uploads/Background-Report-Final-November-2014.pdf>.

48 British Columbia Ministry of Natural Gas Development, *Summary of Shale Gas Activity in Northeast British Columbia 2014* (British Columbia Government, 2014) <https://www2.gov.bc.ca/assets/gov/farming-natural-resources-and-industry/natural-gas-oil/petroleum-geoscience/oil-gas-reports/oil_and_gas_report_2016-1.pdf>.

49 Energy and Mines Ministers' Conference, *Responsible Shale Development: Enhancing the Knowledge Base on Shale Oil and Gas in Canada* (Natural Resources Canada, 2013) <https://www.nrcan.gc.ca/sites/www.nrcan.gc.ca/files/www/pdf/publications/emmc/Shale_Resources_e.pdf> 17; British Columbia Ministry of Natural

Gas Development, *Summary of Shale Gas Activity in Northeast British Columbia 2014* (British Columbia Government, 2014) <https://www2.gov.bc.ca/assets/gov/farming-natural-resources-and-industry/natural-gas-oil/petroleum-geoscience/oil-gas-reports/oil_and_gas_report_2016-1.pdf> 3.

50 Tracy E. Stobbe, Alison J. Eagle, Geerte Cotteleer and G. Cornelis van Kooten, 'Farmland Preservation Verdicts – Rezoning Agricultural Land in British Columbia' (2011) 59 *Canadian Journal of Agricultural Economics* 555.

51 *An Act to Amend the Agricultural Land Commission Amendment Act*, SBC 2002 (2nd Sess), c 36.

52 *Agricultural Land Commission Act*, SBC 2002, c 36, s 4.2.

53 *An Act to Amend the Agricultural Land Commission Amendment Act*, SBC 2002 (2nd Sess), c 36.

54 Arthur Green, Siobhan McPhee, Aviv Ettya, Britta Rocker and Christina Temenos, *British Columbia in a Global Context* (An Open Education Resource Textbook) (BCcampus OpenEd, 1st ed, 2014) <https://opentextbc.ca/geography/chapter/6-6-case-studies/>.

55 Ibid.

56 *Agricultural Land Commission Act*, SBC 2002, c 36, s 13(4).

57 Ibid, s 44.

58 Ibid, s 4.3.

59 Ibid, s 4.

60 Ibid, s 11.

61 Ibid, s 11.

62 *Agricultural Land Commission Act*, SBC 2002, c 36, s 11.1.

63 Nathalie Chambers, *Saving Farmland: The Fight for Real Food* (Rocky Mountains Books, 2015).

64 *Agricultural Land Commission Act*, SBC 2002, c 36, s 20 (1).

65 Ibid.

66 Agricultural Land Commission, *Oil and Gas Development in the Agricultural Land Reserve: The Non-Farm Use of Agricultural Land* (2013) <http://www.llbc.leg.bc.ca/public/pubdocs/bcdocs2014/538680/history%20of%20oil%20and%20gas%20activities%20in%20the%20alr%20november%202013.pdf> 1.

67 Agricultural Land Commission, *Oil and Gas Development in the Agricultural Land Reserve: The Non-Farm Use of Agricultural Land, An Historical Overview of the Agricultural Land Commission's Position Regarding Oil and Gas Activities in the ALR* (2013) <https://www.alc.gov.bc.ca/assets/alc/assets/about-the-alc/working-with other ministries-and-agencies/history-of_oil_and_gas_activities_in_the_alr_november_2013.pdf>.

68 BCOGC, *Certificate of Restoration Application Manual* (2016) <http://www.bcogc.ca/node/12445/download> 620.

69 British Columbia, Oil and Gas Commission, *Delegation Agreement for Oil and Gas Uses in the Agricultural Land Reserve Peace River Regional District and Northern Rockies Regional Municipality* (2014) <https://www.bcogc.ca/node/11130/download>.

70 *Agricultural Land Commission Act*, SBC 2002, c 36, s 50–55.

71 Agricultural Land Commission, *Chapter 5.3 Completing Application Information Details: Agricultural Land Reserve* (2013) <http://www.bcogc.ca/node/13290/download>.

72 Ibid, 1.

73 Barry Smith, *A Work in Progress – The British Columbia Farmland Preservation Program* (2012) <http://www.alc.gov.bc.ca/assets/alc/assets/library/archived-publications/alr-history/a_work_in_progress_-_farmland_preservation_b_smith_2012.pdf>.

74 Murray Rankin, Sandy Carpenter, Patricia Burchmore and Christopher Jones, 'Regulatory Reform in the British Columbia Petroleum Industry: The Oil and Gas Commission' (2000) 38 *Alberta Law Review* 143.

75 British Columbia, Oil and Gas Commission, *Delegation Agreement for Oil and Gas Uses in the Agricultural Land Reserve Peace River Regional District and Northern Rockies Regional Municipality* (2014) <https://www.bcogc.ca/node/11130/download>.

76 Murray Rankin, Sandy Carpenter, Patricia Burchmore and Christopher Jones, 'Regulatory Reform in the British Columbia Petroleum Industry: The Oil and Gas Commission' (2000) 38 *Alberta Law Review* 143, 146.

77 Ibid, 147.

78 British Columbia Ministry of Energy and Mines, *Natural Gas Strategy: Fuelling B. C.'s Economy for the Next Decade and Beyond* (2012) <http://www.gov.bc.ca/ener/popt/down/natural_gas_strategy.pdf> 11.

79 Charles Bois and Sarah Hansen, 'Regulatory and Legal Issues Respecting Coalbed Methane Development in British Columbia' (2008) 45 *Alberta Law Review* 631.

80 *Oil and Gas Activities Act*, SBC 2008, c 36, s 70.

81 British Columbia Oil and Gas Commission, *Oil and Gas Activity Operations Manual Version 1.14* (British Columbia Government, 2017) <http://www.bcogc.ca/node/13274/download> 2.

82 Ibid.

83 Madeline Taylor and Susanne Taylor, 'Agriculture in a Gas Era: A Comparative Analysis of Queensland and British Columbia's Agricultural Land protection and Unconventional Gas Regimes' (2016) 22(3) *Australian Journal of Regional Studies* 459, 469.

84 *Oil and Gas Activities Act*, SBC 2008, c 36, s 19.

85 Ibid, s 70.

86 *Oil and Gas Activities Act*, SBC 2008, c 36, s 4.

87 Ibid, s 17.

88 *Forest Act*, RSBC 1996, c 157.

89 *Forest Practices Code of British Columbia Act*, RSBC 1996, c 159.

90 *Heritage Conservation Act*, RSBC 1996, c 187.

91 *Land Act*, RSBC 1996, c 245.

92 *Environmental Management Act*, SBC 2003, c 53.

93 *Water Sustainability Act*, SBC 2014, c 15.

94 *Oil and Gas Activities Act*, SBC 2008, c 36, s 11.

95 Ibid, s 22(2).

96 Ibid, s 70.

97 *Agricultural Land Commission Act*, SBC 2002, c 36.

98 Ibid, s 26(1)(b).

99 British Columbia Oil and Gas Commission, *Certificate of Restoration Application Manual* (British Columbia Government, 2016) < http://www.bcogc.ca/node/12445/download> 38.

100 British Columbia Oil and Gas Commission, *ALC–OGC Delegation Agreement* (British Columbia Government, 2013) <https://www.bcogc.ca/node/5759/download>.

101 *Agricultural Land Commission Act*, SBC 2002, c 36.

102 Ibid, s 4.

103 British Columbia Oil and Gas Commission, *ALC–OGC Delegation Agreement* (British Columbia Government, 2013) <https://www.bcogc.ca/node/5759/download>.

104 *Agricultural Land Commission Act*, SBC 2002, c 36, s 4.

105 British Columbia Oil and Gas Commission, 'Chapter 5.3 Completing Application Information Details: Agricultural Land Reserve' in British Columbia Oil and Gas Commission (eds), *Oil and Gas Activity Application Manual* (British Columbia Government, 2013) <http://www.bcogc.ca/node/13290/download>.

106 Madeline Taylor and Susanne Taylor, 'Agriculture in a Gas Era: A Comparative Analysis of Queensland and British Columbia's Agricultural Land protection and

Unconventional Gas Regimes' (2016) 22(3) *Australian Journal of Regional Studies* 459, 469.

107   British Columbia Oil and Gas Commission, *Oil and Gas Activity Application Manual Version 1.15* (British Columbia Government, 2017) <http://www.bcogc. ca/node/13267/download>.

108   Agricultural Land Commission, *Message from the Chair* (British Columbia Government, 2013), <http://www.ceaa.gc.ca/050/documents/p63919/97838E.pdf>.

109   *Agricultural Land Commission Act*, SBC 2002, c 36, s 18.

110   Ibid, s 6(c).

111   British Columbia Oil and Gas Commission, *ALC–OGC Delegation Agreement* (British Columbia Government, 2013) <https://www.bcogc.ca/node/5759/ download>.

7 United Kingdom

Introduction

The relationship between unconventional gas extraction and agriculture is complex, varying both spatially and temporally. For some countries, the relationship between the two land uses centres on the need to provide both food security and energy security for the nation, such as the case in China. For other jurisdictions, such as Poland, the need for energy security, arising from geopolitical insecurity and historical relationships, may override the need for food security. For other States, the development of unconventional gas triggers questions of coexistence regarding both food and energy security. The UK is one such jurisdiction that seeks to have both food security and energy security. However, this raises the question of whether energy security is achievable in the UK through the development of shale gas alone. Equally, there is another question as to whether the UK can actually achieve food security given the agricultural land area, population and imports of food from surrounding countries and trading partners.

The UK faces a conundrum in relation to its agricultural land. Whereas many countries see agricultural land to be used to produce agricultural commodities and the grazing of livestock (i.e. agricultural products), agricultural land in the UK plays a number of complex roles. Firstly, it provides both visual amenity and recreational value. Visual amenity has to some extent been challenged with the development of wind farms, and this has heightened the argument of the visual role of agricultural land. Similarly, agricultural areas play an important recreational role. For example, the visiting of farms to buy farm produce, recreational activities such as picnicking, or mere recreational value in the form of observation and leisure pursuits. In addition, it is important to note that the Scottish Outdoor Access Code provides all users in Scotland with the ability to access agricultural land (although cropped land is excluded), thereby increasing the use and amenity of that land.[1] Secondly, agricultural land plays an important role in national identity. Such identity arises due to the historic production of artisan foods that are closely associated with geographic locations. Examples of this include, Scotch whisky, Plymouth Gin, cheeses (such as Red Leicester), and Aberdeen Angus beef.

The issue of energy security in the UK has, until recently, not been an issue to be considered. However, this has changed with declining production in the North Sea, particularly declining gas production as a result of the closing of several southern North Sea Fields. For several decades the UK has had a domestic gas shortfall, which has been met largely through pipeline imports from Norway (Ormen Lange gas field provides 20 per cent of the UK's gas needs),[2] and LNG from Qatar and North Africa. However, recently, the UK Conservative Government has expressed its concern regarding energy security. Since 2011, the UK Government has been required to report annually to Parliament on the availability of electricity and gas to meet UK energy demands.[3] This concern primarily relates to security of energy supply and the impact on Britain, should supply be reduced. An example of a threat to security of supply occurred in March 2018, when LNG from Russia's Yamal field destined for China was rerouted to the UK in order to provide necessary gas and fill a supply deficit, the first in eight years.[4] Demonstrably, there is a need for energy diversity in the UK. This raises the question as to whether domestic shale gas is not only likely to fulfil that function, but also whether it is a suitable source of energy to fulfil that function.

This chapter explores the dilemma that the UK faces in relation to shale gas exploration and production and agricultural land use issues that surround the UK energy debate. Acknowledging that agricultural land issues in the UK are unique, this chapter commences with an assessment of the value, importance and history of shale gas in the UK by examining policy, energy security and unconventional gas activities to date. Specifically, it examines threats relating to fracture-induced seismicity, water use, disposal and contamination, threats to agricultural land from transport and spills associated with shale gas activities, and issues related to radioactive materials and the disposal of such materials. The chapter then considers the value and importance of agricultural land in the UK. In understanding and in consideration of this, it examines not only the role of agricultural land in food security but also the further roles that agricultural land plays within the UK psyche. Finally, this chapter will place the extraction of shale gas within a governance framework, examining land ownership and land access, and how the UK has embraced and applied the precautionary approach in managing shale gas extraction and its impact on agricultural land. In undertaking that analysis, the 2017 permanent moratorium on shale gas extraction to preserve agriculture in Scotland will be considered.

Value, importance and history of shale gas in the UK

The exploration for, and production of, shale gas in the UK first arose on the back of the success of shale gas in the USA. In the early 2010s, the UK Government set out to actively encourage shale gas development on the back of cheap gas[5] and its role in the recovery of the US economy after the 2008 crisis.[6] At the core of this was the energy security trilemma. That is, the need to secure access to energy that was cheap and low carbon. These three fundamental concerns that comprise energy security have played out in the UK energy policy since 2012. In particular, unconventional gas is seen as a transition fuel – that is an energy source to transition from heavier carbon dioxide (CO_2) sources, such as coal, to renewable resources. What is interesting about this environmental consideration of shale gas is that the production of gas domestically means less reliance on the import of gas, resulting in a net reduction in global CO_2 emissions.[7] However, such net reduction in direct emissions is offset, to some extent, by future emission of methane during the unconventional gas production process.[8] This means that the domestic production of unconventional gas fails to provide a net reduction in CO_2 emissions, becoming more or less carbon neutral.[9]

This need to produce onshore unconventional gas has arisen because of the decline in gas production in the North Sea. Since 2000, gas production from the UK's Continental Shelf (UKCS) has declined by almost two thirds,[10] and at 2015 the UK was importing 45 per cent of its annual gas supply.[11] This decline is likely to continue, with new imports likely to increase to 75 per cent of gas consumed by 2030.[12] Therefore, primary motivations to produce onshore shale gas include making up for the shortfall in the offshore production of gas and to reduce reliance on gas imports. However, whereas the production of gas offshore has little, if any, visual amenity and social pollution issues, the production of gas onshore not only provides visual pollution but is a real threat to, and is in direct contestation with agricultural land, local communities and water. In addition, the location of the shale gas resources has divided the UK to some extent along regional and social differences, North and South.

Shale gas resources in the UK are located in two distinct areas. The first area is that of the North UK Carboniferous shale region. This region is characterised by a number of critical basins including: the Midland Valley; Tweed; Northumberland Trough; the Bowland Basin; the Gainsborough; and the West Lincolnshire Basin.[13] The North UK shale gas region is located in the lowlands of Scotland and the central and northern belts of England. Much of shale gas is located in lands historically used for coal mining and other industrial activities. The other location for shale gas in the UK is in the South UK Jurassic shale region.[14] This area is characterised by two large basins, the Wessex Basin to the west and the Weald Basin to the east. The Weald Basin had been the site of a large and protracted shale drilling protest at Balcombe, where in 2013 the gas company Cuadrilla sought and received consent to drill a horizontal test well. Although consent to drill was approved for Cuadrilla, mass public protest ultimately led to the delay and eventual abandonment of drilling. Residents sought judicial review of the Council's decision to grant planning permission for the well, which was quashed, partly on the basis that the claimant's grounds for review concerned matters relating to hydraulic fracturing, yet the 'application did not seek permission for the hydraulic fracturing activity'.[15] Interestingly, Cuadrilla revived plans for exploration in the Weald Basin in 2017, submitting a new planning application to the West Sussex County Council (WSCC) for consent to drill a horizontal well and to undertake hydraulic fracturing for a flow test (often known as a 'frack to flow').[16] The application for consent to drill was approved by the WSCC in January 2018.

Aside from Cuadrilla's reapplication for the Balcombe test well, there has been little shale gas exploration activity in the South UK shale area. This contrasts with the North UK shale area, where attempts to drill have continued. Initial drilling at Preese Hall in the Bowland Basin resulted in two seismic events,[17] two reports,[18] and a temporary moratorium on shale gas exploration. However, since the lifting of the moratorium in 2012,[19] there have been several other attempts to undertake shale drilling and hydraulic fracturing in the North UK shale region. However, since the seismic events in Preese Hall in 2011 there has been strong community resistance to further fracking activities. This resistance has been strengthened and enforced by the refusal of some local councils to grant planning permission for many fracking applications.[20] After the initial refusal for shale gas exploration at Roseacre Wood and Preston New Road, the Secretary of State for Communities and Local Government allowed an appeal against Lancashire County Council's decision to refuse permission for hydraulic fracturing at the Preston New Road site, and ultimately granted planning permission, subject to certain conditions. Furthermore, the Minister reopened a public inquiry for the Roseacre Wood site. Drilling was undertaken at the Preston New Road site, with the operator, Cuadrilla, reporting 'excellent rock quality for hydraulic fracturing and a high natural gas content',[21] and releasing its plan to undertake the first UK hydraulic fracture at the well in 2018. At Kirby Misperton, planning permission was granted for drilling and hydraulic fracturing of a single shale well.[22] Proposed changes to the planning process were granted in May 2018, allowing the drilling of wells associated with shale gas as a 'permitted development.' This would allow drilling operations to commence once consent to drill from the newly created Shale Environmental Regulator has been obtained, as well as the development of criteria to include shale gas production projects into the *Nationally Significant Infrastructure Projects* regime.[23]

In championing the use of gas as an energy fuel, the UK Liberal Party has provided several reasons for its importance:

> despite the welcome improvements in efficiency and innovation from companies operating in the North Sea, the ongoing decline in our offshore gas production has meant that the UK has gone from being a net exporter of gas in 2003 to importing over half (53%) of gas supplies in 2017 and estimates suggest we could be importing 72% of our gas by 2030. Our current import mix, via pipelines from Norway and Continental Europe and LNG terminals that can source gas from around the world, provides us with stable and secure supplies. However, we believe that it is right to utilise our domestic gas resources to the maximum extent and exploring further the potential for onshore gas production from shale rock formations in the UK, where it is economically efficient, and where environment impacts are robustly regulated.[24]

Other reasons for developing shale gas include job creation, economic recovery, internal security and reduced cost of gas for consumers.[25] Such reasoning has been promoted to the electorate as the availability of cheap gas for consumer households.

However, it is important to note that gas comprises only 34 per cent of total energy consumption in the UK. Of this gas usage, 30 per cent is used for electricity generation, 37 per cent by households and the remainder by industry and other users.[26] As noted previously, the UK's position in relation to shale gas extraction and its importance to the UK economy came on the back of the US shale gas revolution and the UK's desire to access abundant, cheap gas similar to the benefits enjoyed by the USA.[27] Therefore, 'hoping to emulate the US experience, the UK Government is trying to promote shale gas development and has introduced a favourable tax regime to encourage investment.'[28] However, what is important to note here, is that since 2011, shale gas exploration and production has had significant opposition from community stakeholders, primarily based on the fear or threat of environmental impacts related to water and seismicity.

What is also interesting to note is the shale gas reserves available for production in the UK. To place the UK reserves in context, Australia holds over 429 trillion cubic feet (Tcf) the USA holds more than 622 Tcf, and China 1115 Tcf.[29] The USA Energy Information Administration (EIA) estimates UK's shale resources at 25.8 Tcf.[30] Cuadrilla has claimed approximately 200 billion cubic feet (Bcf) in Lincolnshire Irongas, declaring 170 Bcf in the Northwest. The British Geological Survey estimates approximately 1329 Bcf, or approximately 46.93 Tcf of shale gas in the Boland Basin alone,[31] almost double the Energy Information Administration assessment of Britain's reserves at 25.8 Tcf.[32] The UK shale reserves, whilst paling into insignificance behind Australia, the USA and China, are ranked sixth highest in Europe.[33] However, although the UK enjoys large reserves, it is important to note that these reserves only relate to technically recoverable gas from shale gas reservoirs. This actual recovery will depend upon issues such as geology, percentage organic content (for example, the difference between the Marcellus and Eagle Ford in the USA), and other geological factors such as sheering and complexity of the reservoir. The reservoir in which drilling has occurred to date (the Bowland Basin) has demonstrated irregular and young geology.[34]

A conundrum exists in the exploration for, and the development of, shale gas; in order to determine the stimulatability of the shale reservoir for shale gas production, there is necessity to undertake horizontal drilling and hydraulic fracturing (a 'frack to flow' appraisal) of a well. Such an appraisal well can determine the behaviour of the reservoir and its fracturing capacity, porosity and permeability and the likelihood of seismicity. However, in undertaking 'frack to flow' tests, there is a chance that earthquakes will occur, such as the case in Preese Hall. As noted above, a major issue for shale gas development in the UK has been public opposition to shale gas development. This public opposition has been sustained and arose initially as a result of the two low intensity earthquakes in 2011 (2.3 on the Richter scale on 1 April 2011, and 1.5 on the Richter scale on 27 May 2011) during a hydraulic fracture at a shale gas well in Preese Hall, Lancashire.[35] These earthquakes were the result of hydraulic fracturing being undertaken in geology that is highly faulted, which is typical of UK shale rock, particularly in that region.[36]

The occurrence of the earthquakes at Preese Hall reinforced public concerns regarding hydraulic fracturing, with many seeking a permanent moratorium on shale gas development to be implemented.[37] Such a moratorium was put in place in May 2011, following the Preese Hall seismic events. During this period, two reports were undertaken. One was by the Royal Society and the Royal Academy of Engineering (Royal Society Report),[38] focusing on a review of hydraulic fracturing and its regulation. The other was an independent report, relating to seismicity and wells, commissioned by Cuadrilla and written by two experts in the field, Professor Peter Styles and Dr Christopher Green (the Green and Styles Report).[39] The comprehensive Royal Society Report concluded that although the regulation framework was robust, there were changes that could be made in order to improve the petroleum framework. Such recommendations were accepted by the government and changes implemented. As part of the changes implemented, the UK Government has developed a Regulatory Roadmap report, providing an overview of best practice in onshore oil and gas extraction regulation.[40]

Whereas the Royal Society Report considered both regulatory as well as physical aspects of the incident, the report by Styles and Green considered merely the seismicity issue and the threat to the environment related to shale gas extraction. Styles and Green concluded that rather than the threat arising from seismic activity, the real threat lay in a failure to ensure the integrity of wells, stridently recommending that well integrity be the centre of regulation. In order to address and mitigate frack-induced seismicity, the UK Government implemented a 'traffic light' system.[41] Furthermore, Styles and Green noted that this failure to preserve well integrity contributes to underground contamination of water sources,[42] an observation that supports the findings of Davies et al. in relation to well integrity and contamination.[43]

The earthquakes at Preese Hall triggered a community backlash against shale activities on UK land. Such a backlash has not been alleviated by the publication of the Royal Society Report and the Green and Styles Report. The lifting of the moratorium on drilling in December 2012 has not been received well by communities, who continue to fight against shale gas development. The impact of shale extraction activities on communities was the subject of the 2015 Taskforce on Shale Gas. In its final conclusions and recommendations, the Taskforce on Shale Gas found that shale gas can be produced safely and usefully in the UK, where industry-leading standards are compulsory and where the risk from shale gas to the local environment and public health is no greater than comparable extractive industries.[44] Furthermore, Lord Smith noted that much of the negativity surrounding shale gas activities stems from the USA, where operator standards are comparatively lax.[45] These findings did little to alleviate public concern. The Scottish Government, upon acquiring the right to grant licences for onshore activities as part of the Smith Commission's devolution of powers in the wake of 2014 Scottish Independence Referendum, chose to undertake a study of shale gas extraction in Scotland. After placing a temporary moratorium on hydraulic fracturing in 2016, the Scottish Government undertook a comprehensive review of shale gas and its impact in Scotland.[46] In October 2017, the Scottish Government chose to ban shale gas extraction because it would undermine the Government's climate change policy and lead to unjustifiable environmental damage.[47] The ban on shale gas, both the initial temporary moratorium and the now permanent ban, was implemented through the planning system and enforced under the new powers to grant onshore licences.

Although the UK Government has cited cheap gas and reduced cost to consumers, there is a question as to whether the extraction of shale gas in the UK would be economically viable and emulate that of the USA experience. This question arises since the cost of commercialisation may well be prohibitive.[48] In particular, the cost of pipelines, wells and other essential infrastructure is likely to be extremely costly since there is no pre-existing infrastructure unlike that which exists in the USA for conventional gas sources.[49] Regardless of the economic benefits, the environmental and social implications of shale gas extraction make it relatively unpalatable for many UK communities. Such unpalatability is reinforced by the questionable economic potential of shale gas in the UK, even though its contribution to the USA economy is demonstrable.[50] Therefore, it is clear that the economic benefits of the development of shale gas are yet to be proven, so any consideration of such activities should move beyond a focus on economic benefits, toward concerns of energy security.

The issue of energy security in the UK is a matter of debate that has its origins in not only the need for energy, but also the type of energy. This debate is fractured along the lines of renewable versus non-renewable energy sources, with communities and some local governments resistant to the development of shale gas, but not seeking to increase levels and use of renewable resources. In the future, it is likely that such a relationship between shale gas extraction, a reduction of CO_2 emissions, and the types of energy used will continue to be debated. What is clear, however, is that there is a nexus between the extraction of shale gas and the use of agricultural land. Most shale gas in the UK occurs in rural areas, and especially on productive agricultural land.

There are many aspects of the development of shale gas resources that provide consternation to the community. The most obvious is that of hydraulic fracturing, and its environmental impact.[51] The greatest concern to communities and agriculturalists alike is the potential impacts of shale gas operations on the land. In particular, those relating to the impact on water, both in terms of the use of water,[52] as well as the disposal of water after hydraulic fracturing has been undertaken.[53] Another concern regarding shale gas development on farmland in the UK is that much of the land is only accessible by narrow roads and that spacious road access to drilling sites, which are the hallmark of the US shale gas network, is limited. This poses problems in relation to both access and community acceptance, given than shale activities in the UK are likely to occur in close proximity to individuals and communities. One consequence of narrow country roads and limited accessibility is the risk of chemicals spillage leading to the possibility of surface contamination as a result of transport accidents. According to Clancy et al., high tanker movements associated with hydraulic fracturing will require appropriate mitigation strategies to minimise the risk of spills associated with well pad activities and fluid transportation.[54]

Value and importance of agricultural land in the UK

As of June 2017, the UK produces 52 per cent of all food it consumes, thereby importing almost half of its food.[55] Of this imported food, the EU accounts for 70 per cent of all imports, as well as accounting for 60 per cent of all food exports.[56] Whether this poses a threat to food security depends on the types of foods imported. At present in the UK, imports exceed domestic production in all areas except for beverage production where exports exceed imports. This is strongly linked to the Scotch whisky sector, as well as the production of boutique gins and other alcoholic beverages. What is most challenging for the UK is that fruit and vegetable imports exceeds exports by approximately 6 to 1, and meat imports exceed exports by a factor of at least 3 to 1.[57] The UK relies on fruit and vegetable imports primarily from two countries; the Netherlands and Spain, which together account for the import of 69 per cent of fresh vegetables.[58] The UK also relies heavily on the import of animal feed stock for livestock production, particularly in relation to poultry and pig sectors, with the majority of such imports sourced from Argentina and Brazil.[59] At the heart of British food security is the reliance on the EU for food imports. This highlights the challenges to UK food security in a post-Brexit where withdrawal from the EU will require the forging of new international food trade deals and finding a market for UK food exports and agricultural labour. The security of food in relation to these factors will be highly dependent on the negotiated outcome of the UK Brexit Exit Strategy.[60]

Of more universal concern in relation to food security and agricultural land is in relation to distribution of food, agricultural yields and access to markets. Of importance here is the issue of agricultural yields. The Parliamentary Office of Science and Technology notes that while it supports sustainable agricultural intensification in the UK, the yield of the UK staple crop – wheat, has not increased for more than a decade.[61] Furthermore, the House of Commons Committee indicates that the cause of such yield plateau is not well understood, and increases to yields will necessarily include the use of new technologies.[62] However, what is heartening to note is that of the crops that can be produced in the UK, the country is 68 per cent self-sufficient.[63]

In the UK, food security concerns are related to the UK's relationship with the EU and the alteration of that relationship as a result of Brexit. This demonstrates that ongoing food security is not tied to the use of farming land for shale gas production, but rather is part of a larger systemic food security issue. Although UK food security concerns are not related specifically to shale gas production, it is yet to be proven that shale gas activities will not have an impact on agricultural land and productivity. In the UK, agricultural land provides an important role in recreation and visual amenity. There have been many concerns about both of these factors in shale gas production (aside from the environmental impact of hydraulic fracturing). Many of the concerns regarding shale gas activities on agricultural land relate to the impact on local communities, the effect on visual amenity, increases in traffic and reduction in the use and enjoyment of the land.[64] These factors have been echoed in the Balcombe Protests as well as the wider issues related to water use in hydraulic fracturing.

The protection of the use and enjoyment of the land is undertaken through the planning system of the UK (England & Wales, and Scotland), which prevents activities being undertaken in areas where such harm may occur. This has been demonstrated by the refusal to grant permission for drilling in Lancashire in the Bowland Basin (Preston New Road). Such rejection of shale gas continues in 2018, with Ineos' application for shale gas exploration in the Rotherham area of South Yorkshire denied by the local council. Similar to other applications for shale gas exploration drilling that have been denied by councils, the reasoning for the denial of the Rotherham application included concerns regarding noise, water and air pollution and traffic created by movement of high volumes of heavy goods vehicles.[65]

Whereas much of the importance of agricultural land is placed on the use and enjoyment of the land in relation to visual and recreational amenity, agricultural land in the UK also plays an important part in artisanal food production. The UK is renowned for the production of foods such as fresh meats, charcuterie and dairy products, such as Stilton cheese. In addition, agricultural land is used to produce ingredients for perhaps the most important artisanal products from the UK, that of Scotch whisky. The unique flavours and global popularity of Scotch whisky are due to the both the barley grown on agricultural land close to the distilleries as well as the water used in the production of whisky (in particular the Spey River), thus illustrating the importance of both agricultural and natural resources. Fortunately, the water sources for Scotch whisky are outside of shale gas production areas. However, the Scottish Government, in its 2017 analysis of shale exploitation, noted that the natural and agricultural value of the land in Scotland was an important consideration in its decision to permanently ban shale gas production.[66]

Accessing agricultural land for shale gas activities

Although shale gas development in Scotland has been permanently banned, it continues in England. As noted previously, decisions regarding final approval of shale activities are made at the local level under the planning system.[67] It is important to understand the overarching regulation for shale gas activities, and the rights and interests of those affected by such activities, in relation to land access and compensation. Such regulation of land access reflects the fundamental property law aspects of the English common law system, particularly in relation to Crown reservation and ownership of minerals.

Overview of legal framework

Under the common law of England and Wales, the landowner owns all mines and minerals underneath the soil of a land. Minerals are defined in the *Town and Country Planning Act 1990* (UK) as, 'all substances of a kind ordinarily worked for removal by underground or surface working, except that it does not include peat for purposes other than for sale'. There is a presumption that the surface owner of the land owns all the minerals in the underlying strata, with the exception of gold and silver. The royal prerogative to the metals of gold and silver was declared in *R v. Earl of Northumberland* [68] (known as the *Case of Mines*) when Queen Elizabeth I was concerned about Crown revenue at a time of conflict, particularly with Spain.

This common law position regarding ownership of minerals in the UK (which presumably included petroleum) was altered by statute in 1934, when ownership of oil and gas onshore in the United Kingdom was vested in the Crown by the *Petroleum (Production) Act 1934* (UK). Under section 1(1) of this Act, 'the property in petroleum exiting in its natural condition in strata in Great Britain is hereby vested in His Majesty, and His Majesty shall have the exclusive right of searching and boring for and getting such petroleum'. Crown ownership of petroleum that was granted under the *Petroleum (Production) Act 1934* (UK) was retained under the *Petroleum Act 1998* (UK), granting the Crown the exclusive onshore and offshore right of 'searching and boring for and getting petroleum', which ensues today. Under section 7(1) of the *Petroleum Act 1998* (UK), the *Mines (Working Facilities and Support) Act 1966* (UK) applies in England, Wales and Scotland to enable a licensee to acquire ancillary rights to assist with the development of the Petroleum Exploration and Development License (PEDL). Such ancillary rights include access rights that encompass entry onto the land to undertake drilling operations, and drilling under the land. Access to, and use of, water is governed by the *Water Act 2014* (UK). This access and use are part of the environment approval that is required before drilling can occur. The permissioning, volume and conditions of access vary according to location and water source (surface versus groundwater). There is yet to be a clear policy or legal framework regarding water access and use for shale gas extraction due to the infant nature of the industry. In order to undertake any shale gas activities, it is essential that the PEDL holder secures access to the land, both on the surface and underground.

Surface access

The right of a PEDL holder to enter onto private land to undertake drilling operations is obtained by private landholder negotiation. However, the permission to enter onto private land can also be granted by a court if it is not reasonably practicable to obtain it by private negotiation. Aside from securing permission to enter onto private land to undertake shale drilling operations, companies are required to inform local residents in an affected area of their intention to undertake hydraulic fracturing activities prior to seeking planning permission.[69] To drill on land, planning permission is required under the relevant regulation provisions of the *Town and Country Planning Act 1990* (UK) and is required prior to the Department of Energy and Climate Change (DECC) granting consent for drilling or production. The decision to allow drilling is made in accordance with the National Planning Policy Framework (NPPF) and the National Planning Practice Guidance (NPPG).

Recognising concerns in relation to the impacts of shale gas on the community as well as the land, the United Kingdom Onshore Operators Group (UKOOG) launched the *Shale Community Engagement Charter* (the Charter) in June 2013, with a revision in 2015. The Charter outlines the steps the industry will take to address concerns around safety, noise, dust, truck movements and other environmental issues, as well as how titleholders will communicate and engage with the community in all areas of concern. Under the Charter, the UKOOG seeks to identify and proactively address local issues and concerns, facilitate sustainable development of extractive resources and achieve an appropriate balance between safe energy production and community need. Based on these objectives, the Charter promises to engage with individuals and organisations in the community from an early stage and to provide compensation to the local community at a rate of £100,000 per well site where hydraulic fracturing takes place. Furthermore, the Charter promises to share the proceeds of production at a rate of 1 per cent of revenue, split 2/3–1/3 between local community and county council level. In addition to the industry-based Charter, the Office of Unconventional Gas and Oil is responsible for ensuring that the local communities benefit from shale gas development in the area.

Underground access

Under the system of landownership, licenseholders do not have an automatic right to drill under landowners' property and require prior permission to undertake such drilling. If permission is refused, then licenseholders can apply through the Secretary of State and courts to gain access. However, this has been seen as cumbersome, with land access seen as a real barrier to the development of shale gas. The need to reform the land access regime was seen as critical after the landmark case of *Star Energy Weald Basin Ltd & Anor v. Bocardo SA*, [70] which found that Star Energy had committed trespass to land by drilling and installing pipelines under Bocardo's land, even though the wells were 950–2800 feet below the surface. The decision affirmed the view that the landowner owns the land and that any activity under a landholder's land, whatever the depth, will constitute trespass. Therefore, in order to avoid committing trespass, the licenseholder is required to come to an agreement with the landholder or apply to a court for an ancillary right pursuant to the *Petroleum Act 1998* (UK) and the *Mines (Working Facilities and Support) Act 1966* (UK).

Access reform was enacted by the *Infrastructure Act 2015* (UK), which grants the right to use deep-level onshore land below 300 m to exploit petroleum and geothermal energy within s 43. 'Use' is defined by s 44 and includes drilling, boring, fracturing, installing and maintaining infrastructure, passing any substance through the land, and keeping or removing any substance put into deep-level land or infrastructure in deep-level land.

Negotiation of compensation and dispute resolution

For surface land access the interested parties, through private negotiation, negotiate compensation. There are no mandated requirements for this negotiation, levels of compensation, or dispute resolution. Where access consent is unreasonably withheld, there is the capacity to gain access through the court. Section 45 of the *Infrastructure Act 2015* authorises the Secretary of State to direct relevant companies to make payments to owners of relevant land or other persons for the benefit of areas in which the relevant land is situated.[71] Furthermore, the company is required to give notice of any payments and activities.

Governance of shale gas activities in agricultural areas

The governance framework in relation to shale gas activities comprises regulation of petroleum extraction at national level (*Petroleum Act 1998* (UK) and associated regulations, as well as applicable safety law and regulation) as part of a consent to drill application made by the license holder. Once a PEDL has been granted for exploration and development operations, the titleholder has the right to drill. However, permission to drill is not automatically granted. Rather, it is a separate process. This permissioning process is made subject to the relevant laws in relation to petroleum operations in the UK and is illustrated in Figure 7.1.

What is important to note from Figure 7.1 is that UK planning permission is an integral part of the governance framework for shale gas activities. Under Planning Regulations, it is the role of the local authority to approve or deny a planning application for a shale gas drill. When undertaking assessment of a planning application, the Minerals Planning Authority is required to assess the application against the local development plan and to consider whether the proposed development constitutes an acceptable use of the relevant land and whether the impact of those uses is acceptable.[72] A planning application will then be either approved or denied and this will be on the grounds of a consideration of all aspects including its impact on the local population.[73] The consent to drill permission system is overseen by the Oil and Gas Authority, which undertakes a coordinating function. Drilling consent is only granted when the relevant planning authority has granted permission and environmental approvals and permits from the relevant environmental agency have been obtained. The Health and Safety Executive approves the well design, based on the assessment and certification of an independent and component well examiner, and notification of the intent to drill to the British Geological Survey (BGS). This last requirement, that of BGS notification, serves to ensure

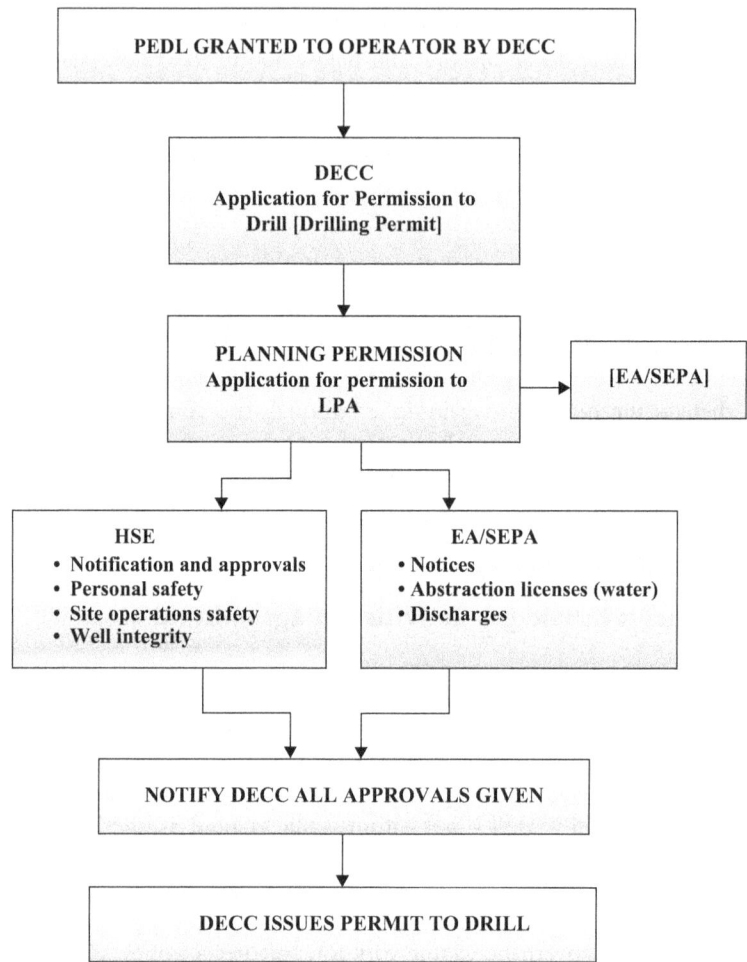

Figure 7.1 **Permissioning process**
Source: John Paterson and Tina Hunter, 'Shale Gas Law and Regulation in the United Kingdom' in Tina Hunter (ed.), *Handbook of Shale Gas Law and Policy* (Intersentia, 2016) 254

that the traffic light system (of regulatory review) implemented after Preese Hall can be implemented and seismology monitored.

As can be seen from the analysis above, since the initial hydraulic fracturing that was undertaken in 2011, the UK has undergone a long process of evaluation, regulatory reform and public concern in relation to shale gas extraction. It is clear that the UK Government seeks to extract shale gas in order to meet its perceived energy security shortfalls. The issue of whether the extraction of shale gas will indeed meet such shortfalls is not the subject of this chapter.

Rather, this chapter analyses whether the approach taken by the UK Government to shale gas activities on agricultural land to date is appropriate.

Since 2011, fear and community consternation has been at the forefront of shale gas activates in the UK. The Government has sought to undertake consultation and provide the public with assurances of reforms to governance and oversight in relation to shale gas extraction, in particular through the establishment of the Task Force on Shale Gas, the Royal Society Report and the Styles and Green Report after Preese Hall. Since the release of these reports, there has been a measured approach to shale gas industry development. However, recent overturning of planning decisions indicate that although the UK Government embraces a precautionary approach, it is keen to develop shale gas. What is missing from the UK's precautionary approach is a detailed, proactive comprehensive report on the impact of shale gas activities on UK land, agriculture and communities. Although the EU has commissioned a number of reports in relation to the impact of shale gas extraction on health, environment and social impact, a comprehensive report for the UK does not yet exist. Perhaps the best international example of such a comprehensive analysis is that of the Pepper Report[74] which considered in comprehensive detail and scope the scientific, social, legal and economic impact of shale gas extraction in the Northern Territory, Australia.

Whilst the precautionary principle is not explicitly stated within the shale gas regulatory framework in the UK, the aim of the precautionary principle has been codified in EU Environmental Law. Indeed, article 191 of the 2010 Treaty on the Functioning of the EU (TFEU) explicitly states that environmental policies should be based on the precautionary principle and on the principles that 'preventative action should be taken'.[75] Article 194 of the TFEU includes a provision on integrating environmental considerations into the EU energy policy as energy policies need to be made 'with regard for the need to preserve and improve the environment'.[76] Such imbued principles have been incorporated into the UK's domestic framework and, unless this framework is altered in the post Brexit era, the principles are likely to remain.

Conclusion

With the decline in off sea gas fields, onshore shale gas has become relatively important in the UK energy mix. Political leaders in the UK believe that shale gas holds many of the answers when viewed through an energy security and economic development lens. While this chapter does not analyse the veracity of such views, it has considered the impact of shale gas development on agricultural land in the UK. Rather than only considering the implications on food security, this chapter has extended the examination of the impact of shale gas activities to incorporate its impact on the use and enjoyment of agricultural land. This has been necessary because of the intrinsic value of agricultural land in the UK. This value is derived not just from activities in relation to food security, but also values founded on food sovereignty principles such as the value of agricultural land held by small-scale artisanal farming, the related lifestyle and recreational values, all of which have been held dearly by the British for generations. The importance of agricultural land for recreation and visual amenity traces its origins to Victorian times, particularly when rural areas provided relief for crowded industrial cities. This social role of agricultural land today has a major bearing on how shale gas activities are to be conducted in the British landscape.

Given the UK Government's belief that its precautionary approach to shale gas activities represents best practice, the Government sees shale gas development as ongoing, and is fully committed to such activities occurring. This commitment is demonstrated by the recent ministerial decision to overturn the denial of planning permission in relation to drill applications. This represents a new paradigm in shale gas extraction, one where the visual and recreational amenity of the agricultural land, rather than food production value alone, is considered when granting permission to drill. However, the UK approach remains imbued in the precautionary approach principle that is integrated into its regulatory framework as a result of EU legislation and competencies. This is likely to remain after Brexit occurs in 2019.

Notes

1 The Scottish Land Access Code was implemented as part of the *Land Reform (Scotland) Act 2004* (Scot).
2 Shell, *Ormen Lange* (Shell Global, 2018) <https://www.shell.com/about-us/major-projects/ormen-lange.html>.
3 This obligation arises under section 172 of the *Energy Act 2004* (UK), as amended by section 80 of the *Energy Act 2011* (UK).
4 See David Sheppard and Andrew Ward, 'Siberian Gas Delivery to UK Offers Relief After Cold Blast: LNG Arrival Highlights Fragile State of British Energy Security', 3 March 2018, *Financial Times*, <https://www.ft.com/content/31e076e2-1e28-11e8-956a-43db76e69936>.The LNG cargo was required after a series of cold weather periods placed extreme pressure on gas supplies in the UK. The gas was sourced and collected by Shell by an at-sea transfer off the coast of western France. Since the LNG cargo was originally set to be delivered to China via the Northern Sea Route, the LNG carrier was an icebreaker. The UK's energy security vulnerability has been linked to the closure of the Rough gas storage facility in the North Sea, which was capable of storing one tenth of the UK's daily peak gas demand.
5 The Henry Hub price for gas hovers around US$3 as a result of the availability of cheap shale gas in the USA.
6 Department of Environment and Climate Change and Department of Communities and Local Government, *Shale Gas and Oil Policy Statement by DECC and DCLG* (United Kingdom Government, 2015) <https://www.gov.uk/government/publications/shale-gas-and-oil-policy-statement-by-decc-and-dclg/shale-gas-and-oil-policy-statement-by-decc-and-dclg>. The national need to explore shale oil and gas resources is set out in this document.
7 Laurence Stamford and Adisa Azapagic, 'Life Cycle Environmental Impacts of UK Shale Gas' (2014) 134 *Applied Energy* 506, 507.
8 Christopher McGlade, Steve Pye, Paul Ekins, Michael Bradshaw and Jim Watson, 'The Future Role of Natural Gas in the UK: A Bridge to Nowhere?' (2018) 113 *Energy Policy* 454, 454–455.
9 Joanne Hawkins, 'Fracking: Minding the Gaps' (2015) 17(1) *Environmental Law Review* 8, 10.

10 Tina Hunter, 'Converging Energy Governance in Mature Petroleum Provinces: Political, Legal, and Economic Dimensions in Governing Mature Petroleum Fields in the North Sea' in Slawomir Raszewski (ed.), *The International Political Economy of Oil and Gas* (Springer, 2017) 168 and Figure 11.1.

11 Department of Environment and Climate Change, *Digest of UK Energy Statistics* (United Kingdom Government, 2015) <https://www.gov.uk/government/collections/digest-of-uk-energy-statistics-dukes#2015>.

12 Ibid.

13 US Energy Information Administration, *Technically Recoverable Shale Oil and Gas Resources: United Kingdom* (United States Government, 2015), XI-1.

14 Ibid.

15 *R (Frack Free Balcombe Residents Association) v West Sussex County Council* [2014] EWHC 4108 (Admin) [131].

16 West Sussex County Council, *Application No WSCC/040/17/BA* (WSCC, 2017) <http://buildings.westsussex.gov.uk/ePlanningOPS/loadFullDetails.do?aplId=2178>.

17 British Geological Survey, *Fracking and Earthquake Hazard* (Natural Environment Research Council, 2018) <http://earthquakes.bgs.ac.uk/research/earthquake_hazard_shale_gas.html>. These events mentioned were two minor earthquakes on 1 April 2011 (2.3 magnitude) and 27 May 2011 (1.5 magnitude) that were induced by hydraulic fracture treatments at the Preese Hall well PH1.

18 Royal Society and Royal Academy of Engineering, *Shale Gas Extraction in the UK: A Review of Hydraulic Fracturing* (Royal Society, 2012) <https://www.raeng.org.uk/publications/reports/shale-gas-extraction-in-the-uk>; Christopher Green, Peter Styles and Brian Baptie, *Preese Hall Shale Gas Fracturing: Review and Recommendations for Induced Seismic Mitigation* (Assets Publishing, 2012) <https://assets.publishing.service.gov.uk/government/uploads/system/uploads/attachment_data/file/15745/5075-preese-hall-shale-gas-fracturing-review.pdf>.

19 Edward Davey, *Written Ministerial Statement by Edward Davey: Exploration for Shale Gas* (United Kingdom Government, 2012) <https://www.gov.uk/government/speeches/written-ministerial-statement-by-edward-davey-exploration-for-shale-gas>.

20 Applications denied planning consent include Roseacre Wood and Preston New Road, Fylde and Grange Road, Singleton.

21 Cuadrilla Resources, *Preston New Road: Latest News* (CR, 2018) <https://cuadrillaresources.com/site/preston-new-road/>.

22 North Yorkshire County Council, *Kirby Misperton Fracking Operations* (NYCC, 2017) <https://www.northyorks.gov.uk/kirby-misperton-fracking-operations>.

23 Greg Clark, *Energy Policy Written Statement – HCWS690 Statement to Parliament 17 May 2018* (United Kingdom Government, 2018) <https://www.parliament.uk/business/publications/written-questions-answers-statements/written-statement/Commons/2018-05-17/HCWS690/>.

24 Ibid.

25 Ibid.

26 Christopher McGlade, Steve Pye, Paul Ekins, Michael Bradshaw and Jim Watson 'The Future Role of Natural Gas in the UK: A Bridge to Nowhere?' (2018) 113 *Energy Policy* 454, 455.

27 Laurence Stamford and Adisa Azapagic, 'Life Cycle Environmental Impacts of UK Shale Gas' (2014) 134 *Applied Energy* 506, 507.

28 Jasmin Cooper, Laurence Stamford and Adisa Azapagic, 'Economic Viability of UK Shale Gas and Potential Impacts on the Energy Market up to 2030' (2018) 215 *Applied Energy* 577, 578.

29 US Energy Information Administration, *World Shale Resource Assessments* (United States Government, 2015) <https://www.eia.gov/analysis/studies/worldshalegas/>.

30 Ibid.

31 British Geological Survey, *Bowland Shale Gas* (BGS, 2016) <http://www.bgs.ac.uk/research/energy/shaleGas/bowlandShaleGas.html>.

32 US Energy Information Administration, *World Shale Resource Assessments* (United States Government, 2015), <https://www.eia.gov/analysis/studies/worldshalegas/>.

33 Other significant reserves include, in descending order, Poland (145.8), France (136.7), Ukraine (127.9), Romania (50.7), and the Netherlands (25.8).

34 US Energy Information Administration, *Technically Recoverable Shale Oil and Gas Resources: United Kingdom* (United States Government, 2015), XI-3.

35 British Geological Survey, *Fracking and Earthquake Hazard* (BGS, 2018) <http://earthquakes.bgs.ac.uk/research/earthquake_hazard_shale_gas.html>.

36 US Energy Information Administration, *Technically Recoverable Shale Oil and Gas Resources: United Kingdom* (United States Government, 2015), XI-3.

37 See, eg, the public protests surrounding the drilling of the Balcombe well since 2013.

38 Royal Society and Royal Academy of Engineering, *Shale Gas Extraction in the UK: A Review of Hydraulic Fracturing* (Royal Society, 2012) <https://www.raeng.org.uk/publications/reports/shale-gas-extraction-in-the-uk>.

39 Christopher Green, Peter Styles and Brian Baptie, *Preese Hall Shale Gas Fracturing: Review and Recommendations for Induced Seismic Mitigation* (Assets Publishing, 2012), <https://assets.publishing.service.gov.uk/government/uploads/system/uploads/attachment_data/file/15745/5075-preese-hall-shale-gas-fracturing-review.pdf>.

40 UK Government, *Regulatory Roadmap – Onshore Oil and Gas Exploration in the UK: Regulation and Best Practice* (United Kingdom Government, 2018) <https://www.gov.uk/government/publications/regulatory-roadmap-onshore-oil-and-gas-exploration-in-the-uk-regulation-and-best-practice>.

41 For a discussion on the 'traffic light' system see, eg, Oil and Gas Authority, *OGA Traffic Light Monitoring Scheme to mitigate induced seismicity* (OGA) <https://www.ogauthority.co.uk/media/3860/traffic-light-system-doc-for-website_final.pdf>. Under this system, where seismicity is in the range 0.00–0.5, fluid injection is to be at a reduced rate, while any seismicity over 0.5 requires a halt to fluid injection.

42 Christopher Green, Peter Styles and Brian Baptie, *Preese Hall Shale Gas Fracturing: Review and Recommendations for Induced Seismic Mitigation* (Assets Publishing, 2012) <https://assets.publishing.service.gov.uk/government/uploads/system/uploads/attachment_data/file/15745/5075-preese-hall-shale-gas-fracturing-review.pdf>.

43 Richard Davies et al., 'Oil and Gas Wells and their Integrity: Implications for Shale and Unconventional Resource Exploitation' (2014) 56 *Marine and Petroleum Geology* 239.

44 Task Force on Shale Gas, *Final Conclusions and Recommendations* (TFSG, 2015) <https://darkroom.taskforceonshalegas.uk/original/d6f5f84dbfecbe9c22bddbc7f93d31bc:cb2ee01d6a9d7a96cd7d10262971d586/task-force-on-shale-gas-final-conclusions-and-recommendations.pdf>.

45 Ibid, 1.

46 Scottish Government, *Onshore Oil and Gas* (2017) <http://www.gov.scot/Topics/Business-Industry/Energy/onshoreoilandgas>.

47 Scottish Government, *Unconventional Oil and Gas – Statement* (Minister for Business, Innovation and Energy, Scottish Government, 2017) <https://news.gov.scot/speeches-and-briefings/unconventional-oil-and-gas-statement>.

48 Jasmin Cooper, Laurence Stamford and Adisa Azapagic, 'Economic Viability of UK Shale Gas and Potential Impacts on the Energy Market up to 2030' (2018) 215 *Applied Energy* 577, 578.

49 Ibid; only six wells have been drilled to date in the UK since 2010.

50 Laurence Stamford and Adisa Azapagic, 'Life Cycle Environmental Impacts of UK Shale Gas' (2014) 134 *Applied Energy* 506, 506.

51 For a detailed discussion on hydraulic fracturing and its impact see Peter Styles, 'Shale, Shale Gas and Hydraulic Fracturing' in Tina Hunter (ed.), *Handbook of Shale*

Gas Law and Policy: Economics: Access, Law and Regulation in Key Jurisdictions (Intersentia, 2016).

52 Jake Hays et al., 'Considerations for the Development of Shale Gas in the United Kingdom' (2015) 512–513 *Science of the Total Environment* 36.

53 For a detailed discussion on the impact of hydraulic fracturing and shale gas operations on water (both use and water contamination) see David Campin, 'Regulating Hydraulic Fracturing' in Tina Hunter (ed.), *Handbook of Shale Gas Law and Policy: Economics: Access, Law and Regulation in Key Jurisdictions* (Intersentia, 2016); Andrew Garnett, 'Regulating Well Integrity' in Tina Hunter (ed.), *Handbook of Shale Gas Law and Policy: Economics: Access, Law and Regulation in Key Jurisdictions* (Intersentia, 2016); Y. Kharaka et al., 'The Energy–Water Nexus: Potential Groundwater-quality Degradation Associated with Production of Shale Gas' (2013) 7 *Procedia Earth and Planetary Science* 417; and Jose Estrada and Rao Bhamidimarri, 'A Review of the Issues and Treatment Options for Wastewater from Shale Gas Extraction by Hydraulic Fracturing' (2016) 182 *Fuel* 292.

54 Sarah Clancy et al., 'The Potential for Spills and Leaks of Contaminated Liquids from Shale Gas Developments' (2018) 626 *Science of the Total Environment* 1463, 1463.

55 House of Parliament, Parliamentary Office of Science and Technology, *Security of UK Food Supply* (United Kingdom Government, 2017), number 556, 1.

56 Ibid.

57 Ibid.

58 Ibid.

59 Ibid, 2.

60 Ibid.

61 Ibid, 3.

62 House of Commons Environment, Food and Rural Affairs Committee, *Food Security: Second Report of Session 2014–15* (United Kingdom Government, 2014) 22–23.

63 House of Parliament, Parliamentary Office of Science and Technology, *Security of UK Food Supply* (United Kingdom Government, 2017), number 556, 3.

64 For a social analysis of shale gas issues see John Whitton et al., 'Shale Gas Governance in the UK and US: Opportunities for Public Participation and the Implications for Social Justice' (2017) 26 *Energy Research and Social Science* 11.

65 Adam Vaughan, 'UK Fracking Backlash: Seven of Eight Plans Rejected in 2018', *The Guardian* (Online) (2018) <https://www.theguardian.com/environment/2018/mar/08/uk-fracking-backlash-seven-out-of-eight-plans-rejected-in-2018>.

66 Scottish Government, *Unconventional Oil and Gas – Statement* 3 October 2017 (Minister for Business, Innovation and Energy, Scottish Government, 2017) <https://news.gov.scot/speeches-and-briefings/unconventional-oil-and-gas-statement>.

67 For a detailed consideration of the role of the planning system in unconventional gas exploitation in the UK, see Tina Hunter, Steven Latta and Greg Gordon, 'Current Practice and Emerging Trends in Regulating Onshore Exploration and Production in Great Britain' in Greg Gordon, John Paterson and Emre Usenmez (eds), *UK Oil and Gas Law: Current Practice and Emerging Trends, Volume I* (EUP, 3rd edn, 2018).

68 *R v Earl of Northumberland* (1568) 1 Plowden 310.

69 Department of Environment and Climate Change, *Onshore Oil and Gas Regulation in the UK: Regulation and Best Practice* (United Kingdom Government, 2013).

70 *Star Energy Weald Basin Ltd & Anor v Bocardo SA* [2010] UKSC 35.

71 *Infrastructure Act 2015* (UK) s 45.

72 *Town and Country Planning Act 1990* (UK) ss 70(2), 71, 71A.

73 For a further consideration of planning system and its impact and control over shale gas regulation, see Tina Hunter, Steven Latta and Greg Gordon, 'Current Practice

and Emerging Trends in Regulating Onshore Exploration and Production in Great Britain' in Greg Gordon, John Paterson and Emre Üsenmez (eds), *UK Oil and Gas Law: Current Practice and Emerging Trends, Volume I* (Edinburgh Press, 3rd edn, 2018).
74 Rachel Pepper et al., *Scientific Inquiry into Hydraulic Fracturing in the Northern Territory: Final Report* (Northern Territory Government, 2018) ('Pepper Report').
75 *Treaty on the Functioning of the European Union*, opened for signature 7 February 1992, [2009] OJ C 115/13 (entered into force 1 November 1993) ('TFEU') art 191.
76 Ibid, art 194.

8 New York State, USA

Introduction

Shale gas extraction and exploration in the USA is longstanding, having been in operation since 1825 in New York State. Thus, legislative and policy issues in the USA, as a 'mature State' for shale gas, are much more developed in some aspects than in other jurisdictions, such as Australia and China. However, it was not until the development of 'slickwater' hydraulic fracturing by George Mitchell in Texas, in the decade from 1997, that shale gas became a 'game changer' to the USA.[1] As of 2017, unconventional gas now equates to 60 per cent of total USA natural gas production.[2] This rapid increase has been attributed to the advent of slickwater hydraulic fracturing techniques, coupled with the injection of chemicals and water to 'fracture' the shale rock and produce unconventional gas.[3]

The expansion of hydraulic fracturing in the USA has led to a plethora of diametric regulatory positions amongst the states. States such as Texas and Pennsylvania have rushed to embrace shale gas through an adaptive management approach. This has encouraged exploitation, positioning shale as a transition fuel during the shift from oil and coal burning to energy generation from carbon-free sources. This is contrary to the position of other states, such as New York, which have chosen to place a permanent ban on hydraulic fracturing. This ban is based on the negative findings associated with hydraulic fracturing on rural communities, the environment and public health implications stemming from potential contamination of the Catskill Water Table.[4] The Marcellus shale formation is the main source of natural gas production in the USA and is located underneath the states of New York, Pennsylvania, West Virginia and Ohio.[5] In 2015, after a 7-year review, the state of New York permanently banned all hydraulic fracturing activities of shale gas in its regional jurisdiction of the Marcellus play. The socio-legal environment that led to the ban in New York State, compared with the adaptive and market-led regulation of neighbouring state Pennsylvania,[6] demonstrates the application of the precautionary principle.

This chapter presents an analysis of application of the precautionary principle to hydraulic fracturing activities in New York State. It explores the regulatory enactment of the state-wide ban on hydraulic fracturing, before analysing the regulatory response of rural communities in the adoption of the Municipal Home Rule. The implementation of the Municipal Home Rule, by banning oil and gas activities, has effectively empowered rural communities to pass self-administered laws for the protection and enhancement of their general welfare.[7] An example of self-regulation is the Joint Landholders Coalition Of New York (JLCNY), which demonstrates the power of collective action as landholder groups collectively negotiate community oil and gas leases. The New York experience provides important and salutary examples of the role played by local communities in influencing the trajectory of state government legislation in response to the challenges of shale gas industrial activity.

The precautionary principle as a regulatory approach

As examined in Chapter 4, the precautionary principle is a regulatory tool applied in the context of scientifically uncertain and potentially harmful activity. Principle 15 of the *Rio Declaration on Environment and Development* contains the well-cited definition of the principle, which states:

> In order to protect the environment, the precautionary approach shall be widely applied by States according to their capabilities. Where there are threats of *serious or irreversible damage*, lack of full scientific certainty shall not be used as a reason for postponing cost effective measures to prevent environmental degradation.[8]

Although the Rio Declaration is only legally binding upon signatory States, Principle 15 requires a precautionary approach to be taken in the face of serious or irreversible damage coupled with scientific uncertainty. The precautionary principle is a fall-back position for States who are concerned about the potential effects of a certain regulatory activity as surmised by Peel, 'In a global "risk society", the continuing relevance of precaution is assured by communities' concern with the possibility of harm in circumstances where science is unable to offer guarantees of safety'.[9]

The precautionary principle has been enshrined within a range of international law instruments.[10] For example, the first inclusion of the precautionary principle in international law was adopted within the 1987 *Declaration of the Second International North Sea Conference on the Protection of the North Sea*, describing the principle in its specific application to the North Sea:

> in order to protect the North Sea from possible damaging effects of the most dangerous substances, a precautionary approach is necessary which may require action to control inputs of such substances even before a causal link has been established by absolute clear scientific evidence.[11]

The Lowell Statement provides an additional dimension to the precautionary principle, advocating the adoption of the principle 'when complex and uncertain threats must be addressed … even if the exact nature and magnitude of the harm are not fully understood'.[12] The Lowell Statement therefore extends the precautionary principle to enable a participatory regulatory process through the 'application of transparent and inclusive decision-making processes that increase participation of all stakeholders and communities, particularly those potentially affected by a policy choice'.[13] Kriebel et al. provides a complementary definition of the precautionary principle in their four element explanation: 'preventive action in the face of uncertainty; shifting the burden of proof to the proponents of an activity; exploring a wide range of alternatives to possibly harmful actions; and increasing public participation in decision making'.[14]

According to Resnik's model, two key questions must be considered by law makers when applying the precautionary principle to determine the appropriate regulatory approach: '(1) Are the threats to be addressed by the precautionary principle plausible? (2) Are the precautionary measures adopted reasonable?'[15] Therefore, lawmakers must consider the 'level and nature of uncertainty' and the 'spatial, magnitude, longevity and manageability of the threat'[16] in determining whether to prohibit an activity on the basis of precaution.

The differential approaches in literature regarding the precautionary approach, as well as its application to natural resource activities, reveal that enacting prohibitive or mitigating measures to avoid irreversible damage and engaging public participation are essential steps for lawmakers considering the effective regulation of new and evolving regulatory challenges. In evaluating the effectiveness of the application of the precautionary principle, a number of practical measures must be taken into account, including 'effectiveness, proportionality, cost-effectiveness, realism and consistency, to determine whether a response to a threat is reasonable' in the application of the principle.[17]

The precautionary principle holds the benefit of regulatory flexibility for a State. When applied effectively, it permits the State to regulate via a variety of regulatory tools,[18] depending on the level of anticipated degradation and the corresponding desired legislative response. These tools range in scale from a regulatory prohibition to mandating activity licensing regimes and development approval schemes. A State, for example, may apply a regulatory prohibition where there is high risk and threat of degradation to agricultural land, as a non-renewable natural resource. However, a licensing and contractual-based authorisation, where the State confers permissory rights to extract subsurface natural resources to private mining companies according to certain conditions, may be an effective application where strong economic incentives are apparent. Therefore, a 'precautionary approach', as described in Article 15 of the Rio Declaration,[19] is applied to decision-making and regulation based on the presence of uncertainty, rather than the precautionary approach as an automatic trigger for protective measures in general.

In order to strike an optimum balance between competing interests in the application of the precautionary principle, a level of acceptable risk must be embraced in conjunction with the economic need to encourage innovation and the imperative to preserve agricultural land and its communities. Peel cautions against the precautionary principle:

In situations where a high degree of value attaches to the aspect of health or the environment potentially affected by an activity, applying ... precaution would lead to stricter precautionary measures than in situations where strong, countervailing social and economic considerations exist.[20]

In order to aid States in determining the level of precaution required, comprehensive scientific and policy reviews, community consultation and stakeholder engagement are additional regulatory tools often used to manage and guide the regulatory-making process. Experience within, and commentary on, the development of the hydraulic fracturing ban in New York State demonstrates that the precautionary principle can be effectively applied where the development of unconventional gas would not be beneficial to communities if it is likely to create irreversible and permanent damage to agricultural areas.

The American legal framework regulating unconventional gas

Land granted in fee simple does not equate to absolute ownership in most Commonwealth jurisdictions such as the UK, Australia and Canada who hold regalian property systems – whereby all land and natural resources are under the exclusive control of the State. This differs to the property law model in the USA, where ownership of subsurface rights in a private ownership framework mirrors the unconditional gas leasing process as a private transaction between landowners and unconventional gas titleholders.[21]

The taxonomy of land and natural resource ownership in the USA allows the surface estate landholder to sever their mineral estate from the surface estate. This severing allows the sale of the mineral estate or allows it to be leased by reservation in order for private oil and gas companies to exploit natural resources.[22] Unlike regalian natural resource ownership, this separation of the surface and natural resource estate does not result in a renouncement of title by the surface estate holder, as outlined by *Del Monte Mining:* 'unquestionably, at common law the owner of the soil might convey his interest in minerals beneath the surface without relinquishing his title to the surface'.[23]

The general framework for land and natural resources ownership in the USA stems from the fundamental common law maxim, *cuius est solum, eius est usque ad coelum et ad inferos*, meaning whoever owns the soil also owns 'up to the sky and down to the depths'.[24] In most states of the USA, the maxim continues to function as the primary and paramount regulatory mechanism for surface estate ownership over *in situ* subsurface minerals such as coal.[25] As the *ad coelum* maxim focuses on surface rights and ownership, when applied to minerals, the principle affords the surface estate owner to retain ownership of minerals on the grounds of 'corporeal proximity'.[26] However, due to the transitionary nature of oil and gas, the *ad coelum* maxim could not easily be applied. For this reason, the 'rule of capture' principle was adopted as the legal ownership of unconventional gas in the USA.[27]

Since 1886,[28] the American judiciary has upheld the common law principle of 'the rule of capture' as the fundamental ownership principle for unconventional gas.[29] The rule of capture, originally founded in relation to the capture of wild animals and the common law groundwater doctrine,[30] in the oil and gas context provides that whoever 'captures' hydrocarbons can claim legal ownership. Thus, a private landholder may claim ownership and reap the benefits of a neighbouring landowner's oil or natural gas that has migrated from a neighbouring legal estate. Once shale gas is 'captured' under the rule, the holder will retain full common law ownership of the gas subject only to public policy obligations and regulatory restrictions.[31] Consequently, until oil and gas is produced in commercial form, under the rule of capture the surface right holder could not hold a proprietary interest. This led to the problematic outcome of mass drilling, trespass to land and the reduction of gas permeability, leading to a reduction in ultimate gas recovery.

The rule of capture has been qualified by a number of correlative rights to limit the extent of private ownership as summarised by Hepburn:

> (1) the owner only has the right to capture the natural flow of gas; (2) only reasonable means may be used to capture flowing gas and conservation rules may not be breached; (3) there must be no injury caused to the common source of supply; and (4) the common source of supply must not be destroyed through intentional or negligent behaviour.[32]

However, according to Ingleson, 'short of drilling, correlative rights of adjoining lessees could not be protected. The result was that more recently courts in a number of USA jurisdictions have moved to support a principle of absolute ownership. This brought oil and gas ownership doctrine into line with that applied to hard minerals'.[33]

The shale revolution

The 'shale revolution' commenced in Texas in the late 1990s, led by George Mitchell and his development of hydraulic (or 'slickwater') fracturing in the Barnett Shale. This 'revolution', triggered by rising gas prices and the affordability of fracturing techniques generating profitable royalties for private landholders and oil and gas companies, has resulted in unconventional gas becoming one of the 'key components' of the USA energy system.[34] Shale gas, in comparison with CSG, has low levels of permeability and porosity and always requires hydraulic fracturing. Further, shale gas is usually best accessed via horizontal drilling techniques, rather than the vertical drilling techniques used for CSG.[35] Hydraulic fracturing is prolific and widespread in the USA compared with Australia (where extraction of CSG requiring hydraulic fracturing occurs in less than 10 per cent of all wells) and Canada (where the technique is more costly and has arguably met more opposition in some provinces).

Overarching ownership taxonomy for unconventional gas does not exist in the USA, as all natural resources activities including exploration, drilling, production and transportation are regulated by the respective environmental and petroleum regulations, and enforced by penalties. The *Energy Policy Act 2005* affirms that the US states, rather than the federal government, holds municipal jurisdiction of the oil and gas industry.[36] Hydraulic fracturing is exclusively regulated by individual US states rather than federally by the United States Environmental Protection Agency. In drafting regulations relating to the recovery of shale gas, lawmakers in individual states must act 'in the public interest … including landowners and the general public'.[37] Consequently, the variation in state regulatory responses has led to 'fragmented federalism'[38] and considerable variation in state-level policy of unconventional gas extraction.

Further, US states hold constitutionally reserved powers to regulate public health, safety and welfare,[39] in combination with the EPA delegation of federal authority over oil and gas development to the states. State-based regulation of hydraulic fracturing has created a wide range of approaches in the USA from the adaptive management of Texas as one of the most active states in unconventional gas development, to the precautionary approach of New York State in banning hydraulic fracturing as regulated by the Oil & Gas Permitting and Management division of the Department of Environmental Conservation (DEC). Consequently, state lawmakers and local governments play an increasingly important role in regulating oil and gas development in establishing permit conditions and regulation of unconventional gas and its related activities and creating zoning districts in which oil and gas development is allowed.[40]

The precautionary approach of New York State

Article 23 of the *Environmental Conservation Law* 3 NY ENV Law Consol § 23–0301 (ECL) confers regulatory power to the Department of Environmental Conservation (DEC) over oil and natural gas development in New York State. This is comparable with other jurisdictions such as Australia and Canada, which confer oil and gas licensing and regulatory responsibilities to their respective Mining and Natural Resources Departments. The policy of the DEC is to provide for the ultimate recovery of oil and gas while expressly taking into account 'the rights of all persons including landowners and the general public may be fully protected'.[41] Any person or corporation seeking to drill and extract oil or natural gas must obtain a permit from the Department in accordance with Title 5 of Article 23 of the ECL.

In 1992, New York's Department of Environmental Conservation prepared a Generic Environmental Impact Statement (EIS) on the Oil, Gas and Solution Mining Regulatory Program in permitting hydraulic fracturing in the state.[42] An EIS was issued in order to create 'specific conditions or criteria under which future actions will be undertaken or approved, including requirements for any subsequent State Environmental Quality Review Act (SEQRA) compliance'.[43] However, during the period of 1992–2009, the developments in high-volume hydraulic fracturing had altered drastically with the advent of horizontal drilling and multi-well pads creating significantly new aspects in unconventional gas exploitation.

In particular, the new technological development of hydraulic fracturing required 70 to 300 times more water. In addition, multi-well pads now stretched over 7.4 acres, compared with the previous area needed for exploitation of just 4.8 acres.[44] Consequently, under Executive Order No. 41 in 2009, a Draft Supplemental Generic Environmental Impact Statement[45] (SGEIS) was released and the State Environmental Quality Review (SEQR) was activated under the SEQRA, as 'high volume hydraulic fracturing … raises new, potentially significant, adverse impacts'[46] not previously addressed in New York State's 1992 EIS.

The presence of scientific uncertainty and thus, the likelihood of permanent damage to public health due to potential contamination of drinking water, potential environmental degradation during extraction to agricultural land, amongst other issues, led to a comprehensive and systematic review of high-volume hydraulic fracturing (HVHF). The review was triggered to inform the state's decision on a moratorium in light of its mission to 'conserve, improve and protect its natural resources'.[47] Evidentially, the draft EIS, public consultation and moratorium on hydraulic fracturing represents an adoption of the precautionary principle in recognising that under the SEQRA, permits to drill wells would not be issued until completion of the SGEIS process. The DEC sought to examine whether the re-commencement of hydraulic fracturing activities in New York 'may have a significant adverse impact on the environment'. The SEQR established the precedent to 'prevent or eliminate damage to the environment … human and community resources'.[48]

The public consultation process in examining the 2009 draft SGEIS received over 193,000 written comments during a 6-year public comment period. The first SGEIS draft was released in 2009, revised in 2011, with the final SGEIS statement being released in 2015. In so doing, the New York State Department of Environmental Conservation (DEC) released its report assessing the potential environmental impacts of hydraulic fracturing and evaluating whether mitigation measures could eliminate or reduce significant adverse environmental impacts, and if so, whether measures should be imposed consistent with SEQRA and the ECL.

The 2015 Findings Statement found that hydraulic fracturing phases can result in

> adverse environmental impacts which can range in duration from acute impacts during only one phase, to more permanent impacts which can range in duration from acute impacts … to more permanent impacts that could be present for years or decades after a well is reclaimed.[49]

As these scientific uncertainties could not be mitigated effectively, the significant adverse public health and environmental impacts could not be avoided or minimised to an acceptable level in accordance with SEQRA. Consequently, the DEC concluded a No-Action permanent ban, with the goal of prohibition as opposed to a moratorium as a temporary suspension of an activity. This permanent ban was needed in order for New York State to meet its social, economic aims and other essential considerations, such as the protection of agricultural land. The No-Action regulatory decision is a precautionary approach in that it recognises the significant likelihood of serious or irrevocable environmental impacts of hydraulic fracturing. The DEC also recognised the cumulative impacts of natural gas infrastructure including pipelines, gathering lines and compressor stations. In so doing, it found these adverse cumulative impacts would cause permanent damage to agricultural lands.[50]

The DEC also considered the application of the New York State Constitution article XIV s 4 to 'encourage the development and improvement of its agricultural lands for the production of food and other agricultural products [which] … shall include the protection of agricultural lands'.[51] Consequently, DEC concluded permitting hydraulic fracturing is in violation with article XIV in impacting agricultural land quality. This was despite the potential solution proposed to consult with the Department of Agriculture and Markets to develop additional permit conditions and reclamation guidelines for well heads located on agricultural land. Consequently, the DEC found that 'the only means of completely eliminating the risks of impacts to farmlands and livestock is to employ the No-Action alternative'.[52]

The DEC report highlighted the possibility of unconventional gas development in New York State equating to severe impacts of community character, history, demographics and culture for rural communities.[53] In considering the comparative jurisdiction examples of neighbouring states Pennsylvania and West Virginia, the DEC found that large-scale hydraulic fracturing activities had the potential to 'industrialise rural areas of New York … (raising) additional concerns relating to agriculture and socioeconomic impacts'.[54] Therefore, the degree of change in community character, effect on natural physical features, community identification and community character for rural communities are recognised as being extremely important to the state.[55] Community character as a regulatory factor in decision-making is unique and powerful to the New York State context. While planning theory refers to 'identity' and 'place-making' as important considerations – they are just this – considerations rather than hard regulation.[56]

Overall, hydraulic fracturing was classified by the DEC as causing likely permanent and irreversible 'negative impacts on the local communities'[57] exacerbated by the 'significant gaps [which] exist in the knowledge of potential public health impacts from HVHF [high volume hydraulic fracturing]'.[58] This led to the conclusion of the New York State Department of Health (DOH) that 'until the science provides sufficient information to determine the level of risk to public health from HVHF to all New Yorkers and whether the risks can be adequately managed … HVHF should not proceed in New York State'.[59]

In considering a number of potential mitigation measures to avoid cumulative impacts of hydraulic fracturing in New York State, from ecosystem and wildlife protection, to air quality and greenhouse gas (GHG) mitigation, the DEC concluded as follows: '(1) the effectiveness of the mitigation is uncertain; (2) the potential risk and impact from high-volume hydraulic fracturing to the environment and public health cannot be quantified at this time, and (3) there are some impacts that are simply unavoidable'.[60] Therefore, the DEC found the No-Action ruling on hydraulic fracturing the 'only alternative that meets the SEQRA legal mandate because authorizing high-volume hydraulic fracturing under any scenario would not adequately mitigate adverse impacts to ecosystems … community character and public health'.[61] This finding is consistent with the DEC's stated mission to 'conserve, improve and protect natural resources to enhance the health, safety and welfare of New York and the overall economic and social well-being of its citizens'.[62]

The DEC's No-Action ruling in banning HVHF confirmed that individual or site-specific permit applications for wells using HVHF will no longer be processed. Therefore, the DEC states:

> Consistent with the social, economic and other essential considerations from among the reasonable alternatives available, the No-Action alternative avoids adverse environmental impacts to the maximum extent practicable; including impacts disclosed in the supplemental environmental impact statement.[63]

In applying a 'precautionary approach', as stipulated by Principle 15 of the Rio Declaration, through the creation of its draft SGEIS, the DEC adopted the two-fold methodology of Resnik.[64] First, this was to find that the threats to New York State's water system and rural communities were plausible and, indeed, likely. Second, the precautionary mitigation measures that could be adopted were not reasonable, in the face of scientific uncertainty and likely irreversible environmental degradation. Consequently, enacting a permanent ban on HVHF would be an effective adoption of the precautionary principle due to the high 'level and nature of uncertainty' of the activity.[65]

The participation of stakeholders in a transparent and inclusive decision-making process, as required by the Lowell Statement,[66] is also evident within the 160,000 public responses throughout the 7-year EIS review process undertaken by the DEC. Further, at the community level, many communities in New York State exercised their right to oppose hydraulic fracturing.

This ban has now been extended to a state-wide ban of hydraulic fracturing under Assembly Bill A3243, introduced on 27 January 2017, and passed by the New York State Senate and Assembly. Upon signing by the Governor of New York the law took immediate effect, pursuant to section 5 of the Bill. Interestingly, according to section 1, the ban includes not only fracturing using water, but also liquid propane gas. This ban on hydraulic fracturing deviates from the ordinary meaning of hydraulic fracturing utilising water, thus precluding the use of future technologies to create porosity and permeability in shales.

During the moratorium on hydraulic fracturing which commenced in 2009, and as New York State moved to fully assess potential risks of the activity, communities began to self-regulate and prohibit oil and gas activities within their municipal zone. By 2014, some 150 towns had moratoria on the process,[67] led by the precedent of Municipal Home Rule landmark cases *Wallach v. Town of Dryden* [68] and *Cooperstown Holstein Corp. v Town of Middlefield.* [69]

The Municipal Home Rule Law

The Municipal Home Rule (MHR) empowers local governments in New York State to enact laws, both for the protection and enhancement of the physical and visual attributes of a community. In addition, the MHR empowers local governments to enact zoning laws for the purpose of 'fostering health, safety, morals, or the general welfare of the community'.[70] It is enshrined within Article IX of the New York Constitution which states 'every local government shall have power to adopt and amend local laws not inconsistent with the provisions of this constitution or any general law ... except to the extent that the legislature shall restrict the adoption of such a local law'.[71]

Fundamentally, the MHR empowers local communities to enact regulations concerning the 'protection, order, conduct, safety, health and well-being of persons or property therein'. The case of *Wallach v. Town of Dryden* [72] provides an example of community land use powers through the adoption of zoning ordinances, land use restrictions or controls 'for the development of a balanced, cohesive community in consideration of regional needs and requirements'.[73] The Court of Appeals of New York considered whether Dryden, a rural community within Tompkins County, located over the Marcellus Shale region, held the jurisdictional powers to ban oil and gas production activities, including hydraulic fracturing. In 2011, the Norse Energy Corporation acquired oil and gas leases agreed by Dryden landholders. Subsequently, the Dryden Town Board unanimously voted to amend the zoning laws to exclude all oil and gas exploration, extraction and storage activities. The Town Board found unconventional gas extraction's potential in 'endangering the health, safety and general welfare of the community'[74] was too high, and thus would be banned. In alleging that Dryden lacked regulatory authority to prohibit natural gas exploration and extraction, Norse argued that the Oil, Gas and Solution Mining Law (OGSML) supersession clause prohibits communities from enacting local planning laws against oil and gas development.

According to article 23 of the ECL, the code shall 'supersede all local laws or ordinances relating to the regulation of the oil, gas and solution mining industries but shall not supersede local government jurisdiction over local roads or the rights of local governments under the real property tax law'.[75] It was argued by Norse that the ECL therefore intended to pre-empt, and indeed supersede, any community planning laws concerning the regulation of oil and gas activities and therefore the MHR was not applicable in this circumstance. The New York Court of Appeals held Dryden's community municipal zoning regulations were legally valid 'for the development of a balanced, cohesive community' in consideration of 'regional needs and requirements'.[76]

Wallach v. Town of Dryden[77] also considered an action against the town of Middlefield within Ostego County, New York. Similar to Dryden, Middlefield adopted zoning laws to classify a range of heavy industrial uses, including oil and gas, as prohibited uses in the town. As an agricultural community, the Town Board of Middlefield found, after a detailed review on the impacts of unconventional gas on agricultural land, that oil and gas activities and hydraulic fracturing would 'eliminate many … features' (including farms, clean water and rural lifestyle) and 'irreversibly overwhelm the rural character of the town'.[78] The Cooperstown Holstein Corporation, after executing two leases with a Middlefield landholder to undertake hydraulic fracturing activities brought an action to set aside the zoning law in Middlefield, similarly alleging that it was pre-empted by the supersession provision within the OGSML. In *Cooperstown Holstein Corp. v Town of Middlefield*,[79] it was held the zoning law enacted by the Town of Middlefield in 2011, was not pre-empted by the OGSML and was consequently valid.

In both decisions, it was held that municipal law zoning would dictate the technical aspects of oil and gas activities and how they may take place within the municipal. In contrast, a municipal zoning law concerning the location of oil and gas activities would be valid pursuant to the MHR.[80] In confirmation of this principle, the case of *Frew Run* held that zoning laws regulating the location of oil and gas activities are classified as

> incidental control resulting from the municipality's exercise of its right to regulate land use through zoning is not the type of regulatory enactment relating to the [oil, gas and solution mining industries] which the Legislature could have envisioned as being within the prohibition of the statute.[81]

Therefore, Dryden and Middlefield engaged in a reasonable exercise of their zoning authority, taking a precautionary approach, to determine that gas drilling would 'permanently alter and adversely affect the deliberately cultivated, small-town character of their communities'.[82]

The supersession clause within the OGSML was reaffirmed by the Court of Appeals as upholding local zoning laws regulation, and land use prohibition to preserve rural communities. Indeed, it is evident from the cases that the New York legislature did not intend for the OGSML to 'carve out', and indeed encroach on local land use powers. The primary concern of the OGSML was to prevent wasteful oil and gas practices and to ensure that the DEC had the means to regulate the technical operations of the oil and gas industry. As a result of these rulings, communities and local governments within the state may harness the MHR to prevent oil and gas activities within their municipal area via land use zoning regulations.[83]

At the heart of the cases considering the MHR lies the relationship between state and local governments, and their respective exercise of legislative power. It is clear the state retains its jurisdiction to enact policy and state-wide regulation of oil and gas activities. At both the state and local community level, New York has taken a precautionary approach, recognising the serious or irreversible impacts of hydraulic fracturing and empowering rural communities to apply the MHRL to protect community health and well-being and prevent irreversible environmental impacts. Further, the latest State Budget for New York released on March 2018 reinforces and affirms the state's dedication to preserving its agricultural industry with the release of US $1.9 million worth of funding 'for the Farm Viability Institute to help New York's farmers become more profitable and to improve the long-term economic viability and sustainability of farms, the food system, and the communities which they serve'.[84]

The Joint Landholders Coalition Of New York

The success of shale gas landowner coalitions provides an illustration of effective collective action and the empowerment of landholder groups facing mining activities in their community. As noted above, mineral rights within the USA is a private ownership common law legal system, with some aspects of civil law codification. Therefore, private landowners own the minerals, oil and gas beneath their land and negotiate compensation with a corresponding extraction lease with an exploring company, thus transferring their mineral rights to an oil and gas company to develop the shale gas.[85] In the 1953 case of *Benge v Scharbauer*, [86] Texan courts declared that landowners may reserve their mineral rights in oil and gas, thereby enabling the mineral estate to be severed from the surface estate.[87]

In the event of severance, the mineral estate dominates in terms of exploration and extraction and the mineral lessee assumes the same rights owed to the mineral estate owner since the leasing document is perceived as a temporary transference of ownership. According to Timmens,

> the owner of the mineral estate may lease the minerals to third parties for exploration, but law only requires that the lessee (i) notify surface owners of the intent to explore and drill; (ii) have access to as much land as is necessary to explore and drill.[88]

In some American states, landowner coalitions have formed and successfully negotiated favourable contracts protecting their agricultural lands. State governments within the USA and some private entities have exercised the power of 'eminent domain' to take land for public use.[89] What constitutes public use has long been the subject of debate among legal practitioners and academics to ascertain an equitable standard for land acquisitions in the USA.[90]

Despite regulatory differences of petroleum ownership and extraction, the example of rural landowners' collaboration and collective action illustrates the potential use of collective bargaining to gain greater protection for agricultural land. The Marcellus Shale Gas Basin, which lies the Southern Tier Region in New York State, is the site where rural landowners have formed grassroots organisations since 2011, aimed at collective bargaining with natural gas companies over the terms of development leases in their region.[91] These collectives are identified as 'landowner coalitions' .As such, these coalitions exert considerable influence over a substantial portion of the terrain considered attractive to oil and gas drilling. Interestingly, the ban on hydraulic fracturing in New York State was not well received by the JLCNY. A precautionary approach by the state was seen as a violation of private member rights, based on the reasoning that it should be in the hands of individual landholders, as owners of the hydrocarbons, to exercise their right to decide whether or not they enter into a gas lease with a petroleum license holder. It is understandable that the Coalition holds an opposition to an outright permanent ban on hydraulic fracturing, as the landholders have worked to create a model of collectivisation bringing both economic and social benefits to their respective towns.[92] Despite the hydraulic fracturing ban, the Coalition's collective bargaining model continues to operate in relation to oil exploration and remains an instructive case study.

The initial impetus for forming a landowner coalition between members was to secure greater compensation for rural landowners when negotiating oil and gas leases. However, the scope of the landowner coalition has now grown, according to Jacquet and Stedman, to:

> Become the de facto managers of natural resource development across vast and largely contiguous landscape scales. Besides setting rates of compensation, the leases these groups negotiate with energy companies serve as legally-binding operating agreements that can influence environmental and community outcomes.[93]

In this 2011 study, Jacquet and Stedman conducted semi-structured interviews focusing on the timelines, motivations, outcomes and organisational structures of the landholder coalitions and their members with 16 principal leaders of each of the 12 landowner coalitions in the Southern Tier Region of New York State.[94] The two largest landholder coalitions are informal and volunteer-led organisations found in Steuben and Tioga Counties, claiming approximately 162,000 acres with 5,000 owners and 113,000 acres with 1,700 owners, respectively.[95]

Both groups have a leader or spokesperson and a central committee of volunteers that coordinates membership and activities. In comparison, other coalitions are formed by a handful of neighbouring property owners or a leasing consultant who is typically paid a per-acre fee upon successful negotiation. The first more structured coalition model is arguably suited to a number of jurisdictions, such as Queensland, as agricultural landholders forming a collective bargaining group would require an audit of membership, central committee and a structured approach to negotiations while reporting to the ACCC. However, unlike collective bargaining groups, coalitions do require a small per-acre fee to offset legal costs upon negotiating a leasehold agreement.

Emerging from Jacquet and Stedman's 2011 study was the finding that the financial compensation incentive for members represented the primary motivation among members' interest in collective action to include the protection of private property and environmental protection. An interviewee in the study stated, 'you can almost put money as number two, now. The biggest thing is the protection of private assets and private property, and just the knowledge. Not being taken advantage of and protecting yourself'.[96] Consequently, it appears landholder coalition negotiations have the potential to influence natural resource management across a large area of the New York State. For example, some coalitions require environmental protections above those required by the New York State DEC regulations such as additional water testing.[97]

A state-wide peak body, the JLCNY, comprises leaders of individual coalitions and actively produces information for dissemination, negotiation strategies, lobbying and advocacy at the state and federal government levels. The JLCNY's mission statement is 'To foster, promote, advance and protect the common interest of the people as it pertains to natural gas development through education and best environmental practices'.[98] The JLCNY is a registered organisation founded in 2010 and consists of 77,000 landholders in control of one million acres across 14 counties. The JLCNY filed a lawsuit in 2014, objecting to a lack of political objectiveness during a hydraulic fracturing study conducted by the New York State DEC and DOH to allow hydraulic fracturing to begin in the Marcellus Shale region. The Court dismissed the lawsuit on 14 July 2014, based on the arguments that the group lacked standing to sue the state and that the review process had no legally mandated timeline.[99]

The JLCNY appealed the ruling on 24 July 2014, prompting New York Governor Cuomo ultimately to ban hydraulic fracturing across the state in December 2014. The final review of the environmental impact study was issued in June 2015. Although the JLCNY does not collectively sign leases on behalf of the 35 coalitions in New York, it provides legal advice, resources, collective community knowledge and lobbying outreach activities. Landowner coalitions distribute information to their members on common leasing violations, catalogue reported violations and provide legal advice on how to best respond to the violations.[100]

The benefits from collective bargaining are similar to US landholder coalitions, as collective bargaining offers lowered transaction costs for unconventional gas targets through negotiating multiple standardised leases. However, it is noted the American civil law system arguably favours the inclusion of landholder collectivisation as they are direct resource rights holders and financial beneficiaries to oil and gas leases. For example, the concept of 'good faith' bargaining is found in the US *Capper-Volstead Act of 1922*, [101] which provides basic protection for agricultural growers to collectively negotiate on product price.[102] A number of American states have adopted legislation requiring 'good faith' bargaining, meaning that intermediaries must negotiate with an agricultural association.[103] 'Good faith' negotiations with a collective body of agricultural landholders to create a land access agreement may arguably provide a number of advantages to landholders seeking to bolster their bargaining position, improve their knowledge through landholder networks and ultimately, facilitate an effective and equitable land access framework.

Conclusion

According to the DEC, permitting hydraulic fracturing in New York State would create unacceptable adverse and irreversible environmental impacts. The DEC particularly acknowledges that rural communities and agricultural land uses would be impacted due to the pervasive nature of the activity, as well as cumulative social and environmental impacts extending beyond shale gas well pads. This is confirmed in the legal power of the Municipal Home Rule for local governments to enact zoning and municipal development laws to defend community character and exclude oil and gas activities.

The recent New York Court of Appeals decision in the matters of *Wallach v. Town of Dryden* and *Cooperstown Holstein Corp. v. Town of Middlefield* found that ECL Section 23–0303(2) does not disqualify community adopted zoning laws from prohibiting the use of land for HVHF drilling. It is expected these rulings will lead to numerous rural municipalities across New York State to prohibit hydraulic fracturing in addition to the state-wide ban.

A precautionary approach does not imply that states reach the 'right' statutory decision. Rather, the precautionary approach, as demonstrated in the case of New York State is effective when it allows regulation to prevent serious and irreversible harm to agricultural land and rural communities.

Notes

1 According to Hunter, slickwater hydraulic fracturing was developed 'in response to the need to stimulate shale formations. It is often referred to as "slickwater fracking" and is characterised by the use of high volumes of water for each fracture compared to conventional hydraulic fracturing (70–300 times more water). This means that for each fracture, often several million gallons of water are used. Given that each well often requires several hydraulic fractures, this in turn means that tens of millions of gallons of water are used in each well'. In Tina Hunter, 'All Hydraulic Fracturing is Equal, But Some Is More Equal Than Others: An Overview of the Types of Hydraulic Fracturing and the Environmental Impacts' (2014) 19(3) *Australian Environment Review* 66, 68.

2 US Energy Information Administration, *How Much Shale Gas is Produced in the United States?* (EIA, 2018) <https://www.eia.gov/tools/faqs/faq.php?id=907&t=8>.

3 Ibid.

4 Matthew Castelli, 'Fracking and the Rural Poor: Negative Externalities, Failing Remedies, and Federal Legislation' (2015) 3(2) *Indiana Journal of Law and Social Equality* 281.

5 Production from the Eagle Ford, Utica play, and Haynesville plays in the Gulf Coast region is a secondary source to domestic dry natural gas, with production largely leveling off after 2028. For more information, see US Energy Information Administration, *Annual Energy Outlook 2018 with Projections to 2050* (EIA, 2018) 34 <https://www.eia.gov/outlooks/aeo/pdf/AEO2018.pdf>.

6 Stephanie Malin, 'There's No Real Choice But to Sign: Neoliberalization and Normalization of Hydraulic Fracturing on Pennsylvania Farmland' (2014) 4(1) *Journal of Environmental Studies and Sciences* 17; William Brady and James Crannell, 'Hydraulic Fracturing Regulation in the United States: The Laissez-Faire Approach of the Federal Government and Varying State Regulations' (2012) 14 *Vermont Journal of Environmental Law* 39.

7   NY *Municipal Home Rule Law* – MHR § 36.§ 10(1)(ii)(a)(11); NY *Town Law* § 261.
8   Report of the United Nations Conference on the Human Environment, *Rio Declaration on Environment and Development*, UN Doc A/CONF,151/26, (vol. I) / 31 ILM 874 (1992) <https://cil.nus.edu.sg/rp/il/pdf/1992%20Rio%20Declaration%20on%20Environment%20and%20Development-pdf.pdf>.
9   Jacqueline Peel, 'Precaution – A Matter of Principle, Approach or Process?' (2004) 5 *Melbourne Journal of International Law* 483, 484.
10 Cass R. Sunstein, 'Beyond the Precautionary Principle', John M. Olin Law & Economics Working Paper No. 149 (2D SERIES) (2003).
11 *Second International Conference on the Protection of the North Sea.* London 24–25 November 1987, Ministerial Declaration. Note that the nature of this Declaration is that of 'soft law', in that it is non-binding.
12 International Summit on Science and the Precautionary Principle, *Lowell Statement on Science and the Precautionary Principle* (17 December 2001).
13 Ibid.
14 David Kriebel et al., 'The Precautionary Principle in Environmental Science' (2001) 109(9) *Environmental Health Perspectives* 871, 973.
15 D. Resnik, 'Is the Precautionary Principle Unscientific?' (2003) 34 *Studies in History and Philosophy of Science* 329.
16 Sylvester Oscar Nliam, 'International Oil and Gas Environmental Legal Framework and the Precautionary Principle: The Implications for the Niger Delta' (2014) 22(1) *African Journal of International and Comparative Law* 23, 25.
17 D. Resnik, 'Is the Precautionary Principle Unscientific?' (2003) 34 *Studies in History and Philosophy of Science* 329, 342.
18 The 'Regulatory Toolbox', applied to a functional legal methodology, is referred to where regulation occurs through the combination of a number of differing tools and methods, rather than relying upon a single instrument. Bronwen Morgan and Karen Yeung, *An Introduction to Law and Regulation: Text and Materials* (Cambridge University Press, 2007) 9; Arie Freiberg, *Regulation in Australia* (The Federation Press, 2017).
19 Report of the United Nations Conference on the Human Environment, *Rio Declaration on Environment and Development*, UN Doc A/CONF,151/26, (vol. I) / 31 ILM 874 (1992) <https://cil.nus.edu.sg/rp/il/pdf/1992%20Rio%20Declaration%20on%20Environment%20and%20Development-pdf.pdf>.
20 Jacqueline Peel, 'Precaution – A Matter of Principle, Approach or Process?' (2004) 5 *Melbourne Journal of International Law* 483, 492.
21 Alexandra Klass and Hannah Wiseman, *Energy Law* (Foundation Press, 2016).
22 Hannah Wiseman, *Hydraulic Fracturing and Legal Frameworks* (Oxford Handbooks Online, 2017); Gabriel Eckstein, *The International Law of Transboundary Groundwater Resources* (Earthscan, 2017).
23 *Del Monte Mining & Milling Co. v. Last Chance Mining & Milling Co.*, 171 U.S. 55, 60 (1898).
24 John Sprankling, 'Owning the Centre of the Earth' (2008) 55 *UCLA Law Review* 979.
25 Owen L. Anderson, 'Lord Coke, the Restatement and Modern Subsurface Trespass Law' (2011) 6(2) *Texas Journal of Oil, Gas, and Energy Law* 203.
26 Samantha Hepburn, 'Does Unconventional Gas Require Unconventional Ownership: An Analysis of the Functionality of Ownership Frameworks for Unconventional Gas Development' (2013) 8(1) *Journal of Environmental and Public Health Law* 1, 9.
27 Terence Daintith, *Finders Keepers?: How the Law of Capture Shaped the World Oil Industry* (RFF Press, 2010).
28 *Wood Cnty. Petroleum Co. v. W. Va. Transp. Co.*, 28 W. Va. 210 (W. Va. 1886).
29 Terence Daintith, *Finders Keepers?: How the Law of Capture Shaped the World Oil Industry* (RFF Press, 2010); Joel B. Eisen and Jim Rossi, *Energy, Economics, and the Environment: Cases and Materials* (Foundation Press, 2010).

30 *Acton v. Blundell* (1843) 12 M & W 324, 152 ER 1223.
31 Terence Daintith, *Finders Keepers?: How the Law of Capture Shaped the World Oil Industry* (Routledge, 2010).
32 Samantha Hepburn, 'Does Unconventional Gas Require Unconventional Ownership: An Analysis of the Functionality of Ownership Frameworks for Unconventional Gas Development' (2013) 8(1) *Journal of Environmental and Public Health Law* 1, 13.
33 Alastair R. Lucas and Simone Fraser, 'Granting of Shale Gas Licences, Land Access and Property Rights in North America' in Tina Hunter (ed.), *Shale Gas Handbook of Law and Policy* (Intersentia, 2016) 137–138.
34 Alexandra Klass and Hannah Wiseman, *Energy Law* (Foundation Press, 2016) 41.
35 Jay Rutovitz, Stephen Harris, Natasha Kuruppu and Chris Dunstan, 'Drilling Down. Coal Seam Gas: A Background Paper' (Prepared by Institute for Sustainable Futures UTS for the City of Sydney Council, November 2011) 3.
36 Similar to the USA, the regulation of oil and gas activities in Australia is a matter for the states since there is no enumerated power for such regulation in the Australian constitution. For Australian territories, there is the right to regulate activities, but this is subject to the Federal government as a result of s122 of the Australian Constitution. Such intervention was demonstrated in the development of uranium resources in the Northern Territory. See Gerard Carney, 'Constitutional Framework for the Regulation of the Australian Uranium Industry' (2007) 26(3) *Australian Resources and Energy Law Journal* 47.
37 As stated by the Oil, Gas and Solution Mining Law Declaration of Policy – *ECL Article 23 regulating oil and gas activities in New York.*
38 Barbara Warner and Jennifer Shapiro, 'Fractured, Fragmented Federalism: A Study in Fracking Regulatory Policy' 43(3) *Publius The Journal of Federalism* 474.
39 *United States Constitution* amend X.
40 Alexandra Klass and Hannah Wiseman, *Energy Law* (Foundation Press, 2016) 47.
41 NY CLS ECL § 23–0301, art 23.
42 New York Department of Environmental Conservation, *1992 Findings Statement for Oil and Gas GEIS* (1992) <https://www.dec.ny.gov/energy/45912.html>.
43 6 NYCRR §617.10(c).
44 New York Department of Environmental Conservation, *1992 Findings Statement for Oil and Gas GEIS* (1992) <https://www.dec.ny.gov/energy/45912.html>. Department of Environmental Conservation, *2015 Summary & Trends – Oil, Gas and Solution Mining* (2015) <https://www.dec.ny.gov/energy/92904.html.
45 New York Department of Environmental Conservation, *Draft SGEIS* (2009) <ftp://ftp.dec.state.ny.us/dmn/download/OGdSGEISFull.pdf>.
46 New York State Department of Health, *A Public Health Review of High Volume Hydraulic Fracturing for Shale Gas Development* (2014) <http://www.health.ny.gov/press/reports/docs/high_volume_hydraulic_fracturing.pdf>13.
47 New York Department of Environmental Conservation, Final Supplemental Generic Environmental Impact Statement on the Oil, Gas and Solution Mining Regulatory Program, *Findings Statement* (2015) 4.
48 Ibid, 3
49 Ibid, 9.
50 The DEC recognises 'This ancillary activity has the potential to create adverse impacts to state-owned lands, freshwater wetlands, forests and other habitat due to fragmentation, streams where pipelines cross, air resources (from compressor stations), visual resources, agricultural lands, threatened and endangered species, and the spread of invasive species'. New York Department of Environmental Conservation, Final Supplemental Generic Environmental Impact Statement on the Oil, Gas and Solution Mining Regulatory Program, *Findings Statement* (2015) 9.
51 Ibid, 23.

52   Ibid, 175.
53   Ibid, 23.
54   Ibid, 4.
55   Ibid, 23.
56   Derek Thomas, *Placemaking: An Urban Design Methodology* (Routledge, 2016).
57   New York Department of Environmental Conservation, Final Supplemental Generic Environmental Impact Statement on the Oil, Gas and Solution Mining Regulatory Program, *Findings Statement* (2015) 24.
58   Ibid, 26.
59   New York State Department of Health, *A Public Health Review of High Volume Hydraulic Fracturing for Shale Gas Development* (2014) <http://www.health.ny.gov/press/reports/docs/high_volume_hydraulic_fracturing.pdf>.
60   New York Department of Environmental Conservation, Final Supplemental Generic Environmental Impact Statement on the Oil, Gas and Solution Mining Regulatory Program, *Findings Statement* (2015) 34.
61   Ibid.
62   New York Department of Environmental Conservation, *About DEC* (2018) <https://www.dec.ny.gov/24.html>.
63   New York Department of Environmental Conservation, Final Supplemental Generic Environmental Impact Statement on the Oil, Gas and Solution Mining Regulatory Program, *Findings Statement* (2015) 42.
64   D. Resnik, 'Is the Precautionary Principle Unscientific?' (2003) 34 *Studies in History and Philosophy of Science* 329.
65   Sylvester Oscar Nliam, 'International Oil and Gas Environmental Legal Framework and the Precautionary Principle: The Implications for the Niger Delta' (2014) 22(1) *African Journal of International and Comparative Law* 23, 25.
66   International Summit on Science and the Precautionary Principle, *Lowell Statement on Science and the Precautionary Principle* (17 December 2001).
67   Miriam R. Aczel and Karen E. Makuch, 'Environmental Impact Assessments and Hydraulic Fracturing: Lessons from Two U.S. States' (2017) *Case Studies in the Environment* 1.
68   23 N.Y.3d 728 (2014).
69   106 AD3d 1170, 964 NYS2d 431 (N.Y. 2013).
70   *Municipal Home Rule Law* § 10(1)(ii)(a)(11); *Town Law* § 261.
71   *New York Constitution*, art IX, § 2 [c].
72   23 N.Y.3d 728 (2014).
73   'The provisions of this article shall supersede all local laws or ordinances relating to the regulation of the oil, gas and solution mining industries; but shall not supersede local government jurisdiction over local roads or the rights of local governments under the real property tax law' ECL Art 23–0303.
74   *Wallach v. Town of Dryden* 23 N.Y.3d 728 at 715.
75   Ibid.
76   *New York Consolidated Laws, Environmental Conservation Law* – ENV § 23–0303. Administration of article; *Matter of Gernatt Asphalt Prods. v Town of Sardinia*, 87 NY2d 668, 683, 664 NE2d 1226, 642 NYS2d 164 [1996]; *Udell v Haas*, 21 NY2d 463, 469, 235 NE2d 897, 288 NYS2d 888 [1968].
77   16 N.E.3d 1188 (N.Y. 2014) [741].
78   Ibid.
79   *Cooperstown Holstein Corp. v Town of Middlefield*, 106 AD3d 1170, 964 NYS2d 431.
80   *Matter of Frew Run Gravel Prods. v Town of Carroll* 524 NYS2d 25 [1987] 13. This case concerned the Town of Carroll's zoning ordinance which successfully established a zoning district where sand and gravel operations were not permitted.
81   Ibid.

82   *Wallach v. Town of Dryden* 23 N.U.3d 728 at 754.
83   Shaun A. Goho, 'Municipalities and Hydraulic Fracturing: Trends in State Pre-emption' (2012) 64(7) *Planning and Environmental Law* 3.
84   The New York State Senate, *The 2018–2019 State Budget* <https://www.nyse-nate.gov/issues/2018-19-budget> 2.
85   John C. Dernbach and James R. May, *Shale Gas and the Future of Energy: Law and Policy for Sustainability* (Edward Elgar, 2016).
86   259 SW 2d 166 (Tex 1953).
87   *Acker v Guinn*, 464 SW 2d 348, 352 (Tex, 1971).
88   Christopher Timmens and Ashley Vissing, *Shale Gas Leases: Is Bargaining Efficient and What Are the Implications for Homeowners If It Is Not?* (2015) <http://public.econ.duke.edu/~timmins/Timmins_Vissing_11_15.pdf> 15.
89   Dwight H. Merriam and Mary Massaron Ross, *Eminent Domain Use and Abuse: Kelo in Context* (American Bar Association, 2006).
90   *Calder v Bull*, 3 US 386, 388 (1798).
91   Jeffrey Jacquet and Richard C. Stedman, 'Natural Gas Landholder Coalitions in New York State: Emerging Benefits of Collective Natural Resource Manage-ment' (2011) 26(1) *Journal of Rural Social Sciences* 62.
92   JLCNY, Letter to President Trump 4–24–17 (2017) <https://www.jlcny.org/site/index.php/press-room/jlcny-press-releases/2632-letter-to-president-trum-4-24-17>.
93   Jeffrey Jacquet and Richard C. Stedman, 'Natural Gas Landholder Coalitions in New York State: Emerging Benefits of Collective Natural Resource Manage-ment' (2011) 26(1) *Journal of Rural Social Sciences* 62, 63.
94   Ibid, 62.
95   Steuben County Landowners Coalition (SCLC), *Homepage* (2010) <http://mysite.verizon.net/reszcmsk/>; Tioga County Landowners Group (TCLG), Homepage (2016) < http://www.tiogagaslease.org/>.
96   Jeffrey Jacquet and Richard C. Stedman, 'Natural Gas Landholder Coalitions in New York State: Emerging Benefits of Collective Natural Resource Manage-ment' (2011) 26(1) *Journal of Rural Social Sciences* 62, 77.
97   Department of Environmental Conservation, *Regulations and Enforcement* (2016) <http://www.dec.ny.gov/65.html>.
98   *Internal Review Code of 1986*, 26 USC § 501(c)(6).
99   Tom Wilber, *Under the Surface: Fracking, Fortunes, and the Fate of the Marcellus Shale, Updated Edition* (Cornell University Press, 2015).
100 Alex Prud'homme, *Hydrofracking: What Everyone Needs to Know* (OUP USA, 2014).
101 *Capper-Volstead Act of 1922*, 67, PL 67–146.
102 Donald A. Frederick, *Antitrust Status of Farmer Cooperatives: The Story of the Capper-Volstead Act (2002) Cooperative Information Report* 59 <http://www.uwcc.wisc.edu/pdf/CIR59.pdf>.
103 Donald A. Frederick, 'Legal Rights of Producers to Collectively Negotiate' (1993) 19(2) *William Mitchell Law Review* 433.

9 France

Introduction

After the 'shale gas revolution' was heralded in the USA, Europe began to look to unconventional gas as the solution to the dependency from a single external supplier, given that 40 per cent of all gas supplies are imported from Russia.[1] Shale gas holds potential as being the largest unconventional fossil fuel in Europe with estimates from the European Commission's Joint Research Centre totalling 16 trillion cubic metres (tcm) of technically recoverable gas.[2] Although the energy security debate has been topical for many decades, the drive to initiate major change has recently become more urgent and strident as evident in the Ukraine–Russia dispute. In 2006 and 2009, respectively, Russia cut off its gas deliveries to Ukraine, an important transit country for European imports of Russian gas, creating a 'wake up' call[3] for the EU to establish a common energy policy. Therefore, the conundrum of coexistence of two distinct regulatory aims exist within the EU – both of energy security and environmental and land protection.

To address the mounting concern of Russia's evident power over gas supply[4] and to meet estimated increase in energy demand by 27 per cent by 2020, the EU released the European Energy Security Strategy in 2014. Shale gas became the focus to address whether, and to what extent, it can contribute to European energy security. The European Commission (EC) recognised the strategic importance of unconventional gas to 'compensate for declining conventional gas production provided issues of public acceptance and environmental impact are adequately addressed'.[5] Consequently, the EC ordered the streamlining of national administrative procedures, Strategic Impact Assessments and a one-stop shop for hydrocarbon projects in accordance with environmental assessment procedures for energy infrastructure, Projects of Common Interest, and Environmental Impact Assessments (EIAs) for large-scale trans-boundary projects.

Despite the clear need to exploit hydrocarbons to reduce energy dependency on Russia, the position on shale gas exploitation varies considerably amongst EU Member States. Therefore, the role of regulation of unconventional gas activities and application of the precautionary principle within an EU Member State depends on questioning whether 'the law should promote the development of a technology without unduly compromising the protection of society and the environment from consequences of uncertainties and potential threats of an associated technology'.[6] With its shale gas reserves located within the Paris Shale, France could potentially reduce its dependence on gas imports by up to 40 per cent by 2050.[7] Notwithstanding this, France has taken a precautionary regulatory approach as the first EU country to ban shale gas activities permanently.

Law No. 2011–835 of 13 July 2011 was passed by the French National Assembly to ban the exploration and mining of hydrocarbon liquids and gases by hydraulic fracturing and repealed the exclusive licenses to include projects using this technique. This chapter focuses on the precautionary shale gas regulation of France, with the second-largest proven shale gas deposit in the EU, and the first country to enact a legislative ban on the hydraulic fracturing of shale deposits.

Shale gas resources in France and their exploitation in agricultural regions

A study by the US Energy Information Administration in 2013 identified the shale gas resource potential of France to be high, being the second in Western Europe.[8] The two primary areas where shale gas is contained is that of the Paris Basin in the northeast and the South East Basin. Together, they contain approximately 156 tcf, with the Paris Basin comprising the lion's share of gas at 129 tcf and the South East Basin the remainder at 27 tcf.[9] In these basins, shale gas is primarily located in deep relatively stable shales, located approximately 3000–45000 metres below the surface. In addition, the US Energy Information Administration has identified the presence of shale oil. Like other jurisdictions examined within this book, these shale reservoirs occur underneath farmland, as a function of the geological characteristics of the formation of shale in depositional environments, with modern farmland occurring in these depositional environments.[10]

During 2009–2010 France granted licences for the exploration into shale gas resources, primarily to large US oil and gas companies. On the 1 March 2010, the Ministry of Ecology, Energy and Sustainable Development approved the request for three exploration authorisations granting exclusive research capacity for shale gas in south-eastern France. Under the Mining Code in force at that time, public consultation was not required prior to issuing the authorisation. Rather, authorisation was only required if and when exploration indicated that drilling could be profitable; at which time a company must require an exploitation authorisation in the form of a concession. This requirement for public consultation had been eliminated in 1994 to facilitate efforts to identify and understand the potential value of natural resources. In the intervening period, hundreds of research permits were authorised in such a fashion.[11] It was this act of granting a research permit without public consultation that was to form the fundamental basis of public consternation about shale gas extraction, and force a review of the Mining Code.[12] Another flaw of the Mining Code at that time was that the applications of research permits were not required to disclose the techniques they intended to use for such research. Rather, applications were assessed on the basis of financial and technical capacity. Although Keiller notes[13] that the technocrats of the ministry should have recognised the potential problem related to fracking, this is unreasonable given that it was only in September 2010, several months after the grant of the research permits, that the world was made aware of the potential risk and dangers of shale gas by Josh Fox's landmark documentary *Gasland*.

In late December 2010 and January 2011, social media and local activists became aware of the possible risks associated with shale gas activities and the people's revolt against the extraction of shale gas 'began to capture the attention of elected officials in the country side and in Paris'.[14] Perhaps the greatest reasoning for the eagerness to develop shale gas fields was the low hydrocarbon potential in France at that time. In 2012, France had only 60 oil and gas fields, providing 1–2 per cent of the national consumption, with the shortfall fulfilled by imports.[15] Cournil notes that because of this high dependency on imports, it is 'understandable that the potential offered by shale, oil and gas represents an energy El Dorado for France in terms of independence and energy security'.[16] In addition, the shale gas permits were granted under Article L100–1 of the 2015 Energy Code which states that French energy policy must guarantee

> the strategic independence and economic competitiveness of France. This policy aims to: ensure energy security, maintain a competitive energy price, protect human health and the environment, protect aggregation of the greenhouse effect, guarantee social and territorial cohesion by ensuring energy access for all.[17]

Given this policy opposition, it was clear that the legislator needed to balance the need for energy security and protection of the environment as well as social cohesiveness through energy access.

Shale gas and agricultural land

One of the reasons for the profound interest in shale gas and its environmental implications in France is the role that agricultural land plays in both French culture and perhaps more importantly the French economy. Value of agriculture and its land in France is vast[18] – 3.5 per cent of the French population is employed in farming, with France comprising over 49,000 farms.[19] It is the highest ranked agricultural power in the EU, and comprises the largest agricultural country by area. The utilised land for agriculture in France is almost 30 million hectares, equivalent to over half of the French territory. Of that area, 63 per cent is cropping, 34 per cent grasslands and 3 per cent is perennial crops, including vineyards orchids and the like. The French agricultural area provides 90 per cent of all of Europe's agricultural goods. However, there are two types of agriculture in France. The first is that of agriculture as an industrial complex, typified by the activities in the North overlying the Paris Basin and is dominated by mono cropping, particularly cereal crops, fodder crops and industrial crops that together account for 61 per cent of France's total agricultural area in 2010. Permanent grassland is utilised for the raising of livestock, with approximately 22.5 million livestock units recorded in 2010.[20] Organic agriculture has increased in value in the French utilised agricultural area, accounting for almost 2 per cent of the country in 2010.

The second type of agriculture is that of artisanal agricultural and specialist cropping and husbandry that produce culturally significant commodities, some of which have been trademarked according to Geographical Indications of Origin.[21] Water is also an important issue in France, given the need for water for irrigation and livestock purposes. In 2010, approximately 2711 million cubic metres of water irrigated agriculture in France. This amounted to 1712 cubic metres of water per hectare of agricultural land. Importantly, water availability varies significantly in different regions, with lower than national average water use recorded in the Franche-Comté, Corsica, and Nord-Pas-de-Calais areas. Significantly higher water use occurs in the Provence–Alps–Côte d'azur, Réunion and Lorraine regions.[22] This is noteworthy, given that shale gas exploration relies heavily on the use of water for hydraulic fracturing and this has had an important bearing on public rejection of shale gas.

The value of shale gas in France's energy security framework

In order to assess the importance of shale gas exploration and its role in France, it is necessary to examine the place of shale gas and natural gas in France. At the time when shale gas exploration was to commence in France, gas production was only occurring from the Lacq gas field in southwest France. However, since its closure in 2013, no natural gas production has occurred. Therefore, contribution of hydrocarbons in France is the lowest in Europe, with France being the leading low carbon energy producer in the EU. As of 2015, the energy mix of France comprised 46.6 per cent fossil fuel use (28.8 per cent oil, 14.3 gas and 3.6 coal), 46.4 per cent nuclear and the remainder in low carbon options.[23] France sees renewable energy as the future of its energy mix, seeking to increase renewable energy to 40 per cent by 2030 from its 2016 level of 16 per cent. Concomitantly, it seeks to decrease its use of nuclear power for electricity generation from 78 per cent in 2015 to 50 per cent by 2025.[24] The increase in renewable energy has been promulgated under the 2015 low carbon energy framework enacted by the Energy Transition for Green Growth Act of 2015.[25] Focussing on gas consumption, France consumes 36.7 bcm of gas per annum. Of this consumption, 33.4 per cent is industry, 33.1 per cent is residential, 19.8 per cent is commercial and public services and agriculture, 9.3 per cent power generation, 4.1 per cent other energy and transport comprises the remaining 0.3 per cent. Given that no natural gas is produced in France, all gas is imported; totalling 44.4 bcm in 2015. Primary suppliers included Norway at 46.8 per cent, Russia 12.7 per cent, the Netherlands 12 per cent, Algeria 10.5 per cent, Nigeria 3 per cent and other suppliers 15 per cent.[26] The route of supply for these imports is primarily by pipeline which transports 85 per cent of all gas imported, with the remainder supplied through LNG terminals.[27] It is interesting that initially France was supportive of shale gas resources.

Despite the closure of the Lacq field in 2013, which left France with virtually no domestic production and exploration of gas, France is the fourth-largest natural gas consumption market in the EU, with natural gas accounting for 26 per cent of national domestic consumption (and comprising 14.2 per cent of overall energy consumption). In addition, it has EU cross-border pipelines equating to 37,500 km, with a total import capacity of 187.5 mcm/day, with four LNG import terminals in operation in France: Fos Cavaou; Fos Tonkin; Mointoir-de-Bretagne; and Dunkerque.[28]

Banning of shale gas in France and the extension to oil and gas – application of the precautionary principle

The politicisation of shale gas exploration commenced as early as December 2010, with citizens' collectives seeking a political moratorium on the exploration for shale gas and the use of hydraulic fracturing. This call for a moratorium was based on the notion of the precautionary principle.[29] The underlying legal premise for the moratorium was Article 5 of the Charter of the Environment which states that:

> When the occurrence of any damage, albeit unpredictable in the current state of scientific knowledge, may seriously and irreversibly harm the environment, public authorities shall, with due respect for the principle of precaution and the areas within their jurisdiction, ensure the implementation of procedures for risk assessment and the adoption of temporary measures commensurate with the risk involved in order to preclude the occurrence of such damage.[30]

A flash-mobilisation was organised in the Ardèche department (southeastern France) in December 2010. José Bové, an activist grounded in local networks against GMOs, organised a public meeting because one of the licenses targeted the Larzac plateau, a historic landmark for the French altermondialiste movement and for resistance to state infrastructures in the 1970s. This meeting gathered together activist networks and linked the negative impacts of hydraulic fracturing in the USA to the French licenses. Thereafter, a negative meaning was attached to hydraulic fracturing. At the same time, a sense of urgency spread, as opponents generated a discourse of imminent threats to the land opened up to the shale gas industry. From this public meeting, an extraordinary snowball effect is observable: public meetings grew exponentially in most of the hundreds of towns in the area concerned by licenses. In Ardèche alone, activists estimated the number of persons who attended these meetings at 45,000 in three months.

The French government responded to these protests and public outrage by the passing of Law 2011–835,[31] which saw a speedy transit through the French Parliament being approved by the National Assembly on the 21 June 2011 and the Senate on the 30 June 2011. There were two main flaws with Law No. 2011–835, as analysed by Fleming.[32] The first identified was that of pre-existing licences.[33] The law intended to target pre-existing licences since it sought the repeal of exclusive licences using this technique. The second issue was the use of term hydraulic fracturing in the Act. Fleming notes that since only the term 'hydraulic fracturing' was stipulated in the Act, it was possible for shale gas extraction by other forms, such as pneumatic fracturing, to occur and shale gas to be produced in this manner.[34]

In 2017, a new bill and ultimately Act, entitled Law No. 2017–1839 of December 30 2017 Act to end research and exploitation of conventional and unconventional hydrocarbons and to introduce various provisions concerning energy and the environment was passed. This Act sought to rectify the problems identified by Fleming, banning all oil and gas activities by 2040. This meant there was a move from banning a specific extraction technique, where Law 2011–835 identifies hydraulic fracturing, to the banning of the production of the resource itself under Law 2017–1839. Under Law 2017–1839, all license and concessions will not be renewed beyond 2040, and no new exploration permits or renewal of permits will occur after 2020. Also, Law 2017–1839 bans the production of shale gas whatever the method used, thereby correcting the gaps in the 2011–835 Law.

In an extract from the minutes of the Council of Ministers dated 6 December 2017, the Minister of State and the Minister for Ecology and Solidarity Transition introduced the Bill[35] to ensure the coherence of the management policy of French hydrocarbons with the Paris Agreement on Climate. The Bill provides for the goal of achieving carbon neutrality by 2050 and also includes provisions to ensure security of gas supply. Interestingly, under Public Vote number 193 on the Bill as a whole, the French National Assembly voted overwhelmingly for the adoption of the Act.[36] Of the 480 votes cast, 388 votes were for the adoption of the banning of hydrocarbons, 92 against, with 70 absentees.[37]

The French decision to firstly ban the use of hydraulic fracturing and revoke the licenses under Law 2011–835, and then to ultimately ban hydrocarbon extraction under Law 2017–1839, represents a utilisation of the precautionary principle by the parliament to protect the environment from harm arising from an unknown or unproven technique. In seeking to make a decision, the French weighed up options in relation to how to manage extraction of a resource with a high potential for environmental harm. Some jurisdictions, such as Queensland, have undertaken unconventional gas extraction without understanding the possible harm or consequences, seeking to address the damage and impact of the activity as they unfold. This has been the adaptive management approach of the Queensland government in developing CSG and where major environmental impacts of CSG extraction, particularly in relation to water, land use and land access, have been seen. Each of these impacts in Queensland has been remedied in a reactive manner utilising adaptive management as a regulatory approach rather than an environmental approach. This is demonstrated by the implementation of legislation relating to water, particularly the make good provisions when it was discovered that the dewatering of coal seams creates aquifer draw and results in higher than expected water table losses.[38]

When shale gas exploration first commenced in France, the reaction of the French public, particularly the Ardèche movement, in response to proposed activities, was informed by the experience of the USA. At that time, there had been concerns regarding aquifer contamination in the Marcellus Basin and Pavilion, Wyoming,[39] thus invoking the precautionary principle and prompting the introduction and ultimately passing of the Bill for Law 2011–835. This law was implemented by the French parliament to protect farmland and to prevent unknown consequences of an activity that contributes very little to French energy security given the dominance of nuclear energy and the rising importance of renewables.

Value in agricultural land: Food security, food sovereignty and food culture in France: The importance of provenance and Geographical Indicators

France's largest shale gas basin, the Paris Basin, representing 2.09 billion tcf of unconventional gas,[40] lies directly underneath the famed champagne wine-growing region. Given the importance of the champagne appellation and significance in maintaining the value of smallholder grape growers and the rural community of champagne, it is unsurprising the thought of extracting shale gas in the champagne region is extremely unpalatable to the French. It is clear that values of food security, food sovereignty and agrarian artisanal value lies behind the decision to enact a ban on hydraulic fracturing activities in France. The anti-fracking movement of 2011, led by the Ardèche movement, was extraordinary in scope and effect, in arguing hydraulic fracturing represented a threat to the rural and regional environment. The movement accused the then French government that the approval to explore for shale gas was done in an '*en catimini*' (in secret) manner and against the wishes of its citizens.[41] Further, according to the European Parliament, opposition arose to the Paris Basin being fracked to avoid negative impacts to the 'vine rich' area of champagne.[42] In 2018, France released its latest food policy, structured around three key pillars to: guarantee France's food sovereignty; promote healthy food choices that respect the environment; and reduce inequality in accessing good quality, sustainable food. Clearly, the sentiment following the release of the roadmap after the French National Food Conference (EGAlim) in December 2017, was the safeguarding of food sovereignty to 'Pay fair prices to producers, allowing them to earn a decent living from their work; Reinforce the health, environmental and nutritional quality of products; and Promote healthy, safe and sustainable food for all'.[43] France's stance on prioritising the preservation of agricultural land, over that of shale gas and broadly energy security, is evident in its objective in preserving its agricultural production capacity. This is manifested in its policy aim to ensure at least 50 per cent of all produce sourced by the public sector in collective catering is to be originally grown, locally produced or from quality-certified farmers by January 2020.

The importance of Geographical Indications of Origin of agricultural produce for French products cannot be over stated. Article 22.1 of TRIPS defines Geographical Indications of Origin as 'indications, which identify a good as originating in the territory of a Member, or a region or locality in that territory, where a given quality, reputation or other characteristic of the good is essentially attributable to its geographical origin'.[44] Registration of Geographical Indications of Origin provides authorisation of the authenticity of origin and the artisan method of production of the protected product. Historically, the impetus for the creation of the Geographical Indications of Origin system stemmed from the 'need to protect intellectual property related to traditional cultures, geographical diversity and production methods, and thus to protect the product name from misuse'.[45] The system reinforces the credibility and strict legal methods, marketing and production standards in which the protected product must conform to, and seeks to prohibit the 'passing off' of illegitimate products claiming to originate in a protected region. Protection of Geographical Indications of Origin in Europe dates back to medieval times when guilds were granted monopolies by the government over certain industries for regulatory purposes. As stipulated by Bowen,

> Because GIs root production in particular places and protect the unique environmental and cultural resources that have developed over time in these places, scholars and development practitioners have framed them as a means of localizing production within the framework of globalization.[46]

Therefore, Geographical Indications emphasise a statist and prescriptive socio-legal approach of public and regulatory certification conducted by public authorities and producer-led organisations. By granting the authority to utilise a specific artisanal name, grounded in its origin, a protected product carries with it the proprietary rights signalling value, prestige and historic artisanal methods. This value allows farmers and craftsman alike to command a higher price and reach consumers beyond that of an unprotected generic product. This collective interest of registered farmers being certified to utilise an indication of origin is then protected on the basis of greater state involvement in order to protect product's reputation and associated production methods generated by the collective.

Champagne is one of the most famous and fiercely guarded Geographical Indications of Origin in France. Indeed, the wine has been linked to France's 'national identity'.[47] The symbolic status of champagne has, since its creation by Benedictine Monks more than 350 years ago,[48] been attributed to a specific and clearly defined region of the champagne wine-producing region in the northeast of France. The traditional French champagne houses operate from champagne vineyards, such as Moët et Chandon, Veuve Clicquot and Pommery, who make two-thirds of all wines and rely on champagne growers to provide grapes. Champagne production is grounded in attributes of food security and food sovereignty in protecting artisanal and small-scale farmers, in contrast to mono cropping and liberalised agribusiness production. For example, 90 per cent of all champagne is attributed to just 15,000 small grape growers averaging a farm holding of two hectares.[49] The legal protection of the appellation of origin in the label champagne is protected by the European Commission, and thus no other wine, sparkling or still, may carry the name 'champagne', excluding the USA and Russia. The *Comité interprofessionnel du vin de Champagne* (CIVC) acts as the territorial brand manager, and holds extensive quasi-governmental and regulatory powers to ensure the effective management and protection of the brand. Culturally, champagne in France comes with the recognition of the life of artisanal growers and engagement with their communities. It is the protection of such areas from shale gas activities, and to preserve the products produced, that provided much of the impetus for the ban on shale gas extraction.

Another important factor linked to the permanent ban on hydrocarbon extraction in France is related to the intimate connection that French agriculturalists, particularly vintners, have with the soil. The concept of *Terroir* is uniquely French, and is linked with the importance of local, historically significant artisanal products. Terroir represents the agrarian, environmental, social and gastronomic values linked to a specific artisanal locale. It is linked to the *appellation d'origine contrôlée* system in France, which jealously guards its farmers' right to safeguard the 'heightened'[50] Geographical Indication of Origin legal protection in the country. Guided by sensory acuities developed over time, skilled artisanal farmers adjust their agricultural methods to work with the unique seasonal and climatic variations that affect fermentation and aging of wine grapes, as well as attributes lending to its colour and flavour.

As stated by Jurca, 'A terroir can have a reputation to produce a grand cru (exceptional wine), and a wine can also be said to have the goût, or taste, of its terroir'.[51] Consequently, agricultural land protection and land use is highly influenced by the concept of Geographical Indication of Origin. In addition, there is a need to maintain terroir for a two-fold purpose, as identified by Lamarque and Lambin; '(1) better environmental stewardship is required to preserve the biophysical attributes of the terroir associated with the unique characteristics of the product, (2) some requirements or specifications relate to land management practice'.[52]

Conclusion

France holds a comparative level of shale gas reserves to Poland of an estimated 133.6 tcf.[53] However, it has banned the extraction of not only shale gas but all hydrocarbon production from 2040. Since the enactment of Law No. 2011–835, France has prevented any exploration works concerning shale gas in as far as it required hydraulic fracturing. The ban on the use of hydraulic fracturing was codified as articles L.111–13 and L.111–14–1 of the

Mining Code. Law No. 2017–1839, dated 30 December 2017, bans all research and exploitation of hydrocarbons in France: These provisions are codified as articles L.111–4 to L.111.12 of the Mining Code. In particular, the Mining Code Law establishes that no new hydrocarbon exploration licence will be granted, which prohibits the search for new oil or gas deposits. Existing operating concessions shall not be renewed beyond 2040. Liquid and gaseous hydrocarbons and coal, except coal mining, fall within the scope of the Law. These provisions aim at limiting France's greenhouse gas emissions by discouraging the use of hydrocarbons.

The other, intended consequence of the ban on hydrocarbon extraction, is the protection of farmland. This protection is not only for the productive, commercial farmland that provides almost one fifth of Europe's cereal crops,[54] but also for the artisanal food industry. The protection of lands that produce agricultural products that are under the Geographical Indications of Origin legal framework, and the associated *Terroir*, were critical in the banning of shale gas extraction. These uniquely French products are valued, not only for their sale, but for the French lifestyle and culture that they represent. In banning shale gas exploitation through the use of the precautionary principle, France has clearly indicated that the importance of food provenance trumps energy security. Such a position is most certainly supported by the capacity of France to generate energy from nuclear rather than hydrocarbon fuel sources, a decision that other jurisdictions do not have the luxury of making.

Notes

1 Currently, Russia, Norway, Algeria and Libya are the primary suppliers of gas through pipelines to EU, whereas LNG is shipped to Europe from Norway, Nigeria, Algeria, Libya, Egypt, Qatar and Trinidad & Tobago.
2 Reuters, *Shale Gas Will Not Cut EU Import Dependence: Study* (2012) <https://www.reuters.com/article/us-eu-shale-gas/shale-gas-will-not-cut-eu-import-dependence-study-idUSBRE8860W22012090'/>.
3 European Commission, *European Energy Security Strategy*, Brussels, 28.5.2014 COM (2014) 330, 2.
4 As stated by the European Commission, 'Six Member States depend from Russia as single external supplier for their entire gas imports and three of them use natural gas for more than a quarter of their total energy needs. In 2013 energy supplies from Russia accounted for 39% of EU natural gas imports or 27% of EU gas consumption; Russia exported 71 % of its gas to Europe with the largest volumes to Germany and Italy'. Ibid, 6.
5 Ibid, 13.
6 Ruven C. Fleming and Leonie Reins, 'Shale Gas Extraction, Precaution and Prevention: A Conversation on Regulation Responses' (2016) 20 *Energy Research & Social Science* 131, 131.
7 US Energy Information Administration, *Technically Recoverable Shale Oil and Shale Gas Resources: Other Western Europe* (2015) <https://www.eia.gov/analysis/studies/worldshalegas/pdf/Northern_Western_Europe_2013.pdf> 10.
8 Ibid, 10.
9 Ibid, 2.
10 Ibid, XIII-4.

11 John Keeler, 'The Politics of Shale Gas and Anti-Fracking Movements in France and the UK' in Y. Wang and W. Hefley (eds), *The Global Impact of Unconventional Shale Gas Development* (Springer, 2016) 44.
12 Nicolas Boring, *France: New Mining Code Under Consideration* (2013) <http://loc. gov/law/foreign-news/article/france-new-mining-code-under-consideration/>.
13 John Keeler, 'The Politics of Shale Gas and Anti-Fracking Movements in France and the UK' in Y. Wang and W. Hefley (eds), *The Global Impact of Unconventional Shale Gas Development* (Springer, 2016) 44.
14 Ibid, 46.
15 Christel Cournil, 'Adoption of Legislation on Shale Gas in France' (2013) 4 *European Energy and Environmental Law Review* 141, 144.
16 Ibid, 142
17 Ibid, 143.
18 Eurostat, *Agricultural Census in France* (2012) <http://ec.europa.eu/eurostat/statistics-explained/index.php/Agricultural_census_in_France>. Note that more up to date agricultural data will not be available until 2022, given that the EU agricultural census is only undertaken every 10 years.
19 Ibid.
20 Ibid.
21 As will be discussed later in this chapter.
22 These are the region names as at 2010.
23 International Energy Agency, Energy Policies of IEA Countries: *France* (2016) <https://www.iea.org/newsroom/news/2017/january/energy-policies-of-iea-countries-france-2016.html> 10.
24 Ibid.
25 Ibid, 11.
26 Ibid, 85.
27 Ibid, 85.
28 Ibid.
29 Ruven Fleming, *Shale Gas, the Environment and Energy Security: A New Framework for Energy Regulation* (Edward Elgar, 2017), 118.
30 National Assembly of France, *Charter for the Environment*, Article 5.
31 National Assembly of France, *Law No. 2011–835 of 13 July 2011 to Ban the Exploration and Exploitation of Oil and Gas by Hydraulic Fracturing and to Repeal the Exclusive Licences of Projects Using This Technique for Mining.*
32 Ruven Fleming, *Shale Gas, the Environment and Energy Security: A New Framework for Energy Regulation* (Edward Elgar, 2017, 117–130).
33 Ibid.
34 Ibid.
35 French National Assembly, *Bill No. 155 to End Research and Exploitation of Conventional and Unconventional Hydrocarbons and to introduce Various Provisions Concerning Energy and The Environment*
36 French National Assembly, *Law No. 2017–1839 of December 30 2017 Act to End Research and Exploitation of Conventional and Unconventional Hydrocarbons and to Introduce Various Provisions Concerning Energy and the Environment.*
37 For statistics relating to the vote see <http://www.assemblee-nationale.fr/dyn/15/dossiers/alt/fin_recherche_exploitation_hydrocarbures>.
38 Refer to Chapter 5, Queensland Australia, for a discussion on CSG extraction.
39 United States Environmental Protection Agency, *Investigation of Ground Water Contamination near Pavilion, Wyoming: Draft Report* (2011) <https://www.epa.gov/sites/production/files/documents/EPA_ReportOnPavillion_Dec-8-2011.pdf>; Stephen Osborn, Avner Vengosh, Nathaniel Warner and Robert Jackson, 'Methane Contamination of Drinking Water Accompanying Gas-well Drilling and Hydraulic Fracturing' (2011) 108(20) *Proceedings of National Academy of Science* 8172–8176.

40 USGS, *Assessment of Undiscovered Oil and Gas Resources of the Paris Basin, France*, 2015 (2015) <https://pubs.usgs.gov/fs/2015/3016/pdf/fs2015-3016.pdf>.

41 John T.S. Keeler, *The Politics of Shale Gas and Anti-Fracking Movements in France and the United Kingdom* Masters Thesis, University of Pittsburgh Graduate School of Public and International Affairs (2014) 7.

42 European Parliament, Directorate-General for Internal Policies, *Impact of Shale Gas and Shale Oil Extraction on the Environment and on Human Health* (2011) <https://europeecologie.eu/IMG/pdf/shale-gas-pe-464-425-final> 14.

43 French Government, *Achieving a Balance in Trade Relations in the Agricultural Sector and Healthy and Sustainable Food* (2018) <https://www.gouvernement.fr/en/achieving-a-balance-in-trade-relations-in-the-agricultural-sector-and-healthy-and-sustainable>.

44 World Trade Organisation (WTO), *Agreement on Trade-Related Aspects of Intellectual Property Rights: Module IV Geographic Indications* (2010).

45 Pénélope Lamarquea and Eric F. Lambin, 'The Effectiveness of Marked-based Instruments to Foster the Conservation of Extensive Land Use: The Case of Geographical Indications in the French Alps' (2015) 4 *Land Use Policy* 706, 707.

46 Sarah Bowen, 'Embedding Local Places in Global Spaces: Geographical Indications as a Territorial Development Strategy' (2010) 75(2) *Rural Sociology* 209, 210.

47 Guy Kolleen, *When Champagne Became French: Wine and the Making of a National Identity* (Johns Hopkins University, 2003).

48 Dom Pierre Pérignon has also been attributed as creating champagne according to some sources. Yvon Dufour and Peter Steane, 'Folding the Future Back into the Present: Lessons from the Past – Dom Pierre Pérignon and the Development of Champagne' (2011) 3(1) *Asia-Pacific Journal of Business Administration* 1.

49 Steve Charters and Nathalie Spielmann, 'Characteristics of Strong Territorial Brands: The Case of Champagne' (2014) 67 *Journal of Business Research* 1461.

50 Article 23 (1) of TRIPS prescribes a 'heightened level' of GI protection for wines and spirits and prohibits the incorrect use of GIs and additionally prohibits the use of an indication of a region in conjunction with words such as 'like', 'style' etc. Susanne Taylor and Madeline Taylor, 'The Aroma of Opportunity: The Potential of Wine Geographical Indications in the Comprehensive Economic Cooperation Agreement' in William van Caenegem and Jen Cleary (eds), *The Importance of Place: Geographical Indications as a Tool for Local and Regional Development* (Springer International, 2017) 81–110.

51 Stephen Jurca, 'What's in a Name?: Geographical Indicators, Legal Protection, and the Vulnerability of Zinfandel' (2013) 20(2) *Indiana Journal of Global Legal Studies* 1445, 1457.

52 Pénélope Lamarque and Eric F. Lambin, 'The Effectiveness of Marked-based Instruments to Foster the Conservation of Extensive Land Use: The Case of Geographical Indications in the French Alps' (2015) 42 *Land Use Policy* 706, 707.

53 Poland holds estimated reserves of 133.6 tcf. See Robert Dodge, 'Unconventional Drilling for Natural Gas in Europe' in Yongsheng Wang and William E. Hefley (eds), *The Global Impact of Unconventional Shale Gas Development Economics, Policy, and Interdependence* (Springer, 2017) 97, 110.

54 Eurostat, *Agricultural Census in France* (2012) <http://ec.europa.eu/eurostat/statistics-explained/index.php/Agricultural_census_in_France>.

10 Poland

Introduction

The story of shale gas in Poland is a tale of two woes. On the one hand, is a country that is rich in unconventional gas resources and it was primed to explore and utilise these to gain independence from its old communist master, Russia. On the other hand, even though Poland has the largest shale gas reserves in Europe, in the 10 years since the gas resource was discovered in 2008, it has been unable to develop this bounty and to date does not have a single producing unconventional gas well. The intervening 10 years, from such hope to such despair (in terms of the development of shale gas), provide an excellent study for the interplay between energy security, agriculture and the role of people and governments.

In 2010 Poland was poised on the cusp of shale greatness. The use of hydraulic fracturing and horizontal drilling to access shale reserves in the USA was just beginning to take hold. The prospective shale gas reserves in Poland, and its eagerness to exploit those reserves, provided an example of how shale gas could provide energy security for a country that did not have conventional gas reservoirs. In the 5 years from 2010–2015, great efforts were made to explore for viable shale gas production. This focus on shale gas arose because of Poland's high dependence on Russian gas. Large petroleum corporations, including Total, Eni, Chevron, and ExxonMobil acquired licences to explore for unconventional gas and were filled with hope in terms of shale gas bounty in Poland. Indeed by 2015, investment in shale gas exploration in Poland totalled $US2 billion.[1] Poland was widely perceived by both companies and energy institutions alike as the next frontier.

This has not come to pass. There have been several factors that have influenced the decline of shale gas in Poland. These have included disappointing exploration results leading to downgraded prospectivity, contraction in the petroleum market as a result of lower oil prices that has seen investment in this sector wane and community resistance to the development of shale gas. What was unwavering(until 2015) was the Polish Government's commitment to the development of shale gas, although this changed with the Law and Justice Party's election victory on 25 October 2015. To be sure, other jurisdictions in

Europe (such as Finland) are as dependent on Russian gas. However, Poland's relationship with Russia and indeed the rest of Europe has seen it try to 'go at it alone'; that is, to gain full energy independence. Shale gas was seen as the viable road to independence.

The change of government in 2015 on the back of disappointing shale gas exploration results and weak sector performance has seen a resurgence in a focus on the agricultural industry. This contrasts sharply to the pervading view of the previous government that shale gas and its exploitation was of most importance in Poland. If this chapter was written 3 years ago, it would focus on the imbalance between the importance of shale gas versus agriculture, where shale gas exploitation trumps agriculture. In 2018, that focus has been almost reversed, with agriculture now seen as an important sectorial component of the Polish economy. Previously, Poland provided an example of an over-whelmingly pro-extraction jurisdiction seeking to exploit its shale gas resources to secure energy independence. As a result, agricultural production was firmly delegated to second place, partly because of the availability of food from other EU States at a reasonable price. Yet Poland still seeks to revitalise, restructure and reassert its agricultural economy.[2]

This chapter provides both an historical and a fresh contextual analysis of shale gas in Poland, examining both the past focus on shale gas as an all or nothing approach to resource development as well as the more recent pro-agriculture anti-shale gas approach. What is interesting in this study is that both of these positions represent a statist approach: the period of 2010–2015 marked by the State's pro-shale gas focus and the period from 2015 onwards, as a consequence of the election of a new government, demonstrating a pro-agri-culture approach. In order to undertake this study, the rest of this chapter will be arranged as follows. Firstly, there will be a consideration of the importance of energy security in Poland, given its unique position and relationship with Russia. It will then consider shale gas resources and their development in Poland, as well as their relationship to energy security. This will be followed by an analysis of agricultural land, food security and its relationship with energy within the period of study. Finally, an analysis will be made of what was a largely Polish issue of having to choose between food or energy. This study of a polarised Polish conundrum will be made by examining the statist regulatory approach in Poland and how the State has altered its focus from shale to agriculture.

The importance of energy security in Poland

Energy (in)security in Poland

In their seminal paper *The Power of Policy Regimes*,[3] Goldthau and Labelle note that shale gas has been coined a 'game changer' for both European and Polish energy security. Furthermore, they speculate that it may enable Europe to diversify its sources of natural gas and lower its dependence on Russian imports.[4] In particular, it would enable Poland to increase its energy diversity which hitherto has escaped policymakers as a result of its location in Europe.[5] The promise of Poland, in relation to shale gas, is that it contains the highest amount of resources in Europe and the greatest impetus to develop those resources given its energy insecurity, which has been dominated by its relationship with its former master, Russia.[6] This relationship was shaped by Poland's dependence on Russian gas, which had commenced in the late 1960s, became deeply entrenched by the 1970s and peaked in the mid-2000s.[7]

Shale gas was seen as Poland's answer to its energy insecurity. This insecurity arises as a result of the 2006–2009 Russia–Ukraine gas crisis,[8] and the annexation of the Crimea in 2014. The disruption in gas supply is particularly pertinent for Poland, given that they have a high dependence on Russian gas. Indeed, 88 per cent of all oil and 72 per cent of gas used in Poland originates from Russia.[9] Complicating this factor is the previous relationship between Russia and Poland, where Poland was a Soviet puppet State until 1989, expected to comply with its Soviet master's wishes. The influence of the Soviet regime passed its zenith with the Solidarity movement in the 1980s, the popularity of Lech Wałęsa and ultimately the overthrow of the Soviet-backed regime in 1989. The importance of the Soviet past on Poland is twofold. Firstly, in a country that exhibits a statist approach to the management of energy security and agricultural land, there is a deep mistrust of the State.[10] In addition, a legacy of the Soviet State is corruption which continues to dog the State. The second major issue from Soviet rule that impacts on shale gas development is that of the system of ownership and use of land, whereby individual peasant farms rather than collectivised State farms dominate the largely agrarian Polish State.

In 2009, after the gas crises that marred the winters of 2005–6 and 2008–9, Poland embarked on a comprehensive reinvigoration of its energy policy. There was no single voice in relation to the development of the energy policy, but rather 'a cacophony of competing interests within Poland's coalition government, within the parties that make up the coalition, and among ministries'.[11] The resulting policy, known as the *Energy Policy in Poland until 2030* [12] prioritised the development of conventional energy sources, yet did not provide a plan or policy direction for unconventional gas resources.[13] However, by 2010, shale gas was seen as the answer to the energy insecurity of Poland. This is because Poland saw the opportunity to develop domestic gas resources for the first time. Furthermore, it had witnessed shale gas provide the USA with a new lease of energy life, allowing the USA to move from a net energy importer since 1975 to energy self-sufficiency and the accompanying economic benefits of being an LNG exporter. Poland wanted the same for itself, given its reliance on Russian gas and its relationship with Russia that had the potential to deteriorate significantly. Furthermore, Poland argued that its dependency on Russian gas supplies was risky for the future of the country.[14] The Polish Government has indicated that securing long term energy supply is a government priority[15] and nuclear will play an important part in that role.[16] Johnson similarly argues that Poland's dependence on Russian gas supply is precarious.[17]

The Polish energy mix

A discussion of Polish energy security necessarily focuses on the relationship of Poland with Russia and its dependence in relation to the supply of gas. Very little discussion occurs on other sources of energy in Poland and other suppliers of energy. What is interesting to note is that even though Russia provides 88 per cent of oil and 72 per cent of its gas, Poland still maintains a relatively high degree of energy independence – a construct of the energy source mix that occurs in Poland. Poland is the largest producer of hard coal in the EU, which plays a significant part in its energy mix. Historically, 92–94 per cent of its electricity was generated by coal.[18] World Bank data indicates that in 1960, 97.5 per cent of electricity was generated by coal and this figure has recently fallen to 80.9 per cent in 2015.[19] However, this coal generation capacity is a legacy of the Soviet era, with 62 per cent of all coal electricity generation capacity greater than 30 years old, and an additional 13 per cent is aged between 26–30 years.[20] Given the age of the assets, Poland has to make a decision as to what its energy mix will comprise in the future.[21] Poland has indicated that nuclear will play a part,[22] although a decision on nuclear is not expected until late 2018 or early 2019. Furthermore, it is likely that renewables will play a significant role in the future Polish energy mix, given that renewables have increased from 4 per cent a decade earlier to 10.4 per cent in 2014. Furthermore, Poland recently introduced the *Renewable Energy Act* in 2015 in line with its obligations under the EU Third Energy Package.[23]

As part of its energy policy, Poland seeks to reduce its dependence on coal and lignite by 40 per cent by 2035,[24] as well as a move towards non-carbon electricity generation. To date, Poland's continued use of coal demonstrates the policy emphasis of energy over the environment. This lack of concern about the environment is supported by Dodge who notes that environmental concerns over unconventional gas drilling have been less significant in Poland than other European countries.[25] In relation to the use of petroleum in the Polish energy mix, in the past three decades the use of oil has increased by 96 per cent and the use of gas by 42 per cent.[26] However, the use of gas for electricity generation remains relatively negligible with no gas utilised in 1980 and only 3.8 per cent of gas utilised for electricity generation in 2015.[27]

Today, gas comprises only 15 per cent of Poland's overall energy demand. This percentage makes natural gas the third largest source of energy, of which two-thirds is imported with one-third produced domestically.[28] The total consumption of gas is only 0.44 trillion cubic feet (Tcf) per annum compared with 27 Tcf per annum in the USA and 5 Tcf per annum in China.[29] The energy mix in Poland is likely to change in the future, given that Polish energy policy is largely driven by EU directives and requirements. In particular, two requirements will have a large impact; the requirement to liberalise its electricity and gas markets in line with EU directives, and the EU 20/20/20 goals as part of the Third Energy Package, which requires Poland to increase the share of renewable energy to 15 per cent.[30]

The International Energy Agency (IEA) notes that in 2015, 88 per cent of oil imports and 72 per cent of natural gas imports came from Russia indicating that Polish energy security is dominated by Russian supply.[31] Consequently, Polish energy policy seeks to reduce its dependency on Russia for supply and to diversify its energy sources and supply routes.[32] Therefore, the energy policy 2030 Plan for Poland is predicated on three concepts which align to EU goals: improvement of energy efficiency; enhancement of security of supply; and diversification of electricity generation (including nuclear). Overall, Poland also seeks to reduce the impact of energy on its environment,[33] with the State continuing to take a strong leadership role in determining the energy mix for the future, which may bring Poland in conflict with the goals of the EU as seen in 2012.[34]

Although natural gas usage as a share of energy use is relatively low comprising 18.3 billion cubic metres (bcm), only 15 per cent of Poland's total primary energy supply, its use has been increasing since the decline of the Soviet Union.[35] The share of gas used for heat and power production is now 4 per cent (up from 0 per cent 20 years away), an extremely low figure compared with the IEA average of 19 per cent.[36] The industrial sector accounts for 39 per cent of gas demand, followed by residential use demanding 24 per cent. The long-term energy contracts between Russia and Poland are due to end in 2022, with Poland indicating that it will not be renewing these contacts and instead be seeking to source gas imports from the Netherlands, Norway and elsewhere.[37] US LNG will play a significant role in this new energy supply due to the construction of the LNG terminal in Świnoujście , providing regasification capacity for one quarter of Poland's natural gas supply and enabling shipment from large LNG suppliers such as Qatar and the USA.

Shale gas resources in Poland and their role in energy security

Shale resource plays and distribution

According to USA Energy Information Agency estimates of shale gas resources in 2013, Poland contained 146 Tcf of shale[38] consisting of the *Baltic Basin* in the North, the *Podlasie Basin* in the east/central east and the *Lublin Basin* in the southeast.[39] Of these basins, it appears that the Baltic Basin is the most prospective.[40] Since 2010, many countries have sought energy independence from Russia. As noted by Goldthau, the US shale gas revolution is viewed as a model for nations that are import dependent and seek to diversify their sources of energy through the utilisation of domestic reserves. At the forefront of such a quest for energy independence was Poland. Reeling from recent gas shortages caused by the Russia–Ukraine gas crisis, Poland sought to obtain energy independence from Russia. The shale gas formations containing unconventional gas beginning to be developed in the USA seemed to present the answer to the energy independence players.[41] In the early 2010s, Poland was considered a leading European country in shale gas development – primarily as a result of its high reserves and enthusiasm attributable to energy insecurity.[42] Indeed, Dodge noted that the exploitation of 'an *in situ* energy source that is qualitatively environmentally friendly and quantitatively commercially viable has accelerated exploration for shale gas deposits'.[43]

Prior to 2010, Poland's shale gas industry was non-existent and remained in an early exploratory, pre-commercial phase. Industry was initially excited by Polish prospects, due to the embrace of shale gas development by the government and the relatively minimal regulatory framework within which the activity was undertaken. However, 8 years on, the results have been disappointing at best, and costly and depressing at worst. To date, less than 100 wells have been drilled in Poland and an examination of the reason for this demonstrates that the shale has not been as productive as initially predicted. As a result of poor exploration results, particularly in the Lublin basin, Poland's shale gas capacity has been downgraded to 146 Tcf in its 2013 report.[44] Although exploration confirms shale gas reserve potential, studies indicate that reservoir conditions are more challenging than the industry initially anticipated.[45]

Although the Polish Government, on the back of its energy policy in 2009, was exceptionally keen for shale gas to be developed, with the change of government in 2015 the shale gas agenda has very much taken a back seat and agriculture has become the focus. This is because the Law and Justice Government sees the reform of the agricultural sector as of paramount importance in rebuilding the Polish economy. The shale gas plays in Poland have been exceptionally disappointing. Currently, all of the major players have withdrawn from shale gas exploration and licenses remain in limbo. Thus, there is effectively little to no activity being undertaken in Poland. The reason for this halt in shale gas exploration can be attributed to three main issues. First, the technical issues surrounding the resources which include the lack of prospectivity and yielding disappointing results. Poland was the first jurisdiction outside of North America which whole heartedly embraced shale gas development. However, because of their complex geology and the low percentage of organics, the deposits have yielded results that were not in line with expectations. Such optimism was replaced with a deep pessimism which ultimately resulted in major players abandoning the prospects. On the back of this activity, the Polish Geological Institute has reduced the estimates of technically recoverable reserves from 5300 bcm to 346–768 bcm – a reduction to about 1/10th of the original estimates.[46]

Second, external factors contributing to the cessation of shale gas exploration in Poland included aging infrastructure (pipelines) coming into Poland and EU regulation and sentiment that was increasingly anti-hydraulic fracking. However, perhaps the greatest limiting factor for shale gas regulation in Poland has been internal forces. At the forefront is the under-developed drilling infrastructure and equipment available to undertake the activities.[47] There is also a poor regulatory framework which includes many complexities, and red tape in relation to the licensing and concession system. Because of the Polish government red tape,[48] foreign companies lost support and patience with the government. This was heightened by the change of government in 2015 with the Law and Justice Government tending towards anti-foreign government and influence. This lack of clear regulation and legal framework may be attributed to the remnants of a Soviet system and a failure to reform the legal framework in due course. Added to this is the lack of an entrepreneurial culture in Poland. Prior to 1990, the country was fully under State control and entrepreneurial spirit was crushed.[49] This has impacted upon homegrown advances and, combined with the little support from foreign companies, had a large bearing on the failure of shale gas drilling. If the reserves were of sufficient quality and quantity, perhaps the internal shortcomings could have been overcome. But with poor confirmation of reserves and a tepid government response, the last 5 years have seen enthusiasm from shale gas development wane from high to its present approach.[50] Attempts at reform were undertaken in 2013–2014, especially in relation to the duration of and approval process for licenses. However, this did little to whet the appetite of foreign companies with the change of government sounding the death knell for exploration. Evidentially, one of the greatest impacts has been lack of a comprehensive approach to the issues of legal regulation of shale gas activity and the shortcomings of Polish governments.[51]

Lastly, but certainly not of least significance, is the lack of profitability in developing shale gas. In the USA, both the regulatory framework and the existing infrastructure contributed significantly to the profitability of shale gas.[52] However, in Poland, the lack of both and particularly that of ageing existing infrastructure, combined with resistance from the government in providing concessions for early development, has meant that profitability of the development of shale gas remains poor. Now that all of the majors have withdrawn from shale gas development in Poland, what remains are State-owned companies seeking to develop the resources with little expertise and support. This does not bode well for the development of shale gas in Poland.

Regulation of shale gas extraction

The regulation of shale gas exploration activities is governed by the general Polish mining law, the *Geological and Mining Law of 2011*, amended in 2014, which regulates the licensing and concession framework to provide access to State-owned petroleum resources as well as the extraction activities. An excellent overview of the licensing and legal framework in found in Wiercinski et al.[53] It is important to note, as outlined in Wiercinski et al., that there have been many changes to the Polish regulatory framework for shale gas extraction. While this chapter does not focussing on the changes in shale gas law, but rather the interplay of shale gas extraction and agricultural land, it is worthwhile to note that major reform to the *Polish Mining Code* to incorporate shale gas extraction occurred in 2014 in an attempt to encourage shale gas exploration and production.[54] However, the implementation of the Code has to date been tepid at best and it remains to be seen whether the law is effective. As is demonstrated in Table 10.1, [55] aside from complexity of Polish legislation, shale gas development in Poland is also subject to a myriad of EU legislation. These include environmental, waste, licensing and chemical directives laws. Such legislation and its application in Polish national legislation, as well as the failure of the Polish Government to effectively and efficiently incorporate these requirements into national legislation, has contributed significantly to the withdrawal of foreign companies.

Table 10.1 Summary of law regulating shale gas licencing, permitting and environmental control

Legislation	Brief description	Links to Shale gas	Relevent concerns/issues
Hydrocarbon Directive 94/22/EC	General principles for granting and authorizing exploration and production of hydrocarbons.	Basic legislation regarding hydrocarbons extraction in the European Union. Member States have sovereign rights over their resource.	– In Poland mining usufruct with State Treasury is required next to relevant permits.
Environmental Impact Assessment Directive 2011/92/EU	Procedure to ensure that projects that are likely to have significant effects on the environment are subject to an assesment, prior to approval.	Overall environmental impacts from basis of ongoing academic disputes, regarding drinking water, methane emissions and seismic activity.	– EIA is not mandatory, only subject to so-called Annex II projects. – Polish Environment Ministry states that generally deep drilling methods require an EIA
Water Framework Directive 2000/60/EC, Ground Water Directive 2006/118/EC and Urban Waste Water Treatment Directive 91/271/EEC	These directions all aim to contribute to water protection in the Member States, transposed in Poland in the Water Law Act and the Environment Protection Act	One of the contested areas of environmental impact of hydraulic fracturing and related activities.	– In Poland, there are no specific requirements regarding the prevention of contamination of ground and surface water.
Mining Waste Directive 2006/21/EC	Waste from extractive industries requires prior permitting.	Links predominantly to the disposal of waste water after fracking has occurred.	– While the more general Waste Directive 2008/98/EC may apply, but has not yet been transposed in Poland.
Directives related to the Emissions Trading Scheme (2009/29/EC and others) and Atmospheric Pollutant Directive 2001/81/EC	Legislation safeguarding air quality and monitoring a wide range of emissions, including carbon, methane, nitrogen oxides, volatile organic compounds and sulfur dioxide.	Methane leakage forms one of the most prominent disputes in academia, influencing the overall carbon footprint of shale gas (that proponents see as a bridging fuel.)	– In case of exceeding the national emissions ceiling, the relevant authorities have to make a plan how to limit or revoke emission permits.

(Continued)

Table 10.1 (Continued)

Legislation	Brief description	Links to Shale gas	Relevant concerns/issues
Environment Protection Act and the Act on Protection of Agriculture and Forest Land	Soil protection is a jurisdiction of the Member States, in Poland arranged through two laws. No specific permits relate to soil protection.	Certain land areas may be designated for other purposes than resource extraction.	– For shale gas extraction, it may be necessary to change the use of land as defined in so-called 'land development plans'.
Habit Directive 92/43/EEC and the Wild Bird Directive 2009/147/EC	Directives form the cornerstone of European nature conservation policy, linked to the Natura 2000 network of protected sites.	With many protected nature conservation areas in Poland, regulation following from these Directives can apply to envisaged shale gas extraction sites.	– Activites in protected areas require prior assessment. In Poland projects could still be authorized 'for imperative reasons overriding public interest only'. Unclear whether and if so how this linked to shale gas extraction.
Habit legislation, i.e., 2009/42/EC and 2000/14/EC	Activities related to shale gas are submitted to noise limitations. In Poland, hydrocarbon related activities fall in the scope of existing regulation.	As all industrial activity, shale gas extraction is submitted to noise limitations. This is expected to apply mostly to the drilling phase.	– Outdoor equipment noise standards are set not by the Ministry of the Environment but their colleagues on Economy.
Existing legislation with respect to the transport of chemicals 2008/68/EC and their registration	Legislation aims to regulate safe transport of dangerous goods in the Members States. REACH regulates chemicals and their safe usage, dealing with registration, evaluation, authorization and restriction of chemicals.	Chemicals are a very small but crucial element of the fracking fluids used when extracting shale gas. Amongst others, chemicals substances have been added to prevent corrosion or enhance viscosity.	– Implementation of directive on inland transport of dangerous goods has not been carried out in Poland.

Legislation	Brief description	Links to Shale gas	Relevent concerns/issues
			- No specific disclosure procedure on hydraulic fractturing fluids is currently in place in Poland. Relevant authorities may however require disclosure. - Operators seem to fall under REACH, yet is it not clear whether used chemicals meet relevant thresholds.

Source: Johnson and Boersma (2013), p. 394.

A study by Uppsala University notes that in order for shale gas to be a panacea for Poland's energy self-sufficiency, at least 400 new wells per annum would need to be drilled.[56] As noted previously, a lack of infrastructure and drilling equipment would retard such drilling activity and make self-sufficiency difficult, certainly in the early years of development. Perhaps more significantly, the Polish regulatory framework represents not only a failure of political support for shale gas development, but its lack of clarity means that the system is confusing and unhelpful for stakeholders. Poland's eager embrace of shale gas exploration and development was not favourable to all States in the EU, particularly France.[57] In the early 2010s this created some conflict within the EU. The Polish approach placed shale gas and hydraulic fracturing, a technique associated with the extraction of shale gas, high on the EU's agenda. In 2014, the EU established the European Science and Technology Network on Unconventional Hydrocarbon Extraction (the Network) to collect, analyse and review results from shale gas exploration projects as well as to assess the development of technologies used in unconventional projects.

The establishment of the Network was part of the Directive *Communication from the Commission on the Exploration and Production of Hydrocarbons, Such as Shale Gas Using High Volume fracturing in the EU* (COM/2014/23).[58] The European Commission has also developed a framework for safe and secure unconventional hydrocarbon regulation in the EU with three main objectives: To ensure that opportunities for energy diversification can be safely and effectively taken up if Member States choose to; to provide clarity and predictability for market operators and citizens; and to fully consider greenhouse gas emission and the management of climate and environmental risks in any shale gas development. It is accompanied by a recommendation providing minimum principles for hydraulic fracturing.[59] Article 3 of the COM/2014/23 lays down the requirement for a strategic environmental assessment to prevent, manage and reduce the impacts on and risks for human health and environment based on the requirements of Environmental Impact Assessment Directive.[60] This implements the precautionary approach to environmental management. Such an approach to strategic planning and environmental impact assessment is necessary in Poland given the absence of a sophisticated planning system and minimal protection of the environment that marks the Polish environmental framework.

Agricultural land, food Security and its relationship with energy

Agricultural land in Poland

There is a high degree of overlap between shale gas basins and agricultural land in Poland. Land utilisation maps demonstrate that the area of the Baltic Basin is particularly characterised by cereal crops and root crops, such as potato and sugar beets, as well as some forest areas.[61] Indeed, all shale gas plays are overlain by important agricultural areas. Agriculture has a significant role in the Polish economy for the production of food and the employment of Polish people.

Poland has yet to transition to a fully modern agricultural system; instead, it is reliant upon an intensive labour force (11 per cent of total employment).[62] The agricultural sector is characterised by diverse development conditions, primarily as a result of historical legacies, exacerbated by population decline and ageing.[63] Significantly, one third of the population is rural, contributing around one quarter of the total gross domestic product (GDP) of Poland's economy.[64]

The development of shale gas provides particular pressures on the environment.[65] EU reports[66] indicate that harm arising from shale gas extraction include: the environment and climate, including noise pollution, air pollution and surface and groundwater contamination. Although environmental harm was identified the EU chose not to impose an EU-wide ban on hydraulic fracturing; instead, it decided to leave the regulation of hydraulic fracturing to each individual Member State based on its internal review and recommendation. The majority of companies which sought to develop shale gas in Poland were of North American origin. Historically, the business and political culture of the USA and Canada has valued the development of oil and gas over the environment and the domestic supply of energy over imports.[67] This meant that those companies seeking to develop shale gas had come from a regulatory environment where the development of the resource was of paramount importance. For this reason, this had the potential to pose a great threat to Polish agricultural land. Indeed, modelling by Barinzelli et al. demonstrates that the land required for shale gas development in north Poland is significant.[68] Importantly, competition with agricultural land may become important at a localised level. Because shale gas extraction is specific, Barinzelli et al. note there is a need to consider environmental impact on a case by case basis.[69] Such an impact may be significant given that agricultural land in Poland has to date is comprised of small to medium holdings with little strong infrastructure coupled with the failure to implement large scale mechanisation. Indeed, the impact and importance of agriculture played a significant role in the success of the Law and Justice Government during the 2015 elections.

In October 2015, the Law and Justice Government was elected in response to a campaign to relax repressive food and hygiene regulations which were introduced after Poland joined the EU in 2004.[70] The election of the Law and Justice Government and their relaxation of repressive food and hygiene regulations came on the back of the International Coalition to Protect the Polish Countryside (ICPPC) agenda in a campaign seeking to provide production freedom to 1.3 million smallholdings in Poland.[71] The *ICPPC Charter for Real Food* campaign outlined key factors for maintaining food security and food sovereignty in Poland. As part of its 10-year anniversary program on food sovereignty and self-sufficiency, the ICPPC undertook a family food mapping program, which noted that Poland has a unique countryside with rich wild nature, strong cultural traditions and many generations of farmers. Protection of this natural wealth and diversity and the food security in Poland are viewed as vital issues in Poland, with the formation of ICPPC in November

2000 (initiated by 41 organisations from 18 countries) and a resurgence of the organisation since the new government came into power in 2015. The ICPPC remains strong and has been instrumental in implementing the Law and Justice Government's agricultural programme.

Legislation and studies relating to shale gas in Poland focus primarily on technical issues related to the extraction of shale gas rather than on the social and economic consequences associated with shale gas development on farmland. One important way of addressing these social and economic consequences is through the planning system, such as in the UK. However, there is little protection of high quality soils or a comprehensive planning system in Poland in place at this time. During the period of exploration, 2010 to 2015, the Polish approach to shale gas was to give it greater priority than that of agricultural farmland. A number of protest actions subsequently emerged during shale gas exploration within this period; however, these actions were quashed and many dissenters jailed. It is likely that had shale gas development occurred in any significant manner in Poland, its development would have overridden agricultural protests. One of the first actions of the Law and Justice Government was to free those that had been jailed for protesting and to focus on the agricultural dimension of the Polish economy.[72]

Today, the Polish Government identifies two main threats to Polish agriculture, neither of which are shale gas. The first is that of genetically modified organism (GMO) cropping which has an impact on the use of traditional agricultural species. The second is that of repressive EU food regulations that have a heavy impact on small–medium farm holdings selling agricultural produce locally. It is these two factors that are much more likely to have an impact on food security and food sovereignty in Poland than shale gas. This is because shale gas development has all but ceased and the Law and Justice Government elected in 2015 has continued to focus its attention on the agricultural sector. Such a statist approach to the agricultural sector is a return to the approach taken during the Soviet era. It seeks to correct the harm done when farmland was converted after the removal of the Soviet influence as a result of privatisation and restitution.[73] In addition, this period saw the closure of State farms and the conversion of farmland to other uses. The Law and Justice party seeks, in many respects, to regain control over agricultural land in Poland. It is interesting to envisage what the policy and strategy of the government would be if shale gas development on agricultural land were to re-emerge.

If shale gas had continued to be developed in Poland, of interest would be the management of water use, reuse and disposal. In particular, it is well known that some of the highest impacts of shale gas development are the use of water for hydraulic fracturing and the disposal of produced water returned to the surface after the fracturing is complete. Given the confusion in the Polish regulatory framework concerning shale gas, licensing the control of water use is yet to be determined. In addition, the regulation of waste water disposal arising from hydraulic fracturing is required in accordance with the EU waste regulation. It is these two facts that are likely to have an impact on agricultural land should shale gas development ever occur.

The decline in the development of shale gas exploration occurred at an important political juncture in Poland's agricultural history. For the first time since the fall of the Soviet Union, the State has focused on the agricultural sector and the importance of reform in agriculture to encourage Polish resurgence in agricultural production to increase employment and entice émigrés to return.[74] Given the failure of shale gas development to take hold in Poland, it would appear that early threats to agricultural land and production from shale gas development have largely subsided. Furthermore, the strongly statist approach to agricultural land that the Law and Justice Government has implemented since 2015 will further strengthen the importance of agricultural land in Poland's economic framework. If shale gas exploration recommences in the future, it is possible that conflict between agricultural land use and shale gas development will likely re-occur.

Polarised Poland: Food or energy?

At first blush, it appeared that Poland was going to have to make a choice between energy security and food security, as promising shale gas reserves occurred in shale deposits which underlay prime agricultural land. However, results of exploration provided disappointing outcomes and prospectively was downgraded to one tenth the original estimates. This posed a great threat to Poland's energy security because it meant that it was likely that Poland would continue to rely on Russian imports of gas. However, in the intervening years, the USA has gone from a net energy importer to an energy exporter. Of importance here, is that the export of LNG from Cheniere's Sabine Pass on the border between Texas and Louisiana in the USA to the newly constructed Świnoujście LNG terminal in the Baltic Basin provides an alternative gas supply route and supplier for Poland. This terminal is now seeking to expand – increasing capacity from 5 bcm to 7.5 bcm by 2022. In addition, the use of gas floating storage and regasification unit (FSRU) as an LNG regasification facility will provide an estimated 4.1–8.1 bcm LNG per annum into the gas network.[75] The source of LNG is diversified in Poland, receiving imports from the USA and Qatar. In addition, an important factor will be the development of connection to other pipelines, such as Nordstream, which will provide Russian gas and also bypass the Ukrainian issue since the source of gas is directly from Russia's north, rather than central and southern areas via Ukraine. Such a connection is also likely to provide Norwegian gas,[76] which will transport energy sources to Poland in line with its energy security policy of diversifying supply.

Conclusion

In 2010, Poland looked to the USA as a major influence in its energy security. Since that time, the USA has indeed had an influence on Poland's energy security, but in a different way to that originally envisaged. Early predications were that Poland would develop its shale gas capacity on the back of the USA's energy revolution seeking to utilise American technology and experience to diversify to domestic sources. This has not come to pass, although in 2018, the USA remains a major exporter to Poland and provides the shale gas that is imported through the LNG terminal. Therefore, shale gas development in the USA has indeed become a game changer for energy security in Poland. But rather than providing an example to emulate, surplus gas in the USA is now exported to Poland to ensure a diversity of energy supply there and allowing it to continue to protect and develop its agricultural resources.

Notes

1 Andreas Goldthau, *The Politics of Shale Gas in Eastern Europe* (Cambridge University Press, 2018) 67.
2 OECD, *OECD Rural Policy Reviews: Poland 2018* (2018).
3 Andreas Goldthau and Michael LaBelle, 'The Power of Policy Regimes: Explaining Shale Gas Policy Divergence in Bulgaria and Poland' (2016) 33(6) *Review of Policy Research* 603–622.
4 Ibid, 603.
5 Ibid.
6 J. Wesley Burnett, Randall W. Jackson and Robert Blobaum, 'The State of Play in Poland's Unconventional Shale and Oil Development' (2015) 33(4) *Development Policy Review* 395–414, 400.
7 Robert Dodge, 'Unconventional Drilling for Natural gas in Europe' in Y. Wang and W. Hefley (eds), *The Global Impact of Unconventional Shale Gas Development* (Springer, 2016) 98.
8 For an excellent consideration of the crisis see Aleksandr Kovacevic, *The Impact of the Russia–Ukraine Gas Crisis in south Eastern Europe* (Oxford Institute of Energy Studies, 2009).
9 International Energy Agency, *Energy Policies of IEA Countries: Poland 2016 Review* (2016) <http://www.iea.org/publications/freepublications/publication/Energy_Policies_of_IEA_Countries_Poland_2016_Review.pdf>.
10 J. Wesley Burnett, Randall W. Jackson and Robert Blobaum, 'The State of Play in Poland's Unconventional Shale and Oil Development' (2015) 33(4) *Development Policy Review* 395–414, 399.
11 Ibid, 398.
12 For information of Poland's *Energy Policy in Poland until 2030* see IEA *Energy Policy in Poland until 2030* <https://www.iea.org/policiesandmeasures/pams/poland/name-24723-en.php>.
13 J. Wesley Burnett, Randall W. Jackson and Robert Blobaum, 'The State of Play in Poland's Unconventional Shale and Oil Development' (2015) 33(4) *Development Policy Review* 395–414, 399.
14 Corey Johnson and Tim Boersma, 'Energy (in)Security in Poland: The Case of Shale Gas' (2013) 53 *Energy Policy* 389–399, 397.
15 International Energy Agency, *Energy Policies of IEA Countries: Poland 2016 Review* (2016) <http://www.iea.org/publications/freepublications/publication/Energy_Policies_of_IEA_Countries_Poland_2016_Review.pdf> 11.
16 Ibid, 12.
17 Corey Johnson and Tim Boersma, 'Energy (in)Security in Poland: The Case of Shale Gas' (2013) 53 *Energy Policy* 389–399, 397.

18 J. Wesley Burnett, Randall W. Jackson and Robert Blobaum, 'The State of Play in Poland's Unconventional Shale and Oil Development' (2015) 33(4) *Development Policy Review* 395–414, 400.

19 World Bank, *Electrical Production from Coal Sources* (2016) <https://data.worldbank.org/indicator/EG.ELC.COAL.ZS?locations=PL>.

20 International Energy Agency, *Energy Policies of IEA Countries: Poland 2016 Review* (2016) <http://www.iea.org/publications/freepublications/publication/Energy_Policies_of_IEA_Countries_Poland_2016_Review.pdf> 11.

21 Ibid.

22 Reuters, *Poland to Decide Later this Year on Building Nuclear Plant* (2018) <https://www.reuters.com/article/us-poland-nuclear/poland-to-decide-later-this-year-on-building-nuclear-plant-idUSKBN1FI1Q8>.

23 International Energy Agency, *Energy Policies of IEA Countries: Poland 2016 Review* (2016) <http://www.iea.org/publications/freepublications/publication/Energy_Policies_of_IEA_Countries_Poland_2016_Review.pdf>, 11.

24 J. Wesley Burnett, Randall W. Jackson and Robert Blobaum, 'The State of Play in Poland's Unconventional Shale and Oil Development' (2015) 33(4) *Development Policy Review* 395–414, 403.

25 Robert Dodge, 'Unconventional Drilling for Natural gas in Europe' in Y. Wang and W. Hefley (eds), *The Global Impact of Unconventional Shale Gas Development* (Springer, 2016) 112.

26 International Energy Agency, *Energy Policies of IEA Countries: Poland 2016 Review* (2016) <http://www.iea.org/publications/freepublications/publication/Energy_Policies_of_IEA_Countries_Poland_2016_Review.pdf> 11.

27 World Bank, *Electrical Production from Coal Sources* (2016) <https://data.worldbank.org/indicator/EG.ELC.NGAS.ZS?locations=PL&view=chart>.

28 International Energy Agency, *Energy Policies of IEA Countries: Poland 2016 Review* (2016) <http://www.iea.org/publications/freepublications/publication/Energy_Policies_of_IEA_Countries_Poland_2016_Review.pdf> 18.

29 Ibid.

30 Ibid, 23.

31 Ibid, 24.

32 Ibid, 24.

33 J. Wesley Burnett, Randall W. Jackson and Robert Blobaum, 'The State of Play in Poland's Unconventional Shale and Oil Development' (2015) 33(4) *Development Policy Review* 395–414, 403.

34 In 2012, the European Commission called upon Poland to comply with The Directive on common rules for the internal market in gas (2009/73/EC) and The Directive on common rules for the internal market in electricity (2009/72/EC) within two months.

35 International Energy Agency, *Energy Policies of IEA Countries: Poland 2016 Review* (2016) <http://www.iea.org/publications/freepublications/publication/Energy_Policies_of_IEA_Countries_Poland_2016_Review.pdf> 140.

36 Ibid.

37 Ibid.

38 Energy Information Agency, *Technically Recoverable Shale Oil and Shale Gas Resources: Poland* (2015) <https://www.eia.gov/analysis/studies/worldshalegas/pdf/Poland_Lithuania_Kaliningrad_2013.pdf> VIII-2.

39 Corey Johnson and Tim Boersma, 'Energy (in)Security in Poland: The Case of Shale Gas' (2013) 53 *Energy Policy* 389–399, 395.

40 Polish Geological Institute, *First Report: Assessment of Shale Gas and Shale Oil Resources of the Lower Palaeozoic Baltic–Podlasie–Lublin Basin in Poland* (2012) <https://www.pgi.gov.pl/en/dokumenty-pig-pib-all/aktualnosci-2012/zasoby-gazu/769-raport-en/file.html>.

41 Andreas Goldthau, *The Politics of Shale Gas in Eastern Europe: Energy Security, Contested Technology, and the Social License to Frack* (CUP, 2018) 1.
42 Ibid, 67.
43 Robert Dodge, 'Unconventional Drilling for Natural gas in Europe' in Y. Wang and W. Hefley (eds), *The Global Impact of Unconventional Shale Gas Development* (Springer, 2016).
44 Energy Information Agency, *Technically Recoverable Shale Oil and Shale Gas Resources: Poland* (2015) <https://www.eia.gov/analysis/studies/worldshalegas/pdf/Poland_Lithuania_Kaliningrad_2013.pdf> VIII-2.
45 Ibid.
46 Polish Geological Institute, *First Report: Assessment of Shale Gas and Shale Oil Resources of the Lower Palaeozoic Baltic–Podlasie–Lublin Basin in Poland* (2012) <https://www.pgi.gov.pl/en/dokumenty-pig-pib-all/aktualnosci-2012/zasoby-gazu/769-raport-en/file.html> 7, 25.
47 J. Wesley Burnett, Randall W. Jackson and Robert Blobaum, 'The State of Play in Poland's Unconventional Shale and Oil Development' (2015) 33(4) *Development Policy Review* 395–414, 405.
48 Robert Dodge, 'Unconventional Drilling for Natural gas in Europe' in Y. Wang and W. Hefley (eds), *The Global Impact of Unconventional Shale Gas Development* (Springer, 2016) 112.
49 J. Wesley Burnett, Randall W. Jackson and Robert Blobaum, 'The State of Play in Poland's Unconventional Shale and Oil Development' (2015) 33(4) *Development Policy Review* 395–414, 406.
50 Robert Dodge, 'Unconventional Drilling for Natural gas in Europe' in Y. Wang and W. Hefley (eds), *The Global Impact of Unconventional Shale Gas Development* (Springer, 2016) 112.
51 Jakub Godzimirski, 'Can the Polish Shale Gas Dog Still Bark? Politics and Policy of Unconventional Gas in Poland' (2016) 20 *Energy Research and Social Science* 158–167, 160.
52 Corey Johnson and Tim Boersma, 'Energy (in)Security in Poland: The Case of Shale Gas' (2013) 53 *Energy Policy* 389–399, 390.
53 Andrej Wiercinski, Jakub Jedrzejak and Maciej Kazmarek, 'Poland: A (legal) Step-ahead: New Licensing Procedures for Exploration and Production of Hydrocarbons: New Provisions Dedicated to Shale Gas' in Cecile Musialski, Mattias Altmann and Stefan Lechtenbohmer, *Shale Gas in Europe: A Multidisciplinary Analysis with a Focus on European Specificities* (Claees and Casteels, 2013).
54 For an excellent consideration of the early legal framework for shale gas licensing see Wojciech Baginski, 'Shale Gas in Poland – The Legal Framework for Granting Concessions for Prospecting and Exploration of Hydrocarbons' (2011) 32 *Energy Law Journal* 145–155.
55 From Corey Johnson and Tim Boersma, 'Energy (in)Security in Poland: The Case of Shale Gas' (2013) 53 *Energy Policy* 389–399, 394.
56 Kjell Alekett, Tadeusz Patzek, Bjorn Svensson and Rafal Jaroz, *Can Hydraulic Fracturing Make Poland Self Sufficient in Natural Gas?* (2014) <http://gaia.pge.utexas.edu/papers/140519_Fracking_in_Poland_submitted_Energy_Technology.pdf> 3.
57 For a discussion of shale gas exploration and extraction in France see Chapter 9.
58 European Commission *Communication from the Commission on the Exploration and Production of Hydrocarbons, such as Shale Gas using High Volume Fracturing in the EU* (2014) COM/2014/23.
59 European Commission, *Commission Recommendation of 22nd of January 2014 on Minimum Principles for the Exploration and Production of Hydrocarbons (such as shale gas) using High Volume Hydraulic Fracturing* (2014) 2014/70/EU.
60 European Commission, *Directive 2001/42/EC of the European Parliament and of the Council of 27 June 2001 on the Assessment of the Effects of Certain Plans and Programmes on the Environment* (2001).

61 Food and Agricultural Organisation, *Fertilizer Use by Crop in Poland* (2003) <http://www.fao.org/docrep/005/Y4620E/y4620e00.htm#Contents>; Dominika Milczarek-Andrzejewska, Katarzyna Zawalińska and Adam Czarnecki, 'Land-use Conflicts and the Common Agricultural Policy: Evidence from Poland' (2018) 73 *Land Use Policy* 423.
62 OECD, *OECD Rural Policy Reviews: Poland 2018* (2018), 15.
63 Ibid, 20–21.
64 Ibid, 42.
65 Corey Johnson and Tim Boersma, 'Energy (in)Security in Poland: The Case of Shale Gas' (2013) 53 *Energy Policy* 389–399, 393.
66 Ibid, 393.
67 Ibid, 390.
68 Claudia Barinzelli, Ine Vandecasteele, Ricardo Ribero Baranco, Innes Maru I. Riviero, Nathan Pelletier, Okke Batelaan and Carlo Lavalle, 'Scenarios for Shale Gas Development and their Related Land Use Impacts in the Baltic Basin, North Poland' (2015) 84 *Energy Policy* 80–95, 92.
69 Ibid, 93.
70 Julian Rose and ICPPC, *Polish Government Backs Small Farmers' and Food Sovereignty* (2016) <https://theecologist.org/2016/jan/25/polish-government-backs-small-farmers-and-food-sovereignty>.
71 Ibid.
72 Julian Rose and ICPPC, *Polish Government Backs Small Farmers' and Food Sovereignty* (2016) <https://theecologist.org/2016/jan/25/polish-government-backs-small-farmers-and-food-sovereignty>.
73 Jerzy Banski, 'Changes in Agricultural Land Ownership in Poland in the Period of the Market Economy' (2011) 57(2) *Agricultural Economics* 93–101, 94.
74 Ibid.
75 Ibid.
76 Corey Johnson and Tim Boersma, 'Energy (in)Security in Poland: The Case of Shale Gas' (2013) 53 *Energy Policy* 389–399, 395.

11 China

Introduction

Whereas the issue of food security versus energy security is important for all countries, it is perhaps most important in China. China has the highest population on earth (approximately 1.4 billion people),[1] as well as its challenging geography, land area and water distribution.[2] Given the size of the population, China naturally has a need to produce food to feed its population. Moreover, increased migration of the rural population into urban areas and the expansion of the manufacturing sector has seen China's energy needs rise substantially.[3] This presents a conundrum for the Chinese Central Government – should the development of new sources of energy, particularly unconventional sources such as shale gas, be developed even if such development impinges on and affects agricultural land?

In addition, if shale gas development does occur on agricultural land, what will be the relationship between shale gas extraction and agricultural activities – will both activities occur harmoniously alongside each other, or will the extraction of shale gas be prioritised and, therefore, agricultural land developed to create shale gas production. Contemplation of the development of shale gas also raises the question of water availability and water use, as China struggles with water contamination and availability, and the use of water in the production of shale gas is likely to place further pressure on this.

The political nature of China, that of a communist State, means that the regulatory approach to the issues of shale gas development and agricultural production can be perceived as a statist approach. However, there are also market-economy interests within China, particularly within the manufacturing sector, as well as a burgeoning urban population that need reliable access to energy, with gas the fuel of choice. Therefore, whilst the regulatory approach is naturally statist, there are also market forces and interests that need to be considered.

This chapter will analyse the development of shale gas in China taking into account the importance of both energy and agriculture for the State. It will firstly consider the food security versus energy security dilemma, placing it within the context of changing demographics and geopolitics. It will then consider the legal framework for the regulation of both agriculture and shale gas development, and the interaction of these legal frameworks. Finally, the chapter will analyse the interrelationship between shale gas development, agriculture and water, to determine if the present regulatory approach favours the development of one activity over the other, and whether shale gas development and agricultural activities are contested or coexisting within the Chinese statist framework.

Food security and energy security in China

As its population has grown, China has strived to improve its rates of food poverty, seeking to increase agricultural production, particularly in relation to rice, wheat and corn, which comprise 90 per cent of China's total crop production.[4] This is reflected in the 1996 *White Paper on Grain*,[5] which established a 95 per cent self-sufficiency target for the State,

> while ensuring a continued increase in grain output, to vigorously promote diversified food production, readjust the food structure, and continue to raise the people's quality of life from the stage of simply having enough food and clothing to leading a relatively well-off and comfortable life.[6]

To meet this target, China has sought to increase its grain crop yield, with production usually meeting demand, resulting in a food production/consumption ratio of 1:1 since the mid-2000s.[7] Since the founding of the People's Republic of China, the national grain yield has risen from 1,035 kg/hectare in 1949,[8] to 1,193 kg/hectare in 1961,[9] and 6,029 kg/hectare in 2016.[10] This six-fold increase in yield has come at an environmental price – China has 'excessively and inefficiently used fertilisers',[11] with a three-fold increase in their use since the mid-1980s,[12] to the detriment of the environment.[13]

During the period 1976–2016, grain consumption has doubled from 125 million tonnes to 261 million tonnes.[14] In the same period, meat consumption has increased ten-fold from 7 million tonnes in 1975 to 75 million tonnes in 2016.[15] This means that arable grazing land has become of equal importance as that of cropping land to meet the increased consumption of meat, which in 2017 was approximately 50 kg per capita per annum, of which approximately two-thirds is pork.[16] This huge increase in the consumption of meat is primarily attributable to urbanisation and a growing middle class, which has resulted in a shift from traditional grain-oriented diet to a meat-centric diet. This urbanisation has also increased the demand for other, agriculturally intensive foods such as dairy products. Such dietary changes have led to a change in food consumption rates, with China now reaching 450 kg/capita consumption in 2015,[17] leading to a national rise in obesity rates; the country contains one fifth of all obese people in the world.[18]

Such a shift in the dietary intake of a large percentage of the Chinese population is also reflected in the rapid urbanisation of land use and the increasing importance of agricultural land in China. Land reform in China in 1981 provided incentives to farmers to increase production, as reflected in the

74 per cent increase in grain output, from 354 million tons in 1982 to 618 million tons in 2017.[19] Increasingly, the Chinese government sees the need to preserve agricultural land in order to meet the growing food needs of China's urban middle class, which is projected to be around 60 per cent of China's population by 2020.[20]

China has experienced a socioeconomic shift from a rural to a largely urban population. This, when combined with an ever-increasing manufacturing economic base, creates an increased demand for energy and places energy security at the centre of China's development focus. China's energy demands are high, which is not surprising given its economic growth and urbanisation over the last quarter of a century. Although China's gross domestic product (GDP) grew at a rate of 9.8 per cent per annum in the period 1985–1995, it is expected to grow at an average rate of 6.6 per cent per annum until 2020.[21] Predictions by the State-owned China National Petroleum Corporation (CNPC) indicate that China's demand for oil and gas continues to rise between 1.2–2.7 per cent per annum through to 2030, peaking at 690 million tonnes/year (or 13.8 million barrels per day (bpd)).[22] Similarly, demand for natural gas is expected to rise substantially, peaking at 620 billion cubic metres (bcm) in 2030,[23] up from 237 bcm in 2017,[24] making China the largest source of global demand for gas.[25]

This rising energy demand has to date been largely fulfilled by the use of coal (70 per cent), with only 4 per cent of energy needs met by gas in recent years.[26] The 12th National Five Year Plan of 2011–2015 (12th FYP) sought to address the energy security issue by establishing a national goal of diversified energy sources.[27] This diversification of the national energy mix, as well as the use of natural gas, is further promoted in the 13th National Five Year Plan of 2016–2020 (13th FYP), where the use and development of coal is reduced and shale gas and CSG are actively exploited.[28] Indeed, the express need for the development of unconventional gas is set out in the 13th FYP, which states:

> We will strengthen onshore and offshore oil and gas exploration and exploitation, take well-ordered measures to relax control over mining rights, and actively exploit natural gas, coal seam gas, and shale oil and gas.[29]

In detailing the need for energy development, the 13th FYP specified the acceleration of exploration and development of shale gas in the Changning-Weiyuan region, Sichuan; Fuling, Chongqing; Zhaotong, Yunnan; Yan'an, Shaanxi; and Zunyi-Tongren region, Guizhou.[30] Furthermore, in order to strengthen its energy security, China has also given priority to the establishment of essential infrastructure for the transport of oil and gas, particularly transmission pipelines. Such priority was first iterated in the 12th FYP, which sought to 'accelerate the construction of the strategic transmission channels and improve the domestic trunk oil and gas pipe network'.[31] The importance of shale gas as an energy source to ensure energy security for the state is demonstrated through the *Shale Gas Development Plan 2016–2020*. The *Shale Gas Development Plan 2016–2020* provides guidance for shale gas development under the 13th FYP, including the production goal of 30 bcm by 2030.[32]

Aside from establishing the necessary infrastructure for the import of gas via pipeline and by LNG importation (primarily from Russia and Australia at present), the 12th FYP sought to establish 'cross-regional trunk gas transmission and distribution networks, and create a gas supply layout in which natural gas, coal-bed gas and coal-based gas are balanced'.[33] The establishment of such public utility infrastructure places further pressure on agricultural land. Gas pipelines are constructed in a manner whereby agricultural activities cannot take place after construction is completed, as pipelines have a large diameter of up to 60 inches and are highly disruptive to cropping, grazing and harvesting activities.[34]

The current and projected food security and energy security requirements present China with a challenge. Whereas China seeks to develop its vast unconventional gas reserves,[35] it also needs to focus on the protection of land and the continued development of agriculture. As recognised by the 13th FYP, 'agriculture is the foundation on which we can build a moderately prosperous society in all respects and achieve modernization'.[36]

The implementation of China's *One Belt One Road Initiative*[37] both increases energy requirements, and also provides diverse energy supplies and sources for China's growing energy demands. It is possible that the initiative will not only open up markets for China's manufactured products, but also provide a source of imported food, but this remains to be seen. What is clear is the *Ice Silk Road*[38] will provide China with a new source of energy in the form of LNG from the Arctic Yamal facility in the Russian Federation. This supply of gas will be distributed throughout China using the newly constructed pipeline network, which is referred to in the 13th FYP. It is likely that the *One Belt One Road Initiative* will have an impact on energy security in China, however the depth and extent of this impact remains to be seen until the initiative's final completion in 2049.

Regulatory framework for shale gas extraction in China

In considering the regulatory framework for shale gas it is essential to not only examine the law relating to the extraction of the gas, but also the law relating to the land ownership, land use and agriculture. This is because shale gas development and agriculture are inextricably tied, as illustrated by all jurisdictions examined within this book. Generally, shale gas deposits are found in ancient low-oxygen lacustrine or marine depositional environments with a high percentage of organic material,[39] which have developed into shale hydrocarbons after millions of years of pressure and heating.[40] In the present day, these same organic-rich land-based depositional environments are utilised for agricultural activities, and are often rich agricultural areas essential for cropping. This is especially true in China's eastern regions (where the Sichuan, North China and Songliao Basins shale gas basins are located). In addition, the Tarim and Junggar Basins dominate the critical grazing areas in the western region of China.

In order to consider the regulation of shale gas exploration, it is necessary to consider the issue of land ownership and resources in China. According to Article 10 of the Constitution of the Peoples Republic of China (PRC), 'Land in the cities is owned by the state. Land in the rural and suburban areas is owned by collectives except for those portions which belong to the state in accordance with the law'.[41] Individuals do, however, have the right to use the land that is owned by the state through the granting of usufructuary rights under Articles 40 and 117 of the Property Rights Law 2007, which grants the usufructuary the right to 'possess, use and benefit from immovable property owned by the State'. Regarding mineral resources (including shale gas as it is classified as a mineral rather than petroleum in China), Article 9 of the Constitution of the PRC stipulates that:

> All mineral resources, waters, forests, mountains, grasslands, unreclaimed land, beaches and other natural resources are owned by the State, that is, by the whole people, with the exception of the forests, mountains, grasslands, unreclaimed land and beaches that are owned by collectives as prescribed by law. The State ensures the rational use of natural resources and protects rare animals and plants. Appropriation or damaging of natural resources by any organization or individual by whatever means is prohibited.[42]

Article 9 of the Constitution is implemented through article 3 of the PRC Mineral Resources Law (MRL)[43] and article 46 of the PRC Property Law,[44] which together stipulate that mineral resources belong to the State, and the State Council exercises state ownership rights. This exclusive State ownership of mineral resources occurs on both State-owned land, including construction land,[45] as well as collective-owned land. In practice, this means that on State-owned land there is no fragmentation of the surface rights and underground mineral ownership, which remains intact and vested entirely in the State. However, there is a fragmentation of ownership rights on collective land, whereby the rural collective owns the surface rights of the land only.

The legal basis for the exploitation of shale gas in China is the MRL, combined with the rules for the implementation of the Mineral Resources Law of the PRC (RIMRL). These laws are administered by the national administrative body the Ministry of Natural Resources (MNR),[46] which replaced the Ministry of Land Resources (MLR) in 2018. The role of the MNR is the establishment of integration of land use and natural resource development. This reform was announced on the 13 March 2018 as part of the *Institutional Reform Program of the State Council*.[47] Also part of this reform was the establishment of the Ministry of Agriculture and Rural Industry[48] and the Ministry of Ecology and Environmental Protection[49] from previous administrative bodies. The reasoning behind the reform of administrative bodies was the need to have a more integrated approach to the regulation and administration of land used for multiple purposes in China, as identified in the 13th FYP. There is also a recognition of the need for an integrated approach to the management of the environment and the ecology of China's agricultural land, particularly as shale gas development continues to increase.

The exploitation of shale gas occurs under a licensing system in China, the requirements of which are set out under the MRL. There are two types of licenses granted in relation to shale gas: an exploration license and production license. Shale gas exploration licenses are issued and registered to licensees by the MNR and are of 3 years in duration. Obligations under the license include scheduled minimum exploration expenditures and mandatory reporting to the local county. If an economic deposit of shale gas is discovered, the licensee may apply for a 2-year renewal or retention of the exploration right within 30 days prior to expiration of the permit term, for a maximum period of 4 years or two extensions. The extension will only apply to the area of the economic deposit. Where a commercial deposit has been discovered, a production license is required. Since 2012, exploration and production licenses can be acquired by both Chinese State-owned companies (such as Chinese National Petroleum Company and Sinopec) and privately owned Chinese companies, made possible by the then-MLR approval of shale gas as an independent mineral resource.

In relation to both an exploration and a production license, the grant of the license does not automatically grant the right to either explore for or produce shale gas. Rather, there are additional requirements in order to gain access to the land to undertake the activity. The requirements will vary depending on the ownership of the land, the land classification and the license type. Given that shale gas occurs in agricultural areas, and is generally not developed in highly urbanised areas, it is assumed that most shale gas development will occur on agricultural land, be it cropping or grazing land. This is important, since the type of ownership of rural land is that of collective ownership,[50] as opposed to State ownership of urban land in China. It is important to note that shale gas production under the MNR can only occur on State-owned land and this has implications for the process of gaining access to land, since most of the agricultural land in China is owned by rural collectives.

The first license, an exploration license, grants the holder of the license the right to explore for but not to produce shale gas. Under an exploration license, the licenseholder needs to gain permission to access to the land, by attaining a right over the use of the land. This will require the licenseholder to obtain a contractual right to enter the land, equivalent to a lease, and consent from the rural collective, in order to explore for shale gas.

It is important to note two things within this land access arrangement. Firstly, the licenseholder only gains the right of access to the land to undertake exploration activities. This does not cover any form of ownership or any form of future interest for the production of shale gas. Secondly, this land access right does not enable the licenseholder to produce gas. An interesting legal conundrum occurs where there is a need to 'frack to flow',[51] in order to test the capacity of the shale reservoir to produce as well as its porosity and permeability characteristics. This will result in shale gas being produced in a test environment when the appraisal well is drilled and fractured. It can however, be assumed, that since the release of shale gas is only for testing purposes, and not captured for production, it will still fall under an exploration license and will therefore be not subject to the license requirements of production. Interestingly, such a conundrum has challenged regulators in several jurisdictions, including Western Australia and the Northern Territory in Australia. These jurisdictions have circumvented the issue by defining appraisal wells which involve 'frack to flow' as appraisal wells within the exploration process.[52]

Under a production license, the licenseholder must apply for access to the land in order to undertake production of shale gas for commercial purposes. All shale gas production must occur on State-owned land. Given that most agricultural land is collectively-owned land and not State-owned land, there is a need for the acquisition of the property by the Chinese State in order to release and rezone land for shale gas production purposes. This acquisition process will transfer ownership of the land from the rural collective to the State and the State will grant rights to the license holder for the use and enjoyment of the land to produce shale. This right is akin to a lease however it does not transfer ownership, which remains with the State after acquisition from the rural collective. In order for the State to acquire the land from the rural collective to enable shale gas production to occur, a two-step process is required.

Firstly, the land that is required for the shale gas production needs to undergo a change in land use. This is known as the 'transfer approval process,' which transfers the land from collective land, where agricultural activities occur, to construction land owned by the State. This enables multiple use of the land, including the production of shale gas. In order for this land use transformation from agricultural to construction land to occur, the transfer approval process requires three plans: a socioeconomic plan, an environment plan and a land use plan. The application for the transfer approval process is submitted to the relevant government authority, with the decision made according to the criteria set out and disseminated by the Chinese Central Government.[53] The criteria is based on the land's classification under the Basic Farmland Protection Regulation 1994 (BFPR),[54] and an assessment of that land under Article 12 of the BFPR, including its yield, cultivars and past productivity. Under the land use plan, assessment is made of the use and importance of the agricultural land. Consideration of the land use is undertaken in relation to the BFPR, which is the regulatory basis of the conversion of land from farming to construction land where the building of 'national projects', including nationally important shale gas activities, is deemed unavoidable.[55] Under the socioeconomic plan, issues such as energy security and the national benefit of the project for the Chinese State, as well as local issues, are considered. The environment plan addresses environmental issues such as water use, water disposal and contribution to environmental harm.

These three plans are considered in the assessment of the land transfer, and land use changes are only granted to projects where there is considerable benefit to the State. The capacity to transfer the land from collective to construction land is complex, and thus it is usually only reserved for projects deemed to be of national significance. The assessment of the land use plan is based on the local land use plan, which is devolved from the central government, therefore it is assessed at the local level according to strict, nationally-set criteria. If the transfer approval process is granted, ownership of the land is transferred from the rural collective to the State and compensation is paid.

The *Land Administration Law 1999* (LAL) regulates the acquisition of land and the payment of compensation to landholders. Where land is compulsorily acquired for an activity that is in the national or public interest, compensation will be paid in accordance with Article 2 of the LAL. Under Article 47 of the LAL, compensation is calculated 'on the basis of its original purpose of use'.[56] Compensation for the expropriation of agricultural land is calculated on the basis of 'six to ten times the average annual output value of the expropriated land, calculated on the basis of three years preceding such requisition'.[57] The multiplier factor varies in accordance with the location of the land being compensated and the finances of the local authority granting the compensation.[58]

Once ownership is transferred to the State, the land is reclassified from agricultural land to construction land. This second step is critical since shale gas can only be extracted from construction land and not from agricultural land. The conversion of collective-owned farmland to construction land available for shale gas production is strictly controlled, since farmland is protected. The land enjoying the greatest production is that referred to as basic farmland (Article 2). Basic farmland is cultivated land that may not be occupied and used for any other purpose except for agricultural product. Therefore, the conversion of basic farmland to construction land for shale gas will be reserved only for the projects of greatest importance since this land is highly protected. Necessarily, the larger the area required then the more difficult it is to get approval for the transfer approval process. This provides some level of protection for highly productive farmland, but at the same time it also means that where there is a shale gas project of national importance, noting its high importance as a current national priority, the policy of unconventional gas development will be paramount, and lead to the loss of important farmland.

The interaction of shale gas extraction, agricultural activities and land access: The food security–energy security nexus

Whereas the discussion in the previous section has focussed on the law in relation to accessing shale gas resources and undertaking shale extraction, particularly the law relating to accessing shale gas on agricultural land, this section discusses the interaction between the process of extracting shale gas and agriculture. In particular, this section analyses whether, beyond the mere conversion of basic farmland to construction land to enable shale gas extraction, the process of extraction creates harm to the natural environment that may negatively impact on agricultural production. The main form of harm relates to the use of water resources to undertake hydraulic fracturing (HF), which is used to release the gas from the shale rocks, and the disposal of fluid produced as a result of the process.

As detailed in Chapter 1, the advent of 'slickwater' HF demands the use of high volumes of water, resulting in environmental impacts associated with water use and impingement of agricultural water allocations. China is characterised by both the competition of the use of water with agricultural activities and low water availability. Lohmar notes that this is 'due to institutional failings including a lack of clearly delineated and enforced water usage rights, where upstream users take a greater share of water than entitled to, leaving shortages for downstream users'.[59] This means that availability of water for HF may be limited in such areas where there has been overuse of water by upstream users such as manufacturing, factories or farming activities.

Water availability in China is problematic, with water distribution and quality variable throughout.[60] Cui and Shoemaker note that China's water supply per capita is only 25 per cent of the world's average, and 60 per cent of that water is used for the inefficient irrigation of rural crops.[61] Furthermore, in some water-scarce regions, excessive amounts of groundwater are diverted to agriculture,[62] further impacting on the availability of water. Therefore, in a country where issues of food security remain relevant, water access and availability to undertake HF for shale gas production to meet energy security needs will continue to dominate the regulatory agenda. One solution for China would be the capacity to carry out HF with lower volumes of water, which would rely on the development of low-cost fracking techniques in arid regions.

Another challenge for shale gas extraction in agricultural regions is the potential for water pollution arising from shale gas extraction. This includes the pollution of surface water,[63] as well as the contamination of ground water caused by poorly-cemented wells.[64] The rapid industrialisation of China, combined with poor environmental regulation, has resulted in the contamination of crops from heavy metals, with almost one sixth of China's land contaminated by toxic runoff.[65] Related to this, almost 60 per cent of underground water is polluted and unfit for drinking,[66] and water scarcity is rife. Water pollution can also arise from the fluid used in HF. These fluids contain proppants (such as sand or ceramic beads), and chemicals for thickening the fluid to reduce friction. After HF is complete, this water migrates to the surface, bringing with it elements such as barium, bromide and radium-228, resulting in contaminated produced water.[67] Although some of this fluid can be reinjected or recycled, it can also lead to the concentration of these pollutants, thus contributing to greater water pollution, which is likely to impact on agricultural activities.[68] Sandiglow et al. see the capacity for China to manage such pollution as limited since China's regulatory infrastructure for protecting water is limited, and much less developed than the USA.[69]

Finally, the disposal of produced water from HF poses a great conundrum to China's agricultural activities. In the USA, much of the produced water is disposed of by reinjection into old underground wells, since shale gas extraction is being undertaken in areas where conventional extraction has previously occurred. Such old wells do not exist in China, preventing the underground injection of produced water. However, this is not necessarily detrimental, since such disposal methods have been identified as inducing seismicity.[70] The challenge for China will be the disposal of these contaminated fluids in such a manner that does not further contribute to already high levels of water and land pollution.

The regulatory conundrum – statist regulation in an energy hungry, market-oriented economy

The above analysis of both the laws relating to shale gas development on agricultural land and the environmental impacts of such development on this precious commodity highlights a conundrum that China will continue to experience as it pushes forward with its economic development. This development challenge is both industrial and agricultural. China seeks to become an even greater developed nation, which requires energy. Furthermore, it seeks to provide energy to a greater proportion of its population as net migration to urban areas from the rural areas occurs. As this migration occurs, pressure on energy sources and land will increase. In its 13th FYP, China committed to the industrialisation and marketisation of its agricultural sector:

> The agricultural growth model must be transformed at a faster pace, industrial, production, and business operation systems that work for modern agriculture must be established, and the quality, returns, and competitiveness of agriculture must be strengthened to allow China to embark on a path of agricultural modernization which ensures high yields and safe products, conserves resources, and is environmentally friendly.[71]

This direction outlined in the 13th FYP demonstrates the tension between a statist approach[72] to date and a targeted desire to move to an increasingly market-based approach. At the centre of such tension is the development of shale gas, which to date has been regulated utilising a statist approach. This statist approach has historically focused on participatory intervention, with Chinese national petroleum companies dominating shale gas extraction. However, there is a shift away from a statist approach to a market-based approach as outlined in the 13th FYP, marking a fundamental shift in both thinking and regulation:

> The underlying issue is how to strike a balance between the role of the government and that of the market, and let the market play the decisive role in allocating resources and let the government play its functions better. It is a general rule of the market economy that the market decides the allocation of resources. We have to follow this rule when we improve the socialist market economy. We should work hard to address the problems of market imperfection, too much government interference and poor oversight.

We must actively and in an orderly manner promote market-oriented reform in width and in depth, greatly reducing the government's role in the direct allocation of resources, and promote resources allocation according to market rules, market prices and market competition, so as to maximize the benefits and optimize the efficiency. The main responsibility and role of the government is to maintain the stability of the macro-economy, strengthen and improve public services, safeguard fair competition, strengthen oversight of the market, maintain market order, promote sustainable development and common prosperity, and intervene in situations where market failure occurs.[73]

This shift to a market-based approach is critical if China is to become a modern developed nation. However, the shift also presents a regulatory puzzle, in that it threatens the statist approach to regulating shale gas development on agricultural land. History demonstrates that a statist approach to the development of petroleum resources in a market economy can be successfully undertaken. The dominant example of this is the development of offshore petroleum resources in Norway since the 1960s. During the period of the 1970s and 1980s, the Norwegian government promulgated the statist approach utilising participatory intervention within its developing market economy. Since the 1990s, as a result of international instruments (EU and International Treaties),[74] Norway has had to alter its regulatory approach to fit in with the market-based approach exerted by external forces. Yet, Norway has retained its participatory intervention approach that was set out in its policy principles of 1971.[75] The commonality between Norway and China is that both have set out principles that direct how they wish to regulate the extraction of their resources for the benefit of the nation, the enshrinement of such principles in law,[76] and the retention of statist regulatory approaches.

For both nations, the statist approach remains central to the governance of resource extraction, with the Norwegian experience demonstrating the capacity to implement such a shift. This does not mean, however, that the shift will be necessarily easy or successful. Indeed, it will present China with a number of challenges that Norway did not face. The greatest of these challenges that is unique to China is balancing the provision of energy resources for the development of the country with the provision of food for the people. Unlike Norway, where the extraction of resources is for export and for the benefit of the future generations, China faces a fundamental decision – sacrifice agricultural land for the development of gas resources or continue to develop unconventional gas resources and retain agricultural land? The indication is that China wishes to preserve both:

We will continue to apply the strictest possible protection system for farmland and will designate permanent basic cropland throughout the country. We will put in place a food crop production strategy that is based on farmland management and the application of technology, and with the focus on major growing areas of grain crop and other staple agricultural products, we will make a large-scale push to see the building of farmland irrigation systems and water conservancy infrastructure, the restoration of rural land, the improvement of low- and medium-yield cropland, and the development of high-quality farmland.[77]

As noted in the quote above from the 13th FYP, it is irrigation and water that lie at the heart of agriculture. It is also water and the disposal of HF-produced water that lies at the heart of shale gas development. This presents China with another conundrum – that of water use, reuse and disposal and how these divergent water uses can be managed. In order for water to be successfully managed to ensure that the activities can coexist, the only option is for a statist approach to water management. Any other approach will place water availability and preservation in a vulnerable position, which to date, China has indicated it wishes to avoid.

The transfer of ownership of agricultural land from the State to rural collectives is necessitated by a social need for rural people's control over the land they farm. A future challenge for China will be how land reforms and the increasing transfer of land type from basic farmland to construction land impacts on rural collectives and other farmland holdings. As further land reforms are considered and implemented, the challenge will be to balance agricultural landholder's rights of ownership over their land and protection of farmland itself and the development of shale gas. This development of shale gas has been identified by China as critical to the economy and the development of the country. However, as Vivoda has identified, China also has external forces impacting on its energy security; for example, competition with Japan to secure energy supplies. This has driven both countries to secure bilateral deals to enhance their energy security, reflecting each country's 'desire to "lock in" access to energy in an increasingly competitive environment,'[78] resulting in 'raised costs of energy imports for both countries by delaying the development of domestic and regional energy infrastructure, arguably exacerbating both states' energy insecurity.'[79]

The development of huge gas resources in the Russian Arctic (the existing Yamal field and the future Shtokman field) and the capacity of the Northern Sea Route to provide China with direct access to LNG from the Russian Arctic presents China with a viable alternative source of energy to meet its growing needs. This has been reiterated by China's recognition of the 'Ice Silk Road'[80] in the *One Road One Belt Initiative*, as well as China's development of icebreaking vessels to utilise the Sea Routes in Artic Waters.[81] China is further reinforcing its energy security through the construction of the 'Power of Siberia' Pipeline, which will connect with China's internal West–East Pipeline.[82]

Clearly, the 13th FYP demonstrates a shift in the development of agriculture from statist approach to market-based. However, like other petroleum producing countries, the State chooses to maintain its control through participatory intervention. This intervention in China, on behalf of the State, is through the participation of State-owned petroleum companies in large shale projects and the high degree of regulatory control over shale gas project approval and operation. As China continues to move towards a market-based economy, whilst maintaining its need to feed its population and meet its growing energy needs, balancing these objectives will require a statist approach. This statist approach is iterated through both the 13th FYP and the relevant national laws, indicating that China seeks to be a *dirigiste* economy, similar to other states such as Norway, France and Russia.

The challenge for China, unlike Norway or Russia, will be balancing the market-based approach with food production for its population, challenges that no other country faces to date. The statist approach to regulation provides China, as a Communist State, with the control it requires to implement its food and energy plans. Whereas the market approach provides the necessary flexibility to diversify energy sources to meet energy needs, coordination is retained by the State. This unique approach by the Chinese State appears to provide a tailored solution to the unique challenges facing China today. However, the success of this approach will be dependent on future land reforms, water management and the preservation of farmland for agricultural purposes. These challenges sit firmly within the realm of State control, reinforcing the need for the statist approach to remain.

Conclusion

The food security–energy security nexus in China is unique. Unlikely any other country in the world, China has a mounting double pressure of providing food as well as energy to a large population. The food security–energy security dilemma is heightened by the lack of farmland for cereal production and the pressure that shale gas production places on valuable arable land. This pressure is currently regulated through legal instruments which seek to protect farmland. However, as the 13th FYP indicates, China is moving towards a more market-based approach, as it seeks to industrialise agriculture and substantially increase yields. At the same time, China also seeks to produce shale gas, with targets set and expected to be met. As examined throughout this chapter, this presents China with a conundrum – how do we preserve farmland, increase agricultural yields whilst at the same time increase production of shale gas?

China's approach to date has been, and will continue to be, a statist regulatory framework, where direction from the FYPs and national law drives and directs the implementation of energy security and food security. Diversification of China's energy supply to external sources, particularly from Russia, may well assist the State in maintaining the balance between food security and energy security. Such security will also be contingent upon both land ownership reforms and the future management of precious water resources. It is only the continuation of the statist approach that will enable China to balance these things. Therefore, it is likely that the statist approach currently favoured by China will remain.

Notes

1 Worldometers, *China Population* (2018) <http://www.worldometers.info/world-population/china-population/> accessed 11 May 2018.
2 Whereas China contains roughly 20 per cent of the world's population, it contains only 7 per cent of the world's farmlands. State Council of the Peoples Republic of China, *White Paper – The Grain Issue in China* (1996) 1; Colin A. Carter, Funing Zhong and Jing Zhu, 'Advances in Chinese Agriculture and its Global Implications' 23(1) *Applied Economic Perspectives and Policy* 1.
3 Vlado Vivoda, 'State–Market Interaction in the Hydrocarbon Sector: The Cases of Australia and Japan' in Andrei V. Belyi and Kim Talus (eds), *States and Markets in Hydrocarbon Sectors* (Palgrave, 2015) 257.
4 Global Yield Atlas, *China* (2014) <http://www.yieldgap.org/china>.
5 State Council of the Peoples Republic of China, *White Paper – The Grain Issue in China* (1996)
6 Ibid, 1.
7 China Power, 'How is China Feeding its Population of 1.4 Billion?' (2017) *China Power* 25 January 2017 <https://chinapower.csis.org/china-food-security/>.
8 State Council of the Peoples Republic of China, *White Paper – The Grain Issue in China* (1996), 1.
9 World Bank Food and Agricultural Organisation, *Cereal Yield (Per Hectare)* (2017) <https://data.worldbank.org/indicator/AG.YLD.CREL.KG>.
10 Ibid.
11 Kai Cui and Sharon Shoemaker, 'A Look at Food Security in China' (2018) 2(4) *Nature Partner Journals: Science of Food* 1–2, 1.
12 Ibid.
13 Ibid.
14 China Power, 'How is China Feeding its Population of 1.4 Billion?' (2017) *China Power* 25 January 2017 <https://chinapower.csis.org/china-food-security/>.
15 Ibid.
16 OECD, *Meat Consumption* (2017) <https://data.oecd.org/agroutput/meat-consumption.htm>.
17 This marks an increase from 350 kg/capita/annum only a decade before. Kai Cui and Sharon Shoemaker, 'A Look at Food Security in China' (2018) 2(4) *Nature Partner Journals: Science of Food* 1–2, 1.
18 Yangfeng Wu, 'Overweight and Obesity in China' (2006) 333 *British Medical Journal* 362–363, 363.
19 Kai Cui and Sharon Shoemaker, 'A Look at Food Security in China' (2018) 2(4) *Nature Partner Journals: Science of Food* 1–2, 1
20 China Power, 'How is China Feeding its Population of 1.4 Billion?' (2017) *China Power* 25 January 2017 <https://chinapower.csis.org/china-food-security/> accessed 11 May 2018.
21 World Bank, *China 2020: Development Challenges in the New Century* (Washington, 1997) 21.
22 Muyo Xu and Josephine Mason, *China's Energy Demand to Peak in 2040 as Transportation Demand Grows: CNPC* (2017) <https://www.reuters.com/article/us-china-cnpc-outlook/chinas-energy-demand-to-peak-in-2040-as-transportation-demand-grows-cnpc-idUSKCN1AW0DF>.
23 Ibid.
24 As reported by China's National Development and Reform Commission. Interfax Global Energy, *Chinese Gas Consumption Hits 237 bcm in 2017* (2017) <http://interfaxenergy.com/gasdaily/article/29373/chinese-gas-consumption-hits-237-bcm-in-2017>.
25 IEA, *World energy Outlook 2017: China* (2017) <https://www.iea.org/weo/china/> accessed 12 May 2018.

26 Steven W. Lewis, *Natural Gas in the People's Republic of China* (James A. Baker III Institute for Public Policy, Rice University, 2013) 7.
27 This diversification includes renewable resources. See People's Republic of China, *12ᵗʰ Five Year Plan 2011–2015* (2011), Part 1, Chapter 11, Section 1(Translated by the Delegation of the European Union in China) <http://cbi.typepad.com/china_direct/2011/05/chinas-twelfth-five-new-plan-the-full-english-version.html>.
28 People's Republic of China, *13ᵗʰ Five Year Plan 2016–2020* (2016), Part VII, Chapter 30, Section 1
29 Ibid.
30 Ibid, Chapter 30, Box 11.
31 People's Republic of China, *12ᵗʰ Five Year Plan 2011–2015* (2011).
32 The Shale Gas Development Plan particularly 'specifies the development priorities for the next five years till 2020 and provides a prospect for 2030. The policy analyses the context for shale gas development in China, the opportunities and challenges, puts forward targets and main pilot projects, and clarifies the political and institutional supports'. Asia Pacific Energy, *China: Shale Gas Development Plan (2016–2020)* (2015) <https://policy.asiapacificenergy.org/node/3050>.
33 People's Republic of China, *12ᵗʰ Five Year Plan 2011–2015* (2011), Part 1, Chapter 11, Section 3 (Translated by the Delegation of the European Union in China) <http://cbi.typepad.com/china_direct/2011/05/chinas-twelfth-five-new-plan-the-full-english-version.html> accessed 11 May 2018.
34 Pipeline construction often requires a clear right-of-way of 50–100 feet wide and the need for 24-hour access to the pipeline by the unconventional gas titleholder. John S. Baen, 'The Impact of Mineral Rights and Oil and Gas Activities on Agricultural Land Values' (1996) 64 *Appraisal Journal* 67.
35 China is reported to have 1,115 Trillion Cubic Feet (Tcf) of recoverable shale gas, making it the largest shale gas play in the world. These reserves are mainly concentrated in four basins: the Sichuan (626 Tcf), Tarim (216 Tcf), Junggar (36 Tcf) and Songliao (16 Tcf) Basins. US EIA, *Technically Recoverable Shale Oil and Shale Gas Resources: China* (2015) XX-2.
36 People's Republic of China, *13ᵗʰ Five Year Plan 2016–2020* (2016), Part IV.
37 This initiative is an extensive infrastructure project that seeks to modernize the ancient 'Silk Road' trading routes, through a combination of linking the 'Maritime Road' and the 'Economic Land' routes. This will see the linking of many countries in Asia, the Middle East and Europe by road, as well as Eastern Africa through maritime routes. See Peter Cai, *Understanding China's Belt and Road Initiative* (2017) Lowy Institute, <https://www.lowyinstitute.org/sites/default/files/documents/Understanding%20China%E2%80%99s%20Belt%20and%20Road%20Initiative_WEB_1.pdf>.
38 The 'Ice Silk Road' refers to the cooperation between Russia and China to strengthen cooperation on the development and utilisation of the Northern Sea Route and to research the possibility of shipping along Arctic Routes.
39 Organic content is expressed as a percentage of Total Organic Carbon (TOC). A percentage of 5 per cent (such as that of the Woodford formation in the USA) is considered excellent.
40 For a discussion on the sedimentary depositional environments for shale gas reservoirs in China see C. Zou, D. Dong, S. Wang, J. Li, X. Li, Y. Wang, D. Li and K. Cheng, 'Formation Mechanism, Geologic Properties and Resource Potential of Shale Gas in China' (2010) 37(6) *Petroleum Exploration and Development* 641–653 (in Chinese with English Abstract).
41 *Constitution of the People's Republic of China* art 10.
42 *Constitution of the People's Republic of China* art 9.
43 *Mineral Resources Law of the People's Republic of China* art 3 'Mineral Resources belong to the State. The rights of State ownership in mineral resources is exercised

by the State council. State ownership of mineral resources, either near the earth's surface or underground, shall not change with the alteration of ownership or right to the use of the land which the mineral resources are attached to'.

44 *Mineral Resources Law of the People's Republic of China* art 46 states that 'all mineral resources, waters and sea areas belong to the State'.

45 Construction land in China refers to State-owned land where the user has the right to use the land for any form of constructed activity, including mining and resource extraction, buildings, and the like.

46 This Ministry was formed at the first meeting of the 13th National People's Congress to unify the exercise of duties of owners of all Natural Resources assets, to exercise all land use control and ecological restoral responsibilities and to realise the overall protection and system of restoration of lakes, grasslands, forests, mountains and rivers.

47 *Explanation of the State Council's Institutional Reform Program* (14 March 2018) Xinhua News.

48 This Ministry seeks to adhere to the priority development of agriculture and rural areas the overall implementation of strategy and rural revitalisation and to speed up the realisation of agricultural and rural modernisation. See section I(3) *Explanation of the State Council's Institutional Reform Program* (14 March 2018) Xinhua News.

49 This Ministry was established to protect China's natural environment by integrating decentralised ecological environmental protection responsibilities, unify in the exercise of ecological and urban and rural pollution discharge supervision, strengthen environmental pollution control and safeguard national ecological security. See section I(i)(b) *Explanation of the State Council's Institutional Reform Program* (14 March 2018) Xinhua News.

50 Collective land ownership denotes land owned by peasant collectives (rather than the State) for the purposes of rural/agricultural activities. Note there is no private ownership of land in China, but private, usufructuary rights (akin to very long-term leases) are granted for home ownership. This entitles the holder to the use and enjoyment of the land but not ownership.

51 A 'frack to flow' is where an appraisal well is hydraulically fractured to induce the flow of gas. It is undertaken to test the properties of the reservoir for production capabilities.

52 For instance, refer to division 2 of the *Petroleum and Geothermal Energy Resources Act 1967* (WA).

53 Where the project is less than 82.4 hectares, the provincial government must approve the conversion. However, where the project is more than 82.4 hectares, the State Government must approve the conversion. If the farmland is classified as basic farmland (high productivity), State Council approval is required.

54 The *Basic Farmland Protection Regulation* 1994 (Republic of China) sets out the classification of farmland. The highest level of farmland (as set out in Article 13) is that of basic farmland that has high productivity, and should be protected at all costs. This land cannot be converted to non-agricultural land unless it is in the national interest. Furthermore, under Article 33 of the *Land Administration Law 1999* (Republic of China), at least 80 per cent of the cultivated land in the stipulated area must be basic farmland. Other forms of land in China include resource land (ie forest or water), or wasteland, which includes low-yielding land and fragile land.

55 Under Article 33 of the *Land Administration Law 1999* (Republic of China), if the conversion of land to constructed land is unavoidable in order to make way for national projects, the same amount of land must be replaced by new farmland elsewhere: 'People's governments of provinces, autonomous regions and municipalities directly under the Central Government should strictly implement the overall plans and annual plans for land utilization and take measures to ensure that the total amount of cultivated land within their administrative area remains unreduced.

Where the total amount of cultivated land is reduced, the State Council shall order the government concerned to reclaim land of the same quality and amount as is reduced with time limit … where individual governments of provinces or municipalities directly under the Central Government, for lack of land reserves, cannot reclaim enough land to make up for the cultivated land the used for additional construction projects, they shall apply to the State Council for approval of their reclaiming less or not land within their own administrative areas but of their reclaiming land in other areas'.

56 *Land Administration Law 1999* (PRC) art 47.
57 Ibid.
58 Brook Wilmsen, 'Expanding Capitalism in Rural China Through Land Acquisition and Land Reforms' (2016) 25(101) *Journal of Contemporary China* 701–717, 703.
59 B. Lohmar, *China's Wheat Economy: Current Trends and Prospects for Imports* (2004) WHS 04D-01, Economic Research Service, US Department of Agriculture.
60 Kai Cui and Sharon Shoemaker, 'A Look at Food Security in China' (2018) (4) *Nature Partner Journals: Science of Food* 1–2, 1
61 Ibid.
62 Ibid.
63 Such pollution may come about during the transport and storage process, as well as other leaks and spills associated with hydraulic fracturing. For a detailed discussion of this issue, see S. A. Clancy, F. Worrall, R. J. Davies and J. G. Gluyas, 'The Potential for Spills and Leaks of Contaminated Liquids from Shale Gas Developments' (2018) 626 *Science of the Total Environment* 1463–1473.
64 For a discussion source of groundwater contamination see Rene Lefebre, 'Mechanisms Leading to Potential Impacts of Shale Gas Development on Groundwater Quality' (2017) 4(1) *WIREs Water* e1188 (online)
65 China Power, 'How is China Feeding its Population of 1.4 Billion?' (2017) *China Power* 25 January 2017 <https://chinapower.csis.org/china-food-security/> accessed 11 May 2018.
66 Ibid.
67 David Sandiglow, Jingchao Wu Qing Yang, Anders Hobe and Junda Lin, *Meeting China's Shale Gas Goals* (2014) Columbia, SIPA, (Working Draft), 37.
68 Ibid.
69 Ibid.
70 Nicholas van der Elst et al., 'Enhanced Remote Earthquake Triggering at Fluid-injection Sites in the Midwestern United States' (2013) 341(6142) *Science* 164–167, 164.
71 People's Republic of China, *13th Five Year Plan 2016–2020* (2016), Part IV.
72 As defined and examined in Chapter 4 of this book.
73 Central Committee of the Communist Party of China, *Decision of the Central Committee of the Communist Party of China on Some Major Issues Concerning Comprehensively Deepening the Reform* (2013) (translation by China.org.cn on 16 January 2014) Third Plenary Session of the 18th Central Committee of the Communist Party of China on November 12 2013 <http://www.china.org.cn/china/third_plenary_session/2014-01/16/content_31212602.htm> accessed 12 May 2018.
74 Primarily these instruments include *Directive 94/22/EC of the European Parliament and of the Council of 20 May 1994 on the Conditions for Granting and Using Authorizations for the Prospection, Exploration and Production of Hydrocarbons* (The Hydrocarbon Directive), which applied to Norway upon its entry into the European Economic Area (EEA), and reflected in the *Petroleum Activities Act 1996* (Norway); and the *Energy Charter Treaty* (2016), particularly Article 10 relating to investment.
75 These policy principles are often referred to as the 10 Oil Commandments, as outlined in the Storting *White Paper* 76 (1970–71) <https://www.regjeringen.no/globalassets/upload/kilde/oed/bro/2002/0006/ddd/pdfv/152184-facts_20.pdf>.

76 For instance, the Norwegian principles are captured in section 1–2 of the *Petroleum Activities Act 1996* (Norway).

77 People's Republic of China, *13th Five Year Plan 2016–2020* (2016), Part IV

78 Vlado Vivoda, 'State–Market Interaction in the Hydrocarbon Sector: The Cases of Australia and Japan' in Andrei V. Belyi and Kim Talus (eds), *States and Markets in Hydrocarbon Sectors* (Palgrave, 2015) 257.

79 Ibid.

80 China identifies Russia's Northern Sea Route as the 'Ice Silk Road' under the One Belt One Road Initiative. See 'China, Russia Team up on Ice Silk Road' (2017) *World Maritime News* 10 July 2017 <https://worldmaritimenews.com/archives/224690/china-russia-team-up-on-ice-silk-road/>

81 Ma Chi, 'China's First Home-Built Icebreaker named *Snow Dragon 2*' (2017) *China Daily* 27 September 2017 < http://www.chinadaily.com.cn/china/2017-09/27/content_32544019.htm>.

82 Reuters, *China to Complete Russia Oil, Gas Pipeline Sections by End-2018: Vice Governor* (2017) <https://www.reuters.com/article/us-china-silkroad-russia-pipelines-idUSKBN18819I>

Part III

Socio-regulatory responses and conclusions

12 Collectivisation and collective bargaining

Introduction

In the establishment of a State's unconventional gas regulatory framework, it is relevant to consider the optimal tool to provide access and use of agricultural land. Typically, once unconventional gas is found on a property, the unconventional gas licenseholder is required to negotiate an agreement governing access with the relevant agricultural landholder. A State may leave the terms of this agreement to private negotiations between parties , by taking a 'hands off' minimal, regulatory approach.[1] Alternatively, the State can exercise its regulatory framework to define and monitor relations between the parties in terms of land access, land use, compensation and rehabilitation of the land.[2] The challenge then for regulators is to create effective regulation that is applicable and effective in changing circumstances to allow licenseholders to define land access compensation in a transparent system that balances the value of the economic gain in extracting unconventional gas resources, while protecting landowner interests.[3]

A land access agreement is entered into prior to unconventional gas exploitation which outlines the terms and conditions for access and use of agricultural land throughout the lifecycle of a project. Therefore, the primary purpose of a land access agreement is to manage the conflicting interests of the licenseholder and the affected landholder. The question that naturally arises is how a State can effectively manage agricultural land to ensure food security for future generations while managing the social impact and interests of agricultural landholders?

Within this book, seven jurisdictions have been analysed and categorised based on their regulatory approach to unconventional gas, namely – adaptive, precautionary, or statist. Each State must grapple with a delicate balancing act of managing competing obligations. Firstly, a State must provide energy security for its citizens and, secondly, it must protect agricultural land to support and feed its citizens and drive the economy. Although this book has focused on the specific 'transition' energy source – unconventional gas – managing the conflicting interests between agricultural land and energy development is a universal and seemingly intractable conundrum in a number of differing national contexts. This chapter examines similar challenges and complications found within each of the jurisdictions analysed, while exploring different mechanisms to provide a solution that may pave the way forward for coexistence.

Trends in agricultural land and unconventional gas contestation

Coexistence is one of, if not the key, policy driver for the development of onshore unconventional gas development. It continues to be the subject of national debate and social tension with increased pressure on agricultural land, evident in public unease relating to hydraulic fracturing. Each State approaches the issue of coexistence between these two land uses differently, but what remains consistent is that contestation will become ever more marked in the coming decades as States seek the often elusive goal of achieving energy security. How can coexistence be achieved? One model of coexistence that was active prior to the ban of hydraulic fracturing is found in the Joint Landowners Coalition of New York, as examined in Chapter 8. At a fundamental level, although different jurisdictions have vastly diverse legal and socio-political systems, the regulatory challenge remains the same – how to manage the coexistence of agricultural land and unconventional gas.

This book does not examine in detail all the possible legal questions concerning unconventional gas regulation, agricultural land protection and landholder relations, as it would be an impossible task to propose a 'one size fits all' model for unconventional gas regulation. Rather, the focus has been to examine the socio-regulatory trends of natural resource governance in a number of differing jurisdictions to achieve coexistence. What is evident throughout this book is the increasing trend towards political consideration of the social and environmental issues relating to agricultural land in the regulation of unconventional gas activities. The focus of all jurisdictions surveyed is a political desire to plan a future that is economically productive, preserves agricultural land and provides energy in an increasingly energy-hungry world. The focus on rural land as an increasingly important national resource is due to the economic contribution of agricultural regions to the national economy, and their often significant reserves of unconventional gas located beneath fertile topsoils.

Another obvious development is the balance between the interests of a State's agricultural sector to preserve food security, and in some more evolved agrarian States, to achieve food sovereignty, whilst encouraging unconventional gas activities. There is a lack of legal consensus on how to manage the complex regulatory issues presented by unconventional gas development on agricultural land, particularly given it is an important present and future economic contributor for resource States such as China, Canada and Australia. Public concern relating to agricultural land preservation in all jurisdictions has had the effect of further polarising political debate on unconventional gas in an effort to ensure that equity between energy companies and agricultural landholders is achieved.

The question of the State's role in unconventional gas development continues to be raised as a trend amongst jurisdictions, ranging from minimal intervention, regulatory intervention or participatory intervention as analysed in Chapter 4.[4] Although each jurisdiction, with the exception of the USA, holds the State as the owner of petroleum resources. However, whether the State intervenes and participates in the exploration and production of unconventional gas remains a polarised debate amongst jurisdictions. Where land is owned and managed by a State, and therefore does not require any form of legal access agreements with private landholders, the State exists as the regulator, arbitrator and participant in its unconventional gas industry. In contrast, the market-based, non-statist approach adopts a minimalist approach to State invention, with a preference for free market forces to direct the exploitation of these resources. This model sees private petroleum companies given *de facto* control over unconventional gas activities with licences for exploration and extraction being awarded by the State using an auction bidding system.

Whatever intervention level a State chooses, the theory of collectivisation provides an interesting socio-legal dimension that could be applied systematically to transform the top-down and linear relationship of unconventional gas exploitation. Typically, this linear relationship starts with the granting of an authorisation to a licenseholder (company) by the State, the company exploring for unconventional gas, and culminates in the landholder authorising the extraction (either under law or negotiation) prior to resource exploitation commencing. The public interest in preserving agricultural land provides a unified platform and need for an alternative vehicle and theory to transform this linear relationship between State, landholder and unconventional gas licenseholder. Collectivisation may provide the impetus for the creation of a 'pyramid-like' relationship between actors rather than the current linear relationships. As an illustration, the pyramid would place the State at the pyramid apex, as the authoriser and owner of the natural resource, with the private landholder as owner of the surface estate engaged with the licenseholder company on equal footing at the pyramid base.

The theory and power of collectivisation

Given that there is increasing pressure on agricultural land, contestation will become the new norm in many resource States. Collectivisation is one model to address the imbalance that currently exists in many States. The rush towards the exploitation of energy sources equates to individual property owners facing enormous pressure to comply with regulation that doesn't necessarily represent or serve their best interests. The theory of collectivisation is founded within unionism and is historically associated with socialist States. However, collective vehicles continue to be evident in market-based economies as a mechanism to balance the interests and power of large oligopolies and workers, for example in agricultural cooperatives.

Minimising the social impact of exploitation activities on communities by seeking a 'social licence to operate' as a way of managing the contestation between natural resource companies and agriculture is a well-cited argument for coexistence.[5] The social licence to operate is a well-established narrative in the corporate world and its uptake has been particularly extensive in the mining sector. As stated by Curran 'a social licence – at least from a corporate perspective – can be seen as "a pragmatic calculation" designed to minimise risk by undertaking strategic forms of community engagement'.[6] While there is no universal definition of social licence to operate, the International Energy Agency has identified seven guiding Golden Rules to create a valid social licence to operate in the gas industry as follows:

> Sufficiency and robustness of baseline studies; careful site selection and monitoring; the prevention of leaks; the responsible treatment of water; minimisation of emissions; commitment to a high level of environmental performance; and focused attention on the cumulative economic, social and environmental impacts.[7]

Obtaining a social licence to operate is evidentially very different from obtaining a legal petroleum or mineral title[8] to explore and extract unconventional gas. A social licence to operate does not include legal ramifications or penalties in the event of non-compliance – it is entirely voluntary. However, commentators identify consent and approval from community stakeholders as being an increasingly important aspect to the sustainability and equitability of natural resource operations.[9] Jenkins maintains that a social licence to operate provides a 'cloak of legitimacy' to natural resource activities.[10] The activist group 'Lock the Gate' in Australia criticises the concept of the social licence to operate, stating that CSG companies adopt the term as a strategy of 'greenwash'[11] to demonstrate they recognise the social impacts of their activities in a region. As stated by Curran:

> The issue of a whole community's perceived disenfranchisement in a supposed democracy where corporate 'bullies' threaten to 'bludgeon' communities into submission, often with compliant governments in tow. Whether accurate or not, this is a powerful narrative that governments, with an eye on their political objectives and political capital, and the modern corporation, with an eye on its reputation and bottom line, will want to avoid.[12]

Some agricultural communities and opponents of the social license concept have indicated that the voluntary nature of the social licence to operate suggests that corporate interests are privileged, and that the aim of profit maximisation for shareholder return dominates community interests. This leads to feelings of policy abandonment and governmental indifference to the plight of regional communities. As cautioned by Kapelus:

> Understanding how to engage in such co-operation is not always easy for an industry that historically has not set participation with diverse actors with different ideologies as a top priority. Mining companies have generally wielded the power to implement community development policies (within limits set by government) that allowed them to get on with their core business, namely the extraction of minerals.[13]

This has led some states, such as Queensland, Australia, to identify the need and importance of community engagement during unconventional gas activities, recognising that 'There is a strong need for governments, communities and the gas industry to work closely together to ensure that adverse impacts on community wellbeing are minimised and benefits to community wellbeing occur in the long term'.[14] The apparent 'disenfranchisement' of regional communities in the face of unconventional gas operations has given rise to demands for more transparency at the grassroots level to support these communities. This trend is increasingly prevalent within Queensland's CSG regions as stated by Cheshire et al.:

> it is particularly profound in mining-intensive regions where the number of actors with a stake in local decisions can be high, but weakly organised, and where mining companies (as powerful corporate actors) are formulating their own sets of rules and expectations about where their own, and others', responsibilities lie. Under such circumstances, the impacts of poor governance have undesirable consequences for the communities concerned.[15]

The social impacts of shale gas development in China are explored by Yu, who highlights the social unease in rural communities in the Sichuan and Chongqing provinces due to water depletion, noise pollution and environmental damages. However, the initial activist stance in these rural communities against shale gas, including public protest, has recently ceased with communities conceding their efforts are fruitless based on the view that their interests will not be represented and a fatalism; 'environmental risks and negative social impacts as inevitable'.[16] At the other end of the social impact spectrum, the New York State ban on fracking is seen as a victory of social activism against unconventional gas activities. New York State's portion of the Marcellus shale basin holds one of the richest and largest deposits of shale in the USA, yet the protests against hydraulic fracturing by the rural communities raised public awareness of its impacts and facilitated social collectivisation leading to the eventual ban of the activity.[17] As a result of this civic action, community regulatory power was established in the Municipal Home Rule Law and 245 rural communities in New York State directly banned hydraulic fracturing in 2013, prior to the state-wide ban being introduced.[18] The power of these rural communities is their rhetoric that hydraulic fracturing should not be undertaken 'in anyone's backyard',[19] and demonstrates the community collectivisation that is partially responsible for the eventual ban in New York State. It is clear throughout the jurisdictions examined in this book that the perceptions and support of rural communities in which unconventional gas operations exist should not be understated. This power can be harnessed to determine the regulatory fate to either ban or sustain unconventional gas exploitation on agricultural lands. Given this background, the collective action of farmers to create a loose affiliation and redress the imbalance between corporate and community interests, plays an important role in delivering public benefit, economic outputs and services in relation to unconventional gas activities.[20]

Collective action can be classified as action taken by a group to achieve a common goal for the benefit of the group as a whole. For the purpose of this chapter, the definition of collective action by Marshall and Scott is adopted as follows: 'the action taken by a group (either directly or on its behalf through an organization) in pursuit of members' perceived shared interests'.[21] Meinzen-Dick et al. observe that collective action involves the following key features: 'the involvement of a group of people, shared interests, common and voluntary actions to pursue those shared interests'.[22] Vanni further distinguishes the theory into two types of collective action: (i) *cooperation*: bottom-up, farmer-to-farmer collective action and (ii) *coordination*: top-down, agency-led collective action.[23] Potential collective action within collective bargaining in negotiating land access agreements between agricultural landholders and unconventional gas titleholders is classified as the first type of collective action, namely cooperation.

Socio-cultural literature recognises that collective negotiation and collaboration requires 'trust, voice, reciprocity and a disposition to collaborate for mutually beneficial ends'.[24] The literature also identifies a four-tier criteria framework to assess whether collective action in relation to natural resources will be successful:

1   Resource system characteristics;
2   Group characteristics;
3   Institutional arrangements; and
4   External environment.[25]

The first requirement necessitates the development of robust information so that parties may access resource system characteristics in a practical and clear manner. The absence of information asymmetry is critical to the implementation of collective action, particularly in the agricultural sector, where collective action has historically been exercised as found by van Caenegem et al.[26] In the case of negotiation with unconventional gas companies, agricultural landholders will need a strong base of information dissemination in order to collaborate efficiently and develop a unified negotiating position.

Secondly, the characteristics of the group must include a strong level of homogeneity for successful collective action. Social capital is defined by Putnam as the sum of 'connections among individuals – social networks and the norms of reciprocity and trustworthiness that arise from them'.[27] Nahapiet and Ghoshal expanded this definition by adding that social capital is:

> The sum of the actual and potential resources embedded within, available through, and derived from the network of relationships possessed by an individual or social unit. Social capital thus comprises both the network and the assets that may be mobilized through that network.[28]

Social capital thus comprises a multi-layered relationship of both the network of landholders and their shared information to be harnessed as assets and mobilised within the collective for its benefit. Social capital also recognises that trust norms and reciprocity must be substituted for pure formal contractual rules in order to generate crucial economic efficiencies.[29] Ahn and Ostrom offer a definition of social capital tailored to the field of collective action. In particular, they define social capital as 'an attribute of individuals and of their relationships that enhances their ability to solve collective-action problems'.[30] As stated by Dahal and Adhikari, studies of collective action have widely adopted the definition offered by Putnam et al. of:[31]

> (i) Social capital in collective action is usually related to meso and collective units, such as associations, communities and regions; (ii) social capital is presented as a solution of the barriers of collective action and (iii) the social capital framework is applied to the study of the performance of institutions, such as regional governments.[32]

Social capital theory is effective in describing collective action to protect agricultural land as it complements the traditional public policy approaches based on regulation to address social and environmental problems, according to the World Bank.[33]

Thirdly, the success of collective action is also dependent on institutional arrangements, which must have locally devised and simple rules for the collective group to generate public objectives via 'clustering'.[34] Porter defines industry clustering as 'a geographically proximate group of interconnected companies and institutions in a particular field, linked by commonalities and complementarities'.[35] 'Active' clustering is defined by interaction and functional relationships, knowledge sharing, and collaborative and competitive forces that drive innovation. Nauwelers and Reid describe regional innovation as 'the set of economic, political, and institutional relationships occurring in a given geographical area that generates a collective learning process leading to the rapid diffusion of knowledge and best practice'.[36] The nature of clustering, as outlined by Davies et al., is defined as 'organizations [that] are both in competition and cooperating with one another simultaneously in different areas of their activities, but overall continuing to develop and reinforce the benefits of coexistence'.[37] Participation in clusters offers competitive advantages based on the characteristics of flexibility, sharing of information and resources and links to other networks and opportunities.

Finally, the external environment must be facilitative of collective action. The concept of 'complementarity conditions', a branch of social capital theory, illustrates the necessity of collective action in reaching equitable levels of economic compensation between parties.[38] A situation where agricultural landholders do not gain reasonable compensation[39] for the use of their prime agricultural land and lose agricultural productivity during land access negotiations demonstrates the necessity of a 'complementarity condition' in the form of collective bargaining to attain more favourable contractual conditions.[40] When this 'complementarity condition' is met, other factors such as collective decision-making and collective information-sharing, emerge as important benefits in reaching higher levels of compensation through collective action and the realisation of the 'complementarity condition'.

A collective solution?

Land access agreements, both those regulated by statute and private contracts, are typically aimed at producing effective long-term contracts. The clear goal is to accommodate the needs of both landholders and unconventional gas licenseholders, sustainably. Land access is required for numerous activities over an unconventional gas project's lifecycle, from geological and seismic surveying to drilling, extraction, inspection and processing. In many cases, a cooperative approach to negotiation relies heavily on the local knowledge of stakeholders, namely the agricultural landholders, and on the possibilities to integrate this knowledge into the decision-making process.[41] Thus, collective action increases the credibility and legitimacy of decision-making, but also allows the collection and sharing of information valuable to both licenseholders and landowners during contract negotiations. Many landholders may also have limited access to legal information and oil and gas governance materials when negotiating land access arrangements. Consequently, forming a cluster is one way to ensure an equitable result for all farmers in a certain geographic area located where unconventional gas exploitation takes place. Therefore, horizontal collaborations can be utilised between landowners, for example within an established local agricultural supply chain and governed by a land access agreement and vertical relationships for private and public unconventional gas actors can develop and reinforce the benefits of coexistence.

As each jurisdiction examined differs in its political, economic and legal environment, this chapter confines its analysis of the jurisdiction to the most marked and contested landholder land access agreement arrangements in Queensland, Australia – the largest and sole Australian unconventional gas producer and exporter outside of the USA and Canada.

Collective bargaining within the Australian competition law framework

Collective bargaining can be seen as a transactional regulatory tool for the negotiation of contracts by a collective group as agreements based on negotiation creating an enforceable undertaking.[42] This is in contrast to a pure authorisation regulatory tool, such as a Conduct and Compensation (CCA) land access agreement in Queensland.[43] Collective bargaining is used in multiple sectors to reduce information asymmetry and promotion of effective contractual outcomes for multiple parties and is historically used by unions representing workers in different sectors. Further, collective bargaining often employs objective conduct standards for negotiation to guide parties or to require them to adhere to behaviours and procedures that are conducive to transparent and effective negotiation procedures.

The *Competition and Consumer Act 2010* (Cth), Australia's primary competition law stature, permits small businesses to seek a form of validation, through either the approval of the Australian Competition and Consumer Commission

(ACCC) via its notification or authorisation regime to create a collective bargaining arrangement.[44] The approval of the ACCC is needed, as presenting a collective front to negotiate pricing and contract terms is typically contrary to market liberalisation as an anti-competitive practice. Collective bargaining thus 'grants protection from liability concerning anti-competitive behaviour to any number of businesses including farms that seek to negotiate agreements as a group'.[45] Any small business is permitted to lodge a collective bargaining notice with the ACCC if the small business has made or proposes to give effect to a contract that does not contain a cartel provision, exclusionary provision or a price-fixing provision and the expected value of the contracts between the parties does not exceed the limit of A\$3 million in any 12 month period or A \$5 million for primary production contracts.[46]

The lack of uptake of the notification option may be due to the inflexibilities associated with identifying individuals, targets and low transaction thresholds between A\$3–5 million.

A 'public benefits' test must be satisfied by the ACCC's evaluation of the potential collective bargaining arrangement for either an authorisation or notification pursuant to ss 90(5A) and 90(5B) of the *Competition and Consumer Act 2010* (Cth), which states that the ACCC shall not authorise a provision of a proposed contract, arrangement or understanding that is or may be a cartel provision, unless it is satisfied in all the circumstances that:

- the provision, in the case of s 90(5A), would result or be likely to result or, in the case of s 90(5B), has resulted or is likely to result, in a benefit to the public
- that benefit, in the case of s 90(5A), would outweigh the detriment to the public constituted by any lessening of competition that would result or be likely to result if the proposed contract or arrangement were made or given effect to, or in the case of s 90(5B), outweighs or would outweigh the detriment to the public constituted by any lessening of competition that has resulted or is likely to result from giving effect to the provision.[47]

To satisfy the above collective bargaining tests the ACCC will analyse:

The relevant market affected by the potential collective bargaining conduct; the counter factual; application of the 'future with-and-without test' whereby the ACCC compares the public benefit and anti-competitive detriment generated by the conduct in the future if the authorisation is granted with those generated if the authorisation is not granted; and finally whether there is a public benefit in granting the collective bargaining arrangement.[48]

If the ACCC is satisfied that any public benefit resulting from the cartel provision, exclusionary provision or price-fixing provision does not, or would not, outweigh the public detriment resulting from the provision, the Commission may issue an objection notice.[49]

The ACCC is also given the power, on the grounds of public benefit, to authorise collective bargaining conduct or other conduct constituting cartel conduct or a misuse of market power under the *Competition and Consumer Act 2010* (Cth). The ACCC must make a determination within the relevant period of six months, beginning on the date on which the application was received by it.[50] If the Commission has not determined the application within that time, the authorisation is deemed to have been granted. Although collective bargaining in Australian competition law has historically been utilised by commercial businesses in competition with another to form a vehicle to equalise bargaining positions, there has been a rise in agricultural groups collectively bargaining to create a level playing field between producers, processors and retailers.

Challenges and opportunities of collective bargaining

Collective bargaining is traditionally founded on a competition law exemption to create an arrangement where two or more competitors come together to negotiate with a supplier or a customer over terms, conditions and prices.[51] A land access arrangement represents a commercial venture whereby homogenous small businesses, being agricultural landholders, may be able to collectivise to negotiate terms of compensation, use and access. This makes the land access arrangement a prime candidate to follow a collective bargaining arrangement. The Australian *Agricultural Competitiveness Green Paper*[52] recognises the issue of financial compensation and mining impacts for farmers:

> Stakeholders expressed a concern that the quality of their agricultural land and their life on the farm were being affected by mining activities adjacent to or on their land. Some stakeholders suggested that farmers get a return from mining activities on their land, through a share of royalties.[53]

Contractual agreements are normally reached between farmers and unconventional gas companies pertaining to the conditions of access (with a view to minimising disruption and loss of amenity) and the compensation payable to the landholder.

According to the Australian Productivity Commission's *Inquiry into Mineral and Energy Resource Exploration*,[54] 'In sparsely stocked grazing areas, land holder concerns about exploration activity on their land are not as great as in areas where land is intensively cropped and irrigated'.[55] For example, from 2011–2014, 4,500 land access agreements were negotiated between landholders and CSG companies across the Surat and Bowen Basins in Queensland.[56] The potential for conflict between exploration and agricultural activities tends to rise with the intensity of land use and the magnitude of the potential impact. According to the Association of Mining and Exploration Companies:

Landholder rights relate to the use of the surface of the land. However access to those mining rights often means infringing on the rights of the landholder. Therefore negotiation between the owner of the mining rights and the landholder rights takes place such that the infringement on the rights is appropriately compensated.[57]

The Australian Productivity Commission's inquiry[58] also noted the disadvantages in farmer and gas companies' negotiations in protecting prime agricultural land while coexisting with unconventional gas exploration and extraction:

> Most rural land holders are at some disadvantage in undertaking negotiations with explorers. There is an asymmetry of experience as most land holders will have little or no previous experience in negotiating access agreements and compensation – such negotiations will most likely be a 'one-off'. There is also an asymmetry of information regarding the potential impact of the exploration activity. The land holder will have limited knowledge and experience from which to evaluate the impact of exploration activities on rural land. Further, there is an imbalance of power due to the involuntary nature of the negotiations.[59]

As previously stated, the test for the ACCC to object to a collective bargaining authorisation or notification application is found in the 'public benefits test':

> If the ACCC is satisfied that any benefit to the public that has resulted or is likely to result or would result or be likely to result from the provision does not or would not outweigh the detriment to the public that has resulted or is likely to result or would result or be likely to result from the provision, the ACCC may give the applicant a written notice (the objection notice) stating that it is so satisfied.[60]

'Public benefit' is not defined in the *Competition and Consumer Act 2010* (Cth), however, the ACCC has stated that the term should be given its widest possible meaning. In particular, it includes, 'anything of value to the community generally, any contribution to the aims pursued by society including as one of its principal elements … the achievement of the economic goals of efficiency and progress'.[61] The ACCC model of collective bargaining may result in claims of subverting the public interest if financial benefits flowing to state governments from mining rents are diverted into compensation to farmers, rather than into social welfare provisioning. However, there is a case for the agricultural landholder community to be given special consideration, given the statements of the Australian Senate Select Committee on Unconventional Gas that 'the lack of power and support landholders feel in relation to land access … indicates the overall level of complexity associated with land access involving unconventional gas mining'.[62]

Analysis of ACCC case law based on agricultural collective bargaining authorisations reveals the ACCC's analysis is based on a three-tiered framework: voluntary participation in the collective bargaining group; limiting the number of participating businesses, so that the bargaining group only covers a relatively modest market share; and limiting collective boycotts. These conditions would have to be met before a collective bargaining application could be granted by the ACCC. It is likely a small collective bargaining group of agricultural landholders of 10–20 members, for example, and the formation of landowner groups in specific regions (e.g., the Chinchilla agricultural landholders), would not be contradictory to the ACCC requirement that the group hold a small market share without cartel or collective boycott activity and with voluntary membership.

A limitation to the collective bargaining vehicle is the lack of mandatory enforcement of negotiations and outcomes. A collective bargaining authorisation or notification does not force groups to come to an agreement rather, it allows an opportunity to represent landholder interest in a coordinated and managed approach to empower and benefit potential negotiations between agricultural landholders and resource companies.[63] Therefore, key elements are necessary to overcome limitations and to ensure successful collective bargaining in land access contracts: strong social cohesion and trust within the group; information dissemination; and providing training in leadership, bargaining and negotiation skills to agricultural landholders who will represent collective bargaining groups. Secondly, the characteristics of the land access collective bargaining group will most likely be small in size, as members of the collective bargaining group may most likely be part of an agricultural peak body and will be representative of a certain region (e.g., Dalby).[64] This will allow participants to increase their association through trust, norms, reciprocity, obligations, expectations, values and attitudes.[65] According to Wade[66] and Ostrom,[67] locally devised and simple rules can be used to encourage 'institutional thickness' in collective action within a given territory linked with the combination of 'human capital' (knowledge resources), 'social capital' (trust, reciprocity and other social relations) and 'political capital' (capacity for collective action).

Conclusion: How will landholder input create a more effective land access regime?

The intention of all jurisdictions in seeking the development of an unconventional gas industry is to ensure energy security while balancing the needs of agricultural landholders and future food security by protecting arable land. Often, where jurisdictions purport to offer a centralised system of land access, such as Queensland in its CCAs, the devil is in the detail. For example, the provision relating to opt-out agreements and deferral agreements appears to support the primacy of the petroleum titleholder rather than acting to support landholders. States such as Canada, Australia and China continue the rhetoric in support of balancing conflicting interest while simultaneously watering down legislation that offers the opportunity for grievances and landholders' challenges. It has, in effect, moved the goalposts for landholders in favour of titleholders. Regulatory reviews have tended towards recommending a centralised system of land access arrangements. However, this has in part, led to further uncertainties and duplication, adding more layers of regulation and regulatory burden for landholders and does not simplify the land access regulatory regime. There is mounting evidence from case law, regulatory reviews and criticism from bodies involved in the land access regime that it is flawed in a number of States.

A regulatory tool that may offer redress to the regulatory balance is collective bargaining. It is a flexible and transparent vehicle that has been used successfully in different contexts, including the agricultural sector. When monitored and regulated by a national competition law authority, collective bargaining could provide a vehicle for landholders to form a collaborative body and increase 'good faith' bargaining when reaching a land access arrangement. Collective bargaining may be an effective transactional regulatory tool in negotiating land access agreements. The monitoring of collective bargaining may be conducted by a quasi-judicial administrative body, such as the Surface Rights Board model in British Columbia, or an administrative body specifically tied to agriculture, to ensure parties commit to collaborative and transparent negotiation of land access agreements. While collectivisation will require political will, the first step in moving towards collective bargaining in land access arrangements is to gain support for the proposal in agricultural communities located in key unconventional gas regions.

Notes

1 For example, under s15 of the *Petroleum and Geothermal Energy Resources Act 1967* (WA).
2 Such as in Queensland under the *Petroleum and Gas (Production and Safety) Act 2004* (Queensland).
3 Jostein Aarrestad, 'Resource Extraction, Financial Transaction and Compensation in an Open Economy' (1979) 81(4) *Scandinavian Journal of Economics* 552.
4 Brent F. Nelsen, *The State Offshore: Petroleum, Politics, and State Intervention on the British and Norwegian Conventional Shelves* (Praeger Frederick, 1991) 8.
5 Jason Prno and Scott Slocombe, 'Exploring the Origins of "Social License to Operate' in the Mining Sector: Perspectives from Governance and Sustainability Theories' (2012) 37(3) *Resources Policy* 346.
6 Giorel Curran, 'Social Licence, Corporate Social Responsibility and Coal Seam Gas: Framing the New Political Dynamics of Contestation' (2017) 101 *Energy Policy* 427, 430.
7 International Energy Agency, *Golden Rules for a Golden Age of Gas. World Energy Outlook Special Report on Gas* (2012) <http://www.worldenergyoutlook.org/media/weowebsite/2012/goldenrules/WEO2012_GoldenRulesReport.pdf>.
8 For example, in Australia, unconventional gas is classified as petroleum and thus requires a petroleum license, whereas in China unconventional gas is classified as a mineral and requires a mineral license.
9 Robert Boutilier and Ian Thomson, *Modelling and Measuring the Social Licence to Operate: Fruits of a Dialogue Between Theory and Practice* (2011) <http://socialicense.com/publications/Modelling%20and%20Measuring%20the%20SLO.pdf>.

10 Heledd Jenkins, 'Corporate Social Responsibility and the Mining Industry: Conflicts and Constructs' (2004) 11(1) *Corporate Social Responsibility and Environmental Management* 23, 25.
11 Guy Pearse, *Greenwash: Big Brands and Carbon Scams* (Black Inc, 2012).
12 Giorel Curran, 'Social Licence, Corporate Social Responsibility and Coal Seam Gas: Framing the New Political Dynamics of Contestation' (2017) 101 *Energy Policy* 427, 434.
13 Paul Kapelus, 'Mining, Corporate Social Responsibility and the "Community": The Case of Rio Tinto, Richards Bay Minerals and the Mbonambi' (2002) 39 *Journal of Business Ethics* 275, 287.
14 Department of Industry, Innovation and Science (Cth), Office of the Chief Economist, *Review of the Socioeconomic Impacts of Coal Seam Gas in Queensland* (Commonwealth of Australia, 2015) <https://industry.gov.au/Office-of-the-Chief-Economist/Publications/Documents/coal-seam-gas/Socioeconomic-impacts-of-coal-seam-gas-in-Queensland.pdf> 2.
15 Lynda Cheshire, Jo-Anne Everingham and Geoffrey Lawrence, 'Governing the Impacts of Mining and the Impacts of Mining Governance: Challenges for Rural and Regional Local Governments in Australia' (2014) 36 *Journal of Rural Studies* 330, 337.
16 Chloe Sher and Cary Wu, 'Fracking in China: Community Impacts and Public Support of Shale Gas Development' (2018) 27 *Journal of Contemporary China* 626, 626.
17 Fedor A. Dokshin, 'Whose Backyard and What's at Issue? Spatial and Ideological Dynamics of Local Opposition to Fracking in New York State, 2010 to 2013' (2016) 81(5) *American Sociological Review* 921.
18 Ibid, 940.
19 Ibid, 939.
20 Nico Polman, Krijn Poppe, Jan Wilem van der Schans and Jan-Douwe van der Ploeg, 'Nested Markets with Common Pool Resources in Multifunctional Agriculture' (2010) LXV(2) *Rivista di Economia Agraria* 295.
21 Gordon Marshall and John Scott, *A Dictionary of Sociology* (Oxford University Press, 2009) 96.
22 Ruth Meinzen-Dick, Monica Di Gregorio and Nancy McCarthy, 'Methods for Studying Collective Action in Rural Development' (2004) 82(3) *Agricultural Systems* 197, 200.
23 Francesco Vanni, *Agriculture and Public Foods: The Role of Collective Action* (Springer, 2014) 22.
24 M. Landabaso, C. Outhton and K. Morgan, 'Learning Regions in Europe: Theory, Policy and Practice Through the RIS Experience' (Paper presented at 3rd International Conference on Technology and Innovation Policy: "Global Knowledge Partnerships, Creating Value for the 21st Century", Austin, USA August 30 – September 2, 1999) 12; Frank Shankwitz, *The Five Principles of Collaboration: Applying Trust, Respect, Willingness, Empowerment and Effective Communication to Human Relationships* (Ibeh Agbanyim, 2015).
25 Arun Agrawal, 'Common Property Institutions and Sustainable Governance of Resources' (2001) 29(1) *World Development* 1649; Harry Ayer, 'Grass Roots Collective Action: Agricultural Opportunities' (1997) 22(1) *Journal of Agricultural Resource Economics* 1.
26 William van Caenegem, Madeline Taylor, Jen Cleary and Brenda Marshall, *Collective Bargaining in the Agricultural Sector* (2015) <https://rirdc.infoservices.com.au/downloads/15-055>.
27 Robert Putnam, *Bowling Alone: The Collapse and Revival of American Community* (Simon and Schuster, 2000) 19.
28 Janine Nahapiet and Sumantra Ghoshal, 'Social Capital, Intellectual Capital, and the Organizational Advantage' (1998) 23(2) *The Academy of Management Review* 242, 243.
29 Richard Sexton, 'Market Power, Misconceptions, and Modern Agricultural Markets' (2012) 95(2) *American Journal of Agricultural Economics* 209.

30 T.K. Ahn and Elinor Ostrom, *Foundations of Social Capital* (Edward Elgar, 2003) 13.
31 Robert Putnam, *Making Democracy Work: Civic Tradition in Modern Italy* (Princeton University Press, 1993).
32 GR Dahal and KP Adhikari, 'Bridging, Linking, and Bonding Social Capital in Collective Action' (Working Paper No 79, International Food Policy Research Institute, 2008) 3 <http://www.cifor.org/library/2500/bridging-linking-and-bonding-social-capital-in-collective-action-the-case-of-kalahan-forest-reserve-in-the-philippines/?pub=2500>.
33 World Bank, *World Development Report 2010: Development and Climate Change* (2010) <http://siteresources.worldbank.org/INTWDR2010/Resources/5287678-1226014527953/WDR10-Full-Text.pdf>.
34 Elinor Ostrom, *Governing the Commons* (Cambridge University Press, 2015).
35 Michael Porter, *Competitive Advantage, Creating and Sustaining Superior Performance* (Free Press, 1985) 199.
36 Claire Nauwelaers and Alasdair Reid, *Innovative Regions? A Comparative Review of Methods of Evaluating Regional Innovation Potential* (RIDER, 1995) 95.
37 D. Davies, K. Larkin, and B. Wilson, *Cluster Development: From Theory to Practice—Implications for the Food Industry* (Paper presented at The Agricultural Economics Society, University of Wales, Aberystwyth, 10 April, 2002) 23.
38 James Coleman, *Foundations of Social Theory* (Harvard University Press, 1990) 93.
39 For example, the standard in Queensland, Australia for the award of compensation that is litigated in the Land Court is based on a reasonable amount to fairly compensate the landholders in light of the circumstances. Review for compensation of an unconventional gas lease can only be reviewed and possibly increased if 'there was a material change in circumstances', and thus compensation is limited to compensation for the compensatable effects of the material change in circumstances. See *Nothdurft v QGC Pty Limited* [2017] QLC 41; *Petroleum and Gas (Production and Safety) Act* 2004 (Qld) s 3(1)(h), s 120, s 532(1), s 532(2), s 532(4), s 536C, s 537C (1); *Land Court Act 2000* (Qld) s 24.
40 William van Caenegem, Madeline Taylor, Jen Cleary and Brenda Marshall, *Collective Bargaining in the Agricultural Sector* (2015) <https://rirdc.infoservices.com.au/downloads/15-055> 17.
41 Ibid.
42 Collective bargaining is within the transactional categorisation of regulatory tools. Transactional tools include contracts, grants, legislative agreements, agreements and accords, covenants, compliance agreements, negotiation/ arbitration and enforcement undertakings. Arie Freiberg, *Regulation in Australia* (The Federation Press, 2017) 201.
43 As explored in Chapter 5.
44 Dan Svantesson, *Svantesson on the Law of Obligations* (Centre for Commercial Law, 3rd edn, 2012).
45 William van Caenegem, Madeline Taylor, Jen Cleary and Brenda Marshall, *Collective Bargaining in the Agricultural Sector* (2015) <https://rirdc.infoservices.com.au/downloads/15-055> xii.
46 *Competition and Consumer Act 2010* (Cth) s 93AB(1A); Competition and Consumer Regulations 2010 (Cth) s 71D.
47 ACCC, *Chevron Australia Pty Ltd & Ors A91139 & A91140 & A91160 & A91161*, Draft Determination, 27.
48 Allen Consulting Group, *Report to the ACCC: Gorgon Gas Project Joint Venture Application for Authorisation of Joint Marketing* (24 July 2009) <http://www.domgas.com.au/pdf/Subs_pres/Allens%20Consulting%20Group%20report.pdf> 29.
49 *Competition and Consumer Act 2010* (Cth) s 93AC(1).
50 *Competition and Consumer Act 2010* (Cth) ss 90(10), 90(10A).

51 William Breen Creighton and Anthony Forsyth, *Rediscovering Collective Bargaining: Australia's Fair Work Act in International Perspective* (Routledge, 2012).
52 Agricultural Competitiveness Taskforce, Parliament of Australia, *Agricultural Competitiveness Green Paper* (2014).
53 Ibid, 21.
54 Productivity Commission, *Inquiry into Mineral and Energy Resource Exploration*, Commonwealth of Australia, Canberra, 2013.
55 Ibid, 130.
56 APPEA, *Record Number of Land Agreements signed in Queensland* (2013) <http://www. appea.com.au/media_release/record-number-of-land-agreements-signed-between-gas-companies-and-queensland-farmers/>.
57 AMEC, Submission No 34 to Productivity Commission, *Inquiry into the Non-Financial Barriers to Mineral and Energy Resource Exploration*, March 2013, 2.
58 Productivity Commission, *Inquiry into Mineral and Energy Resource Exploration*, Commonwealth of Australia, Canberra, 2013 132.
59 Ibid, 133.
60 *Competition and Consumer Act 2010* (Cth) s 90.
61 Justice Robert French, 'Authorisation and Public Benefit – Playing with Categories of Meaningless Reference?' (2006) 24 *Federal Judicial Scholarship* 1, 3.
62 Senate Select Committee on Unconventional Gas, Parliament of Australia, *Inquiry into Unconventional Gas Interim Report* (2016) 23.
63 ACCC, *Collective Bargaining and Boycotts* (2016) <https://www.accc.gov.au/business/anti-competitive-behaviour/collective-bargaining-boycotts>.
64 For example, the Darling Downs Cotton Growers Inc representing 110 cotton growers as at 2017 (member numbers fluctuate between 90–250 growers) of the 300–500 growers in the area.
65 Francesco Vanni, *Agriculture and Public Foods: The Role of Collective Action* (Springer, 2014) 24.
66 Robert Wade, *Village Republics: Economic Conditions for Collective Action in South India* (Cambridge University Press, 1988).
67 Elinor Ostrom, *Governing the Commons* (Cambridge University Press, 2015).

13 Conclusion

Introduction

This book explores how resource-rich States can meet the challenge of creating an effective regulatory regime to balance the commercial exploitation of unconventional gas activities on agricultural land. Unconventional gas occupies a unique role in the growth of the world's non-carbon energy market. Since the assent of the Paris Agreement,[1] the drive to meet sustainable energy goals has been solidified while creating an intense debate internationally and in individual nation States. Unconventional gas has been hailed as one solution to a non-carbon energy future in a 'Golden Age of Gas'[2] and States, including Queensland, Australia, British Columbia, Canada and China, have rushed to embrace this energy source to foster a transition from the burning of oil and coal to the generation of energy from carbon-free sources.

Some states, such as Queensland and British Columbia, have been particularly enthusiastic in development of its unconventional gas resources across some of the most productive agricultural land, including in the Bowen Basin of Queensland and areas of the Peace-River Region in British Columbia. Both jurisdictions have chosen to support the development of unconventional gas through land use zoning and land access agreement systems by encouraging licenseholders to develop unconventional gas tenements and reap the economic rewards of the international export market. This is contrary to the position of other states, like New York State, the UK and France, who have chosen to ban hydraulic fracturing in a precautionary stance based on public opposition to the real and perceived environmental and agricultural impacts of the technology. An alternative statist approach, as found in Poland and China, reverses the relationship of the State and licenseholders, by granting control and development rights of unconventional gas exploitation to the State itself. Globally, the rapid exploitation of unconventional gas resources has created political, social and economic externalities. Of these externalities, the conflict between resource companies and landholder interests has assumed primacy and created an increasingly complex regulatory challenge. In seeking to solve the nexus between gaining economic benefits of unconventional gas exploitation and the cumulative impacts on agricultural land, States have sought to either balance, disengage or prioritise natural resource companies to create a system of effective regulation. With the exception of States preventing the exploitation of unconventional gas, it is the underlying aim of all jurisdictions to provide a legal landscape promoting coexistence in balancing the interests of both industries. Indeed, legislative reviews highlight the political and economic requirement for effective natural resource governance to consider how the State will navigate the troubled waters of maximising the broader economic benefits of unconventional gas while sponsoring regulatory frameworks to protect the agricultural sector.[3]

This book critically and comparatively analyses the capacity of the current unconventional gas regulatory regimes of seven jurisdictions to examine the benefits and burdens of differing regulatory frameworks for unconventional gas in managing conflicting land interests. In this global analysis, the book has analysed the differing State policies on energy security, food security, and where applicable food sovereignty, as an added dimension and reasoning for the prioritisation of either unconventional gas or agriculture. A number of natural resource governance functions have been examined including: unconventional gas and agricultural policy; regulatory legislative frameworks; land use zoning; and compensation and land access arrangements. In particular, this book focuses on an analysis of the land use, land access and compensation regimes to find a resolution between the competing interests of licenseholder and agricultural landholder. This chapter presents a concluding summary of research findings and recommendations in exploring how unconventional gas and agricultural farmland can, or cannot, coexist.

The intersection of agriculture and shale gas

When examining the interplay between unconventional gas exploitation and agricultural land, research demonstrates that the intersection between food and agriculture in several jurisdictions is complex, with some states seeking to intensify agriculture, others seeking to protect farmland and others not recognising the value of its agricultural land. For example, the UK government supports the concept of sustainable intensification, which is the production of more food on a finite amount of land in a sustainable way. Furthermore, there is demonstrable interest in preserving the agricultural land in the UK for its visual and recreational amenity, as much as for its agricultural productivity. Given the aesthetic and recreational value of the land, there has been sustained public protest and outcry for a number of years over the prospect of shale gas development, to no avail. The UK government is continuing with its policy of shale gas development, and seeks to utilise its discretionary power over the planning system to ensure that shale gas exploration occurs. Such a position contrasts markedly with that of France, which responded swiftly to public outcry over shale gas exploration, firstly by implementing a moratorium on shale gas exploration, and then by permanently banning all forms of hydrocarbon exploitation.

The relationship between food and energy in China is complex. Given its large population, China seeks to ensure that it has enough food to feed its burgeoning population. Therefore, China has established in its 13th FYP that it seeks to undertake the intensification of agriculture on its finite land resources in order to feed their population. This intensification will be through the industrialisation of agricultural land. Such intensification will necessarily rely on the mediation of polluted land and the control of construction on valuable 'basic farmland'. This is where the conundrum exists – in order to develop shale gas, China needs to utilise 'basic farmland' in many instances. In addition, China faces difficulties in developing its shale gas resources due to restrictions on water. Water resources are relatively sparse, in part due to pressure from agricultural activities and partly due to contamination of much of its water. As a result, both land and water will present significant challenges to China in the years ahead as it seeks to develop its shale gas resources to improve energy security.

Similarly, Poland seeks to undertake a reform of its agricultural system, given that it is relatively parochial, having changed little since the Soviet era. The reforms in Poland are driven from a socioeconomic perspective to target small non-industrial farms, in which EU rules relating to food processing and packaging are crippling, who dominate much of the rural sector. The election of the Law and Justice Party to government in 2015 saw a shift from the declining interest in unconventional hydrocarbons to the rural sector. This was partly motivated by a desire to reform the sector in order to encourage Polish nationals to return to Poland.

Approaches to the regulatory conundrum: Energy security or food security?

The adaptive management approach refers to the regulatory framework that has been implemented in the management of conflicting interests in both Queensland and British Columbia, examined in Chapters 5 and 6, respectively. This approach is typified by its minimal State intervention with preference for market forces to respond and direct unconventional gas exploitation. As an approach originating in scientific ecological management, adaptive management tends to be a 'learning by doing' exercise in creating regulation on the run, as and when, socio-legal issues arise. It grants private unconventional gas companies with *de facto* control and autonomy over its licence area and development of the resource, such as well development and dewatering processes.

In the case of Queensland, the new *Regional Planning Interests* Act 2014 (Qld) (RPIA) regime provides an approvals process for CSG activities on Priority Agricultural Areas to manage land use interests and create coexistence. The RPIA regulatory framework provides an important land use tool permitting CSG activities in agricultural areas once an approval is given by the Department of Infrastructure, Local Government and Planning or the landholders' consent is granted and is 'not likely to have a significant' impact on the agricultural area. In the situation where three governmental departments and three administrative authorities hold jurisdiction over CSG operating on agricultural land, the potential for overlap and regulatory gaps is magnified and there is evidence to suggest that this has, and is, occurring in Queensland.

Land use zoning tools in combination with an administrative regulatory body, such as the Agricultural Land Commission model in British Columbia, outlines an adaptive approach combined with a quasi-judicial tribunal framework. This model seeks to allow the state to fulfil its goal of creating value in the shale gas sector and enabling titleholders to maximise their return. The ALC as the single agricultural land protection regulatory administrative authority, and its Delegation Agreement with the Oil and Gas Commission, demonstrates the creation of such an arrangement which serves to provide the forum for collaboration and coexistence of shale gas development in agricultural land zones.

The precautionary approach refers to the regulatory framework adopted by France, the UK and the state of New York, USA, in preventing and banning unconventional gas activities founded upon the precautionary principle. States exemplifying the precautionary approach all have a common focus on opposition to shale gas due to the emphasis of agricultural land protection to protect food security and food sovereignty while maintaining the viability of its agricultural industries. Chapter 7 examines the cultural importance of the visual and community aspects of agricultural land highlighted by the UK, in viewing shale gas as a threat to the environment, water systems and farming. Although initially the UK sought to advance shale gas exploitation in the face of dwindling North Sea oil and gas stocks, this policy stance was quickly abandoned in response to the Pease Hall incident of 2011. The release of the *Community Engagement Charter* and the recent ministerial decision to overturn the denial of planning permissions for shale gas drilling applications represents the embedding of the precautionary principle founded in article 191 of the Treaty on the Functioning of the European Union.

Similarly, in France, the precautionary principle is stridently applied to all shale gas activities, as the first jurisdiction to ban hydraulic fracturing in 2011 via Law 2011–835. Historically, as examined in Chapter 9, significant artisanal products and unique agrarian conditions creates the strong environmental, social and gastronomic values associated with agricultural land protection in France. The *appellation d'origine contrôlée* system, coupled with strong public opposition commencing with the Ardèche movement, has led to the outright banning of oil and gas activities in France in L.111–13 and L.111–14–1 of the Mining Code. Law No. 2017–1839 from 2040 onwards. However, France's energy policy strategy has allowed its staunch approach towards shale gas to ensue, due to its diversified scheme of import of natural gas from Norway, the Netherlands, Russia and Algeria.

Chapter 8 surveys the unmistakably precautionary approach of the state of New York in banning hydraulic fracturing in 2015 via Bill A3243. In undertaking a state-wide Environmental Quality Review coupled with a new Environmental Impact Statement for high-volume hydraulic fracturing, New York found rural communities and agricultural land protection, in consistence with article XIV of the New York State Constitution, outweighed the potential benefits of shale gas exploitation. New York provides an interesting jurisdictional example of a state providing natural resource governance and self-regulation for its rural communities enshrined within its Municipal Home Rule. Further, the rise of the Joint Landholders Coalition Of New York provides the first example of collective action for landholders in collectively negotiating community-wide oil and gas leases for increased economic benefit.

Finally, the statist approach to unconventional gas regulation is underpinned by strong State intervention in the award, regulation and participation in unconventional gas activities. Both Poland and China have adopted this model, examined in Chapters 10 and 11 respectively, with the State taking a strong and influential role in the management of the agricultural sector, but for different reasons. In Poland, as surveyed in Chapter 10, the statist approach has been utilised by the government to reinvigorate the agricultural sector through reforms relating to farm size and level of industrialisation. Such reform is needed to address the inadequacies of the farming system originating from the communist era. The other reason for reform is the impact that EU regulations relating to agriculture have had on the sector to date. Interestingly, this reform has been able to progress with little need to decide whether food security or energy security trumps since initial shale gas exploration results were poor, providing little impetus to companies to continue with development.

On the other hand, China has taken a strong statist approach, as evident within Chapter 11, seeking to equally value and develop shale gas and agriculture. Through the 13th FYP, and other national government instruments, the Chinese government has taken a strong leadership role in the development of shale gas on agriculture land. The Chinese government seeks to develop the resources, stipulating targets for shale gas production in its energy plan. Yet equally, the government has set out a goal of industrialising agriculture and protecting farmland. Such policies will provide many challenges for China in the coming years. However, if any state can succeed in creating a dominant and successful unconventional gas sector, it likely will be China, primarily due to all encompassing role of the government.

Solutions to conflict: Collective action as a new way forward

A key recommendation of this book is collective bargaining as a vehicle to promote effective methods of coexistence and agreement-making between unconventional gas titleholders and landholders. Providing regulation for collective bargaining within unconventional gas regulatory frameworks may improve communication and strategic landholder engagement, while offering a more streamlined and accessible process to manage land access compensation to deliver long-term improvements and coexistence with the onshore unconventional gas industry. These regulatory aims would support the principle of coexistence between both sectors. The information-sharing reciprocity between collective members, and the opportunity to work towards a collective goal may offer greater opportunities to undertake good faith negotiations. This is in stark contrast to the current situation where landholders must work as individuals to comprehend and interact with complex legal provisions, with limited support from State bodies.

Collective bargaining may be embraced by the unconventional gas industry and agricultural sector in some States, such as Queensland, where both industries have previously utilised collective bargaining mechanisms when negotiating long-term contracts. Further, the monitoring of collective bargaining by a quasi-judicial body such as the Gasfields Commission, supported by the ACCC within the existing Conduct and Compensation Agreement framework, may create a regulatory tool which satisfies coexistence of multiple interests for the benefit of its citizens. However, it must be noted that collective bargaining does not represent a complete solution for all States. In particular, collective bargaining may not be a viable model for China, where organised collective movements are not usually permitted. Instead, it offers one option, rather than a 'silver bullet' for jurisdictions to consider in addressing the increasingly strident calls to 'rebalance' the interests of landholders, particularly in relation to land access of agricultural land.

Conclusion

Each jurisdiction in this book grapples with the challenge of securing both energy and food, while selecting a regulatory approach to either ban or embolden the development of unconventional gas. This concluding chapter highlights the key similarities and differences of the regulatory approaches taken by States to manage the conflicting interests of agricultural landholders and unconventional gas licenseholders. Some jurisdictions require the State to be the regulator, administrator and exploiter of unconventional gas resources. Others have taken a preventative approach in banning unconventional gas where it poses a real and uncompromising threat to agricultural lands and its rural communities. Although all States within this book face the conundrum of choosing a transition fuel to move away from non-renewable sources, and eventually wholly embrace renewable energy, not all have chosen unconventional gas for this task. Such is the case in France in its choice of a nuclear transition to renewables. However, each jurisdiction and its experience with unconventional gas to date seeks the ever-elusive goal of coexistence between two fundamentally different industries – agriculture and energy. The adaptive management approach, as in British Columbia and Queensland, attempts to create regulation on the run, constantly adjusting and amending regulation as each new challenge arises and faces the unconventional gas industry. A precautionary approach, found in France, New York State and the UK, avoids the balancing act between the two land uses by prohibiting hydraulic fracturing of shale gas and clearly prioritising agriculture and its uses over energy extraction. Finally, the statist approach adopted by Poland and China creates a new role for the state as both the participant and regulator in its quest for energy security.

The complex web between land use and access for energy industries on agricultural land is neither unique nor isolated to unconventional gas. Coexistence between energy and agriculture will continue to be a reoccurring issue – with renewable energy farms, such as wind, as the latest rising target for agricultural landholder action. This ultimately requires regulation relating to land access agreements that provide accountability and equitable terms for agricultural landholders. Collective bargaining may be used as an approach to give power and equity back into the hands of agricultural landholders. Collective bargaining can provide the impetus to effectively negotiate and execute land access agreements and provides an example of how regulation may encourage collaboration among landholders in finding sustainable, transparent and equitable arrangements to manage coexistence of unconventional gas activities. This vehicle is not on the horizon of any government surveyed within this book to date, as it is a massive regulatory task that requires significant resources, political will and the fundamental acknowledgement that the current regulatory approach, whether it be adaptive, precautionary or statist, has flaws. Until then, farmers will continue to fight to protect one of the most important and fastest disappearing non-renewable resources, arable land, and it is high time States consider joining this fight too.

Notes

1 United Nations Framework Convention on Climate Change, *Adoption of the Paris Agreement, 21st Conference of the Parties*, Paris: United Nations, U.N. Doc. A/CONF. 541/13 (2015). The Paris Agreement's central aim is to strengthen the global response to the threat of climate change by keeping a global temperature rise this century well below 2 degrees Celsius above pre-industrial levels and to pursue efforts to limit the temperature increase even further to 1.5 degrees Celsius.
2 International Energy Agency, *Golden Rules for a Golden Age of Gas. World Energy Outlook Special Report on Gas* (2012) <http://www.iea.org/publications/free-publications/publication/WEO_2012_Special_Report_Golden_Rules_for_a_Golden_Age_of_Gas.pdf>.
3 Queensland Competition Authority (Qld), *Final Report: Coal Seam Gas Review* (January 2014) <http://www.qca.org.au/getattachment/aaaeab4b-519f-4a95-8a65-911bc46cc1d3/CSG-investigation.aspx>.

References

Aarrestad, Jostein, 'Resource Extraction, Financial Transaction and Compensation in an Open Economy' (1979) 81(4) *Scandinavian Journal of Economics* 552

ABC News, 'New Land Access Code Described as a "Con"' (15 January 2014) ABC News (online) <http://www.abc.net.au/news/2014-01-16/new-land-access-agreement-code-described-as-a-27con27/5202442>

ACCC, 'Collective Bargaining and Boycotts' (2016) <https://www.accc.gov.au/business/anti-competitive-behaviour/collective-bargaining-boycotts>

Aczel, Miriam R. and Karen E. Makuch, 'Environmental Impact Assessments and Hydraulic Fracturing: Lessons from Two U.S. States' (2017) *Case Studies in the Environment* 1

Agrawal, Arun, 'Common Property Institutions and Sustainable Governance of Resources' (2001) 29(1) *World Development* 1649

Agricultural Competitiveness Taskforce, Parliament of Australia, *Agricultural Competitiveness Green Paper* (2014)

Agricultural Land Commission, Chapter 5.3 Completing Application Information Details: Agricultural Land Reserve (2013) <http://www.bcogc.ca/node/13290/download>

Agricultural Land Commission, Message from the Chair (British Columbia Government, 2013) <http://www.ceaa.gc.ca/050/documents/p63919/97838E.pdf>

Agricultural Land Commission, Oil and Gas Development in the Agricultural Land Reserve: The Non-Farm Use of Agricultural Land (2013) <http://www.llbc.leg.bc.ca/public/pubdocs/bcdocs2014/538680/history%20of%20oil%20and%20gas%20activities%20in%20the%20alr%20november%202013.pdf> 1

Agricultural Land Commission, Oil and Gas Development in the Agricultural Land Reserve: The Non-Farm Use of Agricultural Land, An Historical Overview of the Agricultural Land Commission's Position Regarding Oil and Gas Activities in the ALR (2013) <https://www.alc.gov.bc.ca/assets/alc/assets/about-the-alc/working-with-other-ministries-and-agencies/history_of_oil_and_gas_activities_in_the_alr_november_2013.pdf>

Agricultural Land Commission, Provincial Land Commission Annual Report 2014/2015 (ALC, 2015) <http://blogs.ubc.ca/alrmap/files/2016/02/annual_report_2014-2015.pdf>

Ahn, T.K. and Elinor Ostrom, *Foundations of Social Capital* (Edward Elgar, 2003)

Alekett, Kjeell, Tadeusz Patzek, Bjorn Svensson and Rafal Jaroz, Can Hydraulic Fracturing Make Poland Self Sufficient in Natural Gas? (2014) <http://gaia.pge.utexas.edu/papers/140519_Fracking_in_Poland_submitted_Energy_Technology.pdf>

Al-Ibrahim, Ali, Vladimir Strezov, Peter Davies and Ian Wright, 'Environmental Impact of Coal Mining and Coal Seam Gas Production on Surface Water Quality in the Sydney Basin, Australia' (2017) 189(9) *Environmental Monitoring Assessment* 408

Allan, Catherine and George Stankey (eds), *Adaptive Environmental Management* (Springer, Netherlands, 2009)

Allan, Catherine, *Adaptive Management of Natural Resources*, Proceedings of the 5th Australian Stream Management Conference. Australian Rivers: Making a Difference. Charles Sturt University, Thurgoona, New South Wales, 26

Allan, John Anthony, Martin Keulertz, Suvi Sojamo and Jeroen Warner (eds), *Handbook of Land and Water Grabs in Africa: Foreign Direct Investment and Food and Water Security* (Routledge, 2013)

Allen Consulting Group, *Report to the ACCC: Gorgon Gas Project Joint Venture Application for Authorisation of Joint Marketing* (24 July 2009) <http://www.domgas.com.au/pdf/Subs_pres/Allens%20Consulting%20Group%20report.pdf> 29

Alterman, Rachelle, 'Land-Use Regulations and Property Values: The "Windfalls Capture" Idea Revisited' in Nancy Brooks, Kieran Donaghy and Gerrit-Jan Knaap (eds), *The Oxford Handbook of Urban Economics and Planning* (ebook, Oxford University Press, 2011)

Althaus, Catherine, Peter Bridgman and Glyn Davis, *The Australian Policy Handbook* (Allen & Unwin, 2012)

AMEC, Submission No 34 to Productivity Commission, Inquiry into the Non-Financial Barriers to Mineral and Energy Resource Exploration, March 2013

Anderson, James E., *Emergence of the Modern Regulatory State* (Public Affairs Press, 1963)

Anderson, Owen L., 'Lord Coke, the Restatement and Modern Subsurface Trespass Law' (2011) 6(2) *Texas Journal of Oil, Gas, and Energy Law* 203

Androkovich, Robert, Ivan Desjardins, Gordon Tarzwell and Peter Tsigaris, 'Land Preservation in British Columbia: An Empirical Analysis of the Factors Underlying Public Support and Willingness to Pay' (2008) 40(3) *Journal of Agricultural and Applied Economics* 999

APPEA (Australian Petroleum Production and Exploration Association), Australia's Upstream Oil and Gas Industry: A Platform for Prosperity, Issues Paper (2006)

APPEA, Record Number of Land Agreements Signed in Queensland (2013) <http://www.appea.com.au/media_release/record-number-of-land-agreements-signed-between-gas-companies-and-queensland-farmers/>

Argent, Robert, 'Components of Adaptive Management' in Catherine Allan and George Stankey (eds), *Adaptive Environmental Management* (Springer, 2009)

Arora-Jonsson, Seema, *Gender, Development and Environmental Governance: Theorizing Connections* (Routledge, 2013)

Arrow Energy, Analysis of Agricultural Production and Issues in the Darling Downs: Surat Gas Project Supplementary Report to the Environmental Impact Statement Report prepared for Arrow Energy Pty Ltd and Coffey Environments Australia Pty Ltd (AEC Group, 2013) <https://www.arrowenergy.com.au/__data/assets/pdf_file/0003/8670/Appendix_14.pdf>

Asia Pacific Energy, China: Shale Gas Development Plan (2016–2020) (2015) <https://policy.asiapacificenergy.org/node/3050>

Atkinson, Giles, Simon Dietz and Eric Neumayer, *Hand Book of Sustainable Development* (Edward Elgar Publishing, 2007)

Australian Competition and Consumer Commission, Gas Inquiry Interim Report 2017–2010 (2017) <https://www.accc.gov.au/system/files/Gas%20inquiry%20December%202017%20interim%20report.pdf>

Australian Government Department of Environment and Energy, Annual Report 2016–2017 (2017)

Australian Government, National Landcare Program (Australian Government Department of Environment and Energy and Department of Agriculture and Water Resources, 2016) <http://www.nrm.gov.au/>

Australian Productivity Commission, Review of Regulatory Burden on the Upstream Petroleum (Oil and Gas) Sector: April 2009 (Productivity Commission, 2009) 209 <https://www.pc.gov.au/inquiries/completed/upstream-petroleum/report/upstream-petroleum.pdf>

Ayer, Harry, 'Grass Roots Collective Action: Agricultural Opportunities' (1997) 22(1) *Journal of Agricultural Resource Economics* 1

Baen, John S., 'The Impact of Mineral Rights and Oil and Gas Activities on Agricultural Land Values' (1996) 64*Appraisal Journal* 67

Baginski, Wojciech, 'Shale Gas in Poland – The Legal Framework for Granting Concessions for Prospecting and Exploration of Hydrocarbons' (2011) 32*Energy Law Journal* 145

Baker, Mark and Jonathan Kusel, *Community Forestry in the United States: Past Practice, Crafting the Future* (Island Press, 2003)

Baldwin, Robert, 'Better Regulation: Tensions Abroad the Enterprise' in Stephen Weatherill, *Better Regulation* (Bloomsbury Publishing, 2007)

Baldwin, Robert, *Rules and Government* (Clarendon Press, 1995)

Ballem, John Bishop, *The Oil and Gas Lease in Canada* (University of Toronto Press, 2008)

Banski, Jerzy, 'Changes in Agricultural Land Ownership in Poland in the Period of the Market Economy' (2011) 57(2) *Agricultural Economics* 93–101

Barinzelli, Claudia, Ine Vandecasteele, Ricardo Ribero Baranco, Innes Maru I. Riviero, Nathan Pelletier, Okke Batelaan and Carlo Lavalle, 'Scenarios for Shale Gas Development and their Related Land Use Impacts in the Baltic Basin, North Poland' (2015) 84*Energy Policy* 80

Baromey, Neth, *Ecotourism as a Tool for Sustainable Rural Community Development and Natural Resources Management in the Tonle Sap Biosphere Reserve* (Kassel University Press, 2008)

BCOGC, Certificate of Restoration Application Manual (2016) <http://www.bcogc.ca/node/12445/download>

Bellamy, J.A., *Federalism and Regionalism in Australia: New Approaches, New Institutions?* (ANU ePress, 2007)

Beuchelt, Tina and Detlef Virchow, 'Food Sovereignty or the Human Right to Adequate Food: Which Concept Serves Better as International Development Policy for Global Hunger and Poverty Reduction?' (2012) 29(2) *Agriculture and Human Values* 259

Beverton, Raymond and Sidney Holt, *On the Dynamics of Exploited Fish Populations* (Chapman and Hall, 1957)

Birol, Faith, US to Overtake Russia as Top Oil Producer by 2019 at Latest: IEA (2018) Reuters 27 February 2018 <https://www.reuters.com/article/us-energy-iea/u-s-to-overtake-russia-as-top-oil-producer-by-2019-at-latest-iea-idUSKCN1GB0C6>

Black, Julia, 'Regulatory Conversations' (2002) 29(1) *Journal of Law and Society* 163

Black, Julia, Learning from Regulatory Disasters, Society and Economics Working Papers 24/2014 (London School of Economics, Department of Law, 2014) 11

Bois, Charles and Sarah Hansen, 'Regulatory and Legal Issues Respecting Coalbed Methane Development in British Columbia' (2008) 45*Alberta Law Review* 631

Boring, Nicolas, France: New Mining Code Under Consideration (2013) <http://loc.gov/law/foreign-news/article/france-new-mining-code-under-consideration/>

Bothwell, Bob, *Penguin History of Canada* (PenguinCanada, 2007)

Boulle, Laurence, Tina Hunter, Michael Weir and Kath Kurnow, 'Negotiating Conduct and Compensation Agreements for Coal Seam Gas Operations: Developing the Queensland Regulatory Framework' (2014) 17(1) *Australasian Journal of Natural Resources Law and Policy* 75

Boutilier, Robert and Ian Thomson, Modelling and Measuring the Social Licence to Operate: Fruits of a Dialogue Between Theory and Practice (2011) <http://socia license.com/publications/Modelling%20and%20Measuring%20the%20SLO.pdf>

Bowen, Sarah, 'Embedding Local Places in Global Spaces: Geographical Indications as a Territorial Development Strategy' (2010) 75(2) *Rural Sociology* 209

Brady, William and James Crannell, 'Hydraulic Fracturing Regulation in the United States: The Laissez-Faire Approach of the Federal Government and Varying State Regulations' (2012) 14 *Vermont Journal of Environmental Law* 39

Braithwaite, John, Cary Coglianese and David Levi-Faur, 'Can Regulation and Governance Make A Difference?' (2007) 1(1) *Regulation and Governance* 1

Braithwaite, John, Neoliberalism or Regulatory Capitalism (RegNet Occasional Paper No. 5, 2005) <https://www.anu.edu.au/fellows/jbraithwaite/_documents/Articles/ Neoliberalism_Regulatory_2005.pdf>

Breyer, Stephen G., Breaking the Vicious Circle: Toward Effective Risk Regulation (Harvard University Press, 1993)

Breyer, Stephen G., *Regulation and its Reform* (Harvard University Press, 2009)

British Columbia Ministry of Energy and Mines, Natural Gas Strategy: Fuelling B.C.'s Economy for the Next Decade and Beyond (2012) <http://www.gov.bc.ca/ener/p opt/down/natural_gas_strategy.pdf>

British Columbia Ministry of Natural Gas Development, Summary of Shale Gas Activity in Northeast British Columbia 2014 (British Columbia Government, 2014) <https:// www2.gov.bc.ca/assets/gov/farming-natural-resources-and-industry/natural-ga s-oil/petroleum-geoscience/oil-gas-reports/oil_and_gas_report_2016-1.pdf>

British Columbia Oil and Gas Commission, 'Chapter 5.3 Completing Application Information Details: Agricultural Land Reserve' in British Columbia Oil and Gas Commission (ed.), *Oil and Gas Activity Application Manual* (British Columbia Government, 2013) <http://www.bcogc.ca/node/13290/download>

British Columbia Oil and Gas Commission, ALC–OGC Delegation Agreement (2013) <https://www.bcogc.ca/node/5759/download>

British Columbia Oil and Gas Commission, Oil and Gas Activity Operations Manual Version 1.14 (British Columbia Government, 2017) <http://www.bcogc.ca/node/ 13274/download> 2

British Columbia Oil and Gas Commission, Delegation Agreement for Oil and Gas Uses in the Agricultural Land Reserve Peace River Regional District and Northern Rockies Regional Municipality (2014) <https://www.bcogc.ca/node/11130/download>

British Geological Survey, Bowland Shale Gas (BGS, 2016) <http://www.bgs.ac.uk/ research/energy/shaleGas/bowlandShaleGas.html>

British Geological Survey, Fracking and Earthquake Hazard (BGS, 2018) <http://ea rthquakes.bgs.ac.uk/research/earthquake_hazard_shale_gas.html>

Bunter, Michael, *The Promotion and Licensing of Petroleum Prospective Acreage* (Kluwer Law International, 2002) xxii

Burchmore, Patricia and Christopher Jones, 'Regulatory Reform in the British Columbia Petroleum Industry: The Oil and Gas Commission' (2000) 38 *Alberta Law Review* 143

Burnett, J.Wesley, Randall W Jackson and Robert Blobaum, 'The State of Play in Poland's Unconventional Shale and Oil Development' (2015) 33(4) *Development Policy Review* 395–414, 406

Butt, Peter and Peter Nygh, *Encyclopaedic Australian Legal Dictionary* (LexisNexis, 2016)

Butterly, Dee and Dr. Ian Fitzpatrick, *A People's Food Plan* (2017)

Cai, Peter, *Understanding China's Belt and Road Initiative* (Lowy Institute, 2017) <https://www.lowyinstitute.org/sites/default/files/documents/Understanding%20China%E2%80%99s%20Belt%20and%20Road%20Initiative_WEB_1.pdf>

Cameron, James and Juli Abouchar, 'The Precautionary Principle: A Fundamental Principle of Law and Policy for the Protection of the Global Environment' (1991) 14 *Boston College International and Comparative Law Review* 1

Campin, David, 'Regulating hydraulic Fracturing' in Tina Hunter (ed.), *Handbook of Shale Gas Law and Policy: Economics: Access, Law and Regulation in Key Jurisdictions* (Intersentia, 2016)

Carney, Gerard, 'Constitutional Framework for the Regulation of the Australian Uranium Industry' (2007) 26(3) *Australian Resources and Energy Law Journal* 47

Carter, Colin A., Funing Zhong and Jing Zhu, 'Advances in Chinese Agriculture and its Global Implications' 23(1) *Applied Economic Perspectives and Policy* 1

Castelli, Matthew, 'Fracking and the Rural Poor: Negative Externalities, Failing Remedies, and Federal Legislation' (2015) 3(2) *Indiana Journal of Law and Social Equality* 281

CEDA, Australia's Unconventional Energy Options (Report, September 2012) <https://www.ceda.com.au/Research-and-policy/All-CEDA-research/Research-catalogue/Australia-s-Unconventional-Energy-Options>

Central Committee of the Communist Party of China, Decision of the Central Committee of the Communist Party of China on Some Major Issues Concerning Comprehensively Deepening the Reform (2013) (translation by China.org.cn on 16 January 2014) Third Plenary Session of the 18th Central Committee of the Communist Party of China on November 12 2013 <http://www.china.org.cn/china/third_plenary_session/2014-01/16/content_31212602.htm> accessed 12 May 2018

Chambers, Nathalie, *Saving Farmland: The Fight for Real Food* (Rocky Mountains Books, 2015)

Charters, Steve and Nathalie Spielmann, 'Characteristics of Strong Territorial Brands: The Case of Champagne' (2014) 67 *Journal of Business Research* 1461

Chen, Cindy and Alan Randall, 'The Economic Contest Between Coal Seam Gas Mining and Agriculture on Prime Farmland: It May Be Closer than We Thought' (2013) 15(3) *Journal of Economic and Social Policy* 1

Cheshire, Lynda, Jo-Anne- Everingham and Geoffrey Lawrence, 'Governing the Impacts of Mining and the Impacts of Mining Governance: Challenges for Rural and Regional Local Governments in Australia' (2014) 36 *Journal of Rural Studies* 330

Chi, Ma, 'China's First Home-Built Icebreaker Named Snow Dragon 2' (2017) *China Daily*, 27 September 2017 <http://www.chinadaily.com.cn/china/2017-09/27/content_32544019.htm>

China Power, 'How is China Feeding its Population of 1.4 billion?' (2017) China Power 25 January 2017 <https://chinapower.csis.org/china-food-security/>

China, Russia Team up on Ice Silk Road (2017) *World Maritime News*, 10 July 2017 <https://worldmaritimenews.com/archives/224690/china-russia-team-up-on-ice-silk-road/>

Clancy, S. A., F. Worrall, R. J. Davies and J. G. Gluyas, 'The Potential for Spills and Leaks of Contaminated Liquids from Shale Gas Developments' (2018) 626 *Science of the Total Environment* 1463

Clark, Greg, Energy Policy Written Statement – HCWS690 Statement to Parliament 17 May 2018 (United Kingdom Government, 2018) <https://www.parliament.uk/business/publications/written-questions-answers-statements/written-statement/Commons/2018-05-17/HCWS690/>

Coleman, James, *Foundations of Social Theory* (Harvard University Press, 1990)

Commonwealth of Australia, Department of Industry, Innovation and Science, Gas Resources and Energy Quarterly June 2017 (2017)

Conference on the Human Environment, Rio Declaration on Environment and Development, UN Doc A/CONF,151/26, (vol. I) / 31 ILM 874(1992) <https://cil.nus.edu.sg/rp/il/pdf/1992%20Rio%20Declaration%20on%20Environment%20and%20Development-pdf.pdf>

Connor, Linda and Phil McManus, 'What's Mine is Mine(d): Contests over Marginalisation of Rural Life in the Upper Hunter, NSW' (2013) 22(2) *Rural Society* 166

Cooper, Jasmin, Laurence Stamford and Adisa Azapagic, 'Economic Viability of UK Shale Gas and Potential Impacts on the Energy Market up to 2030' (2018) 215 *Applied Energy* 577

Council of Australian Governments, *Best Practice Regulation: A Guide for Ministerial Councils and National Standard Setting Bodies* (2007)

Cournil, Christel, 'Adoption of Legislation on Shale Gas in France' (2013) 4 *European Energy and Environmental Law Review* 141

Creighton, William Breen and Anthony Forsyth, *Rediscovering Collective Bargaining: Australia's Fair Work Act in International Perspective* (Routledge, 2012)

Cross, Frank B., 'Paradoxical Perils of the Precautionary Principle' (1996) 53 *Washington & Lee Law Review* 851

Cuadrilla Resources, Preston New Road: Latest News (CR, 2018) <https://cuadrillaresources.com/site/preston-new-road/>

Cui, Kai and Sharon Shoemaker, 'A Look at Food Security in China' (2018) 2(4) *Nature Partner Journals: Science of Food* 1

Curran, Giorel, 'Social Licence, Corporate Social Responsibility and Coal Seam Gas: Framing the New Political Dynamics of Contestation' (2017) 101 *Energy Policy* 427

Dahal, GR and KP Adhikari, 'Bridging, Linking, and Bonding Social Capital in Collective Action' (Working paper No 79, International Food Policy Research Institute, 2008) <http://www.cifor.org/library/2500/bridging-linking-and-bonding-social-capital-in-collective-action-the-case-of-kalahan-forest-reserve-in-the-philippines/?pub=2500>

Daintith, Terence, *Finders Keepers?: How the Law of Capture Shaped the World Oil Industry* (RFF Press, 2010)

Davey, Edward, Written Ministerial Statement by Edward Davey: Exploration for Shale Gas (United Kingdom Government, 2012) <https://www.gov.uk/government/speeches/written-ministerial-statement-by-edward-davey-exploration-for-shale-gas>

Davidson, Alex T., Michael Dence and The Royal Society of Canada, Institute for Research on Public Policy, *The Brundtland Challenge and the Cost of Inaction* (IRPP, 1988)

Davies, D., K. Larkin, and B. Wilson, Cluster Development: From Theory to Practice – Implications for the Food Industry (Paper presented at The Agricultural Economics Society, University of Wales, Aberystwyth, 10 April 2002)

Davies, Richard, et al., 'Oil and Gas Wells and their Integrity: Implications for Shale and Unconventional Resource Exploitation' (2014) 56 *Marine and Petroleum Geology* 239

de Sa, Paulo, 'Mineral Policy: A World Bank Perspective' in E Bastida, T. Walde and J. Warden-Fernandes (eds), *International Comparative Mineral Law and Policy* (Wolters Kluwer, 2005) 492 de Sadeleer, Nicholas, *Environmental Principles – From Political Slogans to Legal Rules* (Oxford University Press, 2002)

de Schutter, Olivier, A Rights Revolution: Implementing the Right to Food in Latin America and the Caribbean (UN Special Rapporteur to the Right to Food, 2012) <http://www.srfood.org/index.php/en/right-to-food>

de Schutter, Olivier, Interim Report of the Special Rapporteur on the Right to Food: A/68/288 (UN Special Rapporteur on the Right to Food, 2013)

Declaration of Nyéléni in Author Unknown, *Food Sovereignty Framework: Concept and Historical Context* (Nyéléni, 2008) <https://nyeleni.org/IMG/pdf/FoodSovereignity Framework.pdf>

Department of Environment and Climate Change and Department of Communities and Local Government, Shale Gas and Oil Policy Statement by DECC and DCLG (United Kingdom Government, 2015) <https://www.gov.uk/government/publica tions/shale-gas-and-oil-policy-statement-by-decc-and-dclg/shale-gas-and-oil-policy-sta tement-by-decc-and-dclg>

Department of Environment and Climate Change, Digest of UK Energy Statistics (United Kingdom Government, 2015) <https://www.gov.uk/government/collec tions/digest-of-uk-energy-statistics-dukes#2015>

Department of Environment and Climate Change, Onshore Oil and Gas Regulation in the UK: Regulation and Best Practice (United Kingdom Government, 2013)

Department of Environmental Conservation, 2015 Summary & Trends – Oil, Gas and Solution Mining (2015) <https://www.dec.ny.gov/energy/92904.html>

Department of Environmental Conservation, Regulations and Enforcement (2016) <http://www.dec.ny.gov/65.html>

Department of Industry, Innovation and Science (Cth), Office of the Chief Economist, Review of the Socioeconomic Impacts of Coal Seam Gas in Queensland (Common-wealth of Australia, 2015) <https://www.industry.gov.au/sites/g/files/net3906/f/June% 202018/document/pdf/review_of_the_socioeconomic_impacts_of_coal_seam_gas_in_ queensland.pdf>

Department of State Development, Gasfields Commission Review (2016) <https:// www.statedevelopment.qld.gov.au/industry-development/gasfields-commission-re view.html>

Dernbach, John C. and James R. May, *Shale Gas and the Future of Energy: Law and Policy for Sustainability* (Edward Elgar, 2016)

Desmarais, Annette Aurelie, Hannah Wittman and Nettie Wiebe, *Food Sovereignty: Reconnecting Food, Nature & Community* (Fernwood, 2010)

Deville, Adrian and Ronnie Harding, *Applying the Precautionary Principle* (Federation Press, 1997)

Dodge, Robert, 'Unconventional Drilling for Natural gas in Europe' in Y. Wang and W. Hefley (eds), *The Global Impact of Unconventional Shale Gas Development* (Springer, 2016)

Dokshin, Fedor A., 'Whose Backyard and What's at Issue? Spatial and Ideological Dynamics of Local Opposition to Fracking in New York State, 2010 to 2013' (2016) 81(5) *American Sociological Review* 921

Drahos, Peter, *Regulatory Theory: Foundations and Applications* (ANU Press, 2017)

Dryzek, John and David Schlosberg, *Debating the Earth: The Environmental Politics Reader* (Oxford University Press, 2005)

Dufour, Yvon and Peter Steane, 'Folding the Future Back into the Present: Lessons from the Past – Dom Pierre Pérignon and the Development of Champagne' (2011) 3(1) *Asia-Pacific Journal of Business Administration* 1

Dunlop, Claire A. and Claudio M. Radaelli, *Handbook of Regulatory Impact Assessment* (Edward Elgar, 2016)

Dworkin, Richard, *Taking Rights Seriously* (Harvard University Press, 1997)

ECD, Meat Consumption (2017) <https://data.oecd.org/agroutput/meat-consumption.htm>

Eckstein, Gabriel, *The International Law of Transboundary Groundwater Resources* (Earthscan, 2017)

Edelman, Marc, 'Food Sovereignty: Forgotten Genealogies and Future Regulatory Challenges' (2014) 41 *The Journal of Peasant Studies* 959, 959

Edgeworth, Brendan, *Butt's Land Law* (Thompsons Reuters, 7th edn, 2017)

Eisen, Joel B. and Jim Rossi, *Energy, Economics, and the Environment: Cases and Materials* (Foundation Press, 2010)

Emanuel, Lisa, 'Australia: Competition Small Business and Collective Bargaining – Will the New Laws Strike the Right Balance?' *Mondaq* (online) 19 July 2004 <http://www.mondaq.com/australia/x/27357/Corporate+Commercial+Law/Competition+Small+Business+And+Collective+Bargaining+Will+The+New+Laws+Strike+The+Right+Balance>

Energy and Mines Ministers' Conference, Responsible Shale Development: Enhancing the Knowledge Base on Shale Oil and Gas in Canada (Natural Resources Canada, 2013) <https://www.nrcan.gc.ca/sites/www.nrcan.gc.ca/files/www/pdf/publications/emmc/Shale_Resources_e.pdf>

Estrada, Jose and Rao Bhamidimarri, 'A Review of the Issues and Treatment Options for Wastewater from Shale Gas Extraction by Hydraulic Fracturing' (2016) 182 *Fuel* 292

European Commission, *The Shale Gas 'Revolution' Challenges and Implications for the EU* (2013)

European Parliament, Directorate-General for Internal Policies, Impact of Shale Gas and Shale Oil Extraction on the Environment and on Human Health (2011) <https://europeecologie.eu/IMG/pdf/shale-gas-pe-464-425-final>

Eurostat, Agricultural Census in France (2012) <http://ec.europa.eu/eurostat/statistics-explained/index.php/Agricultural_census_in_France>

Everingham, Jo-Anne, Nina Collins, Will Rifkin, Daniel Rodriguez, Thomas Baumgartl, Jim Cavaye and Sue Vink, 'How Farmers, Graziers, Miners and Other Gas-Industry Personnel See Their Potential for Coexistence in Rural Queensland' (2014) 6(2) *SPE Economics and Management* 122, 122

Explanation of the State Councils Institutional Reform Program (14 March 2018) Xinhua News

Finkel, Alan, Independent Review into the Future Security of the National Electricity Market – Blueprint for the Future (2017) <https://www.energy.gov.au/publications/independent-review-future-security-national-electricity-market-blueprint-future>

Fleming, Ruven C. and Leonie Reins, 'Shale Gas Extraction, Precaution and Prevention: A Conversation on Regulation Responses' (2016) 20 *Energy Research & Social Science* 131

Fleming, Ruven, *Shale Gas, the Environment and Energy Security: A New Framework for Energy Regulation* (Edward Elgar, 2017)

Food and Agricultural Organisation of the United Nations, *Glossary on Right to Food* (UN FAO, 2018) <http://www.fao.org/right-to-food/resources/glossary/en/>

Food and Agricultural Organisation, *Country Programming Framework 2012–2015 for People's Republic of China* (UN FAO, 2015) <http://www.fao.org/3/a-ax529e.pdf>

Food and Agricultural Organisation, *Fertilizer Use by Crop in Poland* (2003) <http://www.fao.org/docrep/005/Y4620E/y4620e00.htm#Contents>

Food and Agricultural Organisation, *The State of Food Security and Nutrition in the World 2017: Building Resilience for Peace and Food Security* (UN FAO, 2017)

Food and Agricultural Organization, Chapter 2: Food Security: Concepts and Measurement (FAO, 2006) <http://www.fao.org/docrep/005/y4671e/y4671e06.htm>

Food Secure Canada, *Resetting the Table: A People's Food Policy for Canada* (People's Food Policy Project, 2015) 12 <http://foodsecurecanada.org/sites/default/files/fsc-resetting-2015_web.pdf>

Frederick, Donald A., 'Legal Rights of Producers to Collectively Negotiate' (1993) 19 (2) *William Mitchell Law Review* 433

Frederick, Donald A., Antitrust Status of Farmer Cooperatives: The Story of the Capper-Volstead Act (2002) Cooperative Information Report <http://www.uwcc.wisc.edu/pdf/CIR59.pdf>

Freiberg, Arie, *Regulation in Australia* (The Federation Press, 2017)

French Government, Achieving a Balance in Trade Relations in the Agricultural Sector and Healthy and Sustainable Food (2018) <https://www.gouvernement.fr/en/achieving-a-balance-in-trade-relations-in-the-agricultural-sector-and-healthy-and-sustainable>

French, Justice Robert, 'Authorisation and Public Benefit – Playing with Categories of Meaningless Reference?' (2006) 24 *Federal Judicial Scholarship* 1

Fulcher, Jonathan and Martin Klapper, 'Coal Seam Gas Exploration and Production in NSW: The New Access Argument' (2011) 51(2) *The APPEA Journal* 688

Garnett, Andrew, 'Regulating Well Integrity' in Tina Hunter (ed.), *Handbook of Shale Gas Law and Policy: Economics: Access, Law and Regulation in Key Jurisdictions* (Intersentia, 2016)

Gasfields Commission, About Us (2017) <http://www.gasfieldscommissionqld.org.au/about-us/>

Gasfields Commission Queensland, Annual Report 2016–2017 (2017) <http://www.gasfieldscommissionqld.org.au/resources/documents/Annual%20Report%202016-17%20%20FINAL%20-%20ONLINE.pdf>

Gilles, Jere Lee and Keith Jamtgaard, 'The Commons Reconsidered' (1982) 4(2) *Rangelands* 51

Global Yield Atlas, China (2014) <http://www.yieldgap.org/china>

Godden, Lee and Jacqueline Peel, *Environmental Law: Scientific, Policy and Regulatory Dimensions* (Oxford University Press, 2010)

Godek, Wendy, 'The Complexity of Food Sovereignty Policymaking: The Case of Nicaragua's Law 693' Paper presented at Food Sovereignty: A Critical Dialogue International Conference, Yale University, USA, September 14–15, 2013

Godzimirski, Jakub, 'Can the Polish Shale Gas Dog Still Bark? Politics and Policy of Unconventional Gas in Poland' (2016) 20 *Energy Research and Social Science* 158

Goho, Shaun A., 'Municipalities and Hydraulic Fracturing: Trends in State Preemption' (2012) 64(7) *Planning and Environmental Law* 3

Goldthau, Andreas and Michael LaBelle, 'The Power of Policy Regimes: Explaining Shale Gas Policy Divergence in Bulgaria and Poland' (2016) 33(6) *Review of Policy Research* 603

Goldthau, Andreas, *The Politics of Shale Gas in Eastern Europe: Energy Security, Contested Technology, and the Social License to Frack* (Cambridge University Press, 2018)

Gordon, Richard, *Regulation and Economic Analysis: A Critique Over Two Centuries* (Springer, 1994)

Gouvin, Eric J., 'A Square Peg in a Vicious Circle: Stephen Breyer's Optimistic Prescription for the Regulatory Mess' (1995) 32 *Harvard Journal on Legislation* 473

Government of British Columbia, British Columbia's Natural Gas Strategy (GBC, 2012) <http://www.gov.bc.ca/ener/popt/down/natural_gas_strategy.pdf>

Government of British Columbia, Sector Snapshot 2016: B.C. Agrifood & Seafood (GBC, 2017).

Grabosky, Peter, 'Counterproductive Regulation' (1995) 17(3) *Law and Policy* 257

Gray, Ian and Geoffrey Lawrence, *A Future for Regional Australia: Escaping Global Misfortune* (Cambridge University Press, 2001)

Gray, Kevin and Susan Francis Gray, 'The Idea of Property in Land' in Susan Bright and John K Dewar (eds), *Land Law: Themes and Perspectives* (Oxford University Press, 1998)

Green, Arthur, Siobhan McPhee, Aviv Ettya, Britta Rocker and Christina Temenos, *British Columbia in a Global Context* (An Open Education Resource Textbook) (BCcampus OpenEd, 1st edn, 2014) <https://opentextbc.ca/geography/chapter/6-6-case-studies/>

Green, Christopher, Peter Styles and Brian Baptie, *Preese Hall Shale Gas Fracturing: Review and Recommendations for Induced Seismic Mitigation* (Assets Publishing, 2012) <https://assets.publishing.service.gov.uk/government/uploads/system/uploads/attachment_data/file/15745/5075-preese-hall-shale-gas-fracturing-review.pdf>

Green, Ryan, *Case Studies of Agricultural Land Commission Decisions: The Need for Inquiry and Reform* (University of Victoria Environmental Law Centre, 2006) <http://www.elc.uvic.ca/documents/ALR%20Final%20Report%20(FINAL-2).pdf>

Gregory, Robin, Dan Ohlson and Joe Arvai, 'Deconstructing Adaptive Management: Criteria for Applications to Environmental Management' (2006) 16(6) *Ecological Applications* 2411

Griffin, Carol, 'Watershed Councils: An Emerging Form of Public Participation in Natural Resource Management' (1999) 35(3) *Journal of the American Water Resource Association* 505

Gullet, Warwick, The Precautionary Principle in Australia: Policy, Law and Potential Precautionary EIAs (University of Wollongong Research Online, 2000) <http://ro.uow.edu.au/cgi/viewcontent.cgi?article=1138&context=lawpapers>

Hardin, Garrett, 'The Tragedy of the Commons' (1968) 162(3859) *Science* 1243

Harding, Ronnie and Elizabeth Fisher, 'Introducing the Precautionary Principle' in R. Harding and E. Fisher (eds), *Perspectives on the Precautionary Principle* (The Federation Press, 1999)

Harding, Ronnie, *Environmental Decision-Making* (The Federation Press, 1998)

Haugen, Hans Morten, 'Food Sovereignty – An Appropriate Approach to Ensure the Right to Food?' (2009) 78(3) *Nordic Journal of International Law* 263

Hawkins, Joanne, 'Fracking: Minding the Gaps' (2015) 17(1) *Environmental Law Review* 8

Hays, Jake, et al., 'Considerations for the Development of Shale Gas in the United Kingdom' (2015) 512–513 *Science of the Total Environment* 36

Head, Brian and Elaine McCoy, *Deregulation Or Better Regulation?: Issues for the Public Sector* (Macmillan Education, 1991)

Hepburn, Samantha, 'Does Unconventional Gas Require Unconventional Ownership: An Analysis of the Functionality of Ownership Frameworks for Unconventional Gas Development' (2013) 8(1) *Journal of Environmental and Public Health Law* 1

Hepburn, Samantha, 'Public Resource Ownership and Community Engagement in a Modern Energy Landscape' (2017) 34 *Pace Environmental Law Review* 379

Hepburn, Samantha, *Mining and Energy Law* (Cambridge University Press, 2015)

Hodges, Andrew and Tim Goesch, 'Natural Resource Management: Results for 2004–05 from an ABARE Survey of Australian Farmers' (2006) 13(3) *Australian Commodities* 569

Hohmann, Harald, *Precautionary Legal Duties and Principles of Modern International Environmental Law: The Precautionary Principle: International Environmental Law Between Exploitation and Protection* (Springer, 1994)

Holling, Crawford Stanley, *Adaptive Environmental Assessment and Management* (Wiley, 1978)

Holt-Giménez, Eric, Raj Patel and Annie Shattuck, *Food Rebellions!: Crisis and the Hunger for Justice* (Pambazuka, 2009)

Hospes, Otto, *Overcoming Barriers to the Implementation of the Right to Food* (HALSHA, 2010) <https://hal.archives-ouvertes.fr/hal-00650148/document>

House of Commons Environment, Food and Rural Affairs Committee, Food Security: Second Report of Session 2014–15 (United Kingdom Government, 2014)

House of Parliament, Parliamentary Office of Science and Technology, Security of UK Food Supply (United Kingdom Government, 2017), number 556

Hunter, Tina and John Chandler, *Petroleum Law in Australia* (LexisNexis, 2013)

Hunter, Tina, 'All Hydraulic Fracturing is Equal, But Some is More Equal Than Others: An Overview of the Types of Hydraulic Fracturing and the Environmental Impacts' (2014) 29(3) *Australian Environment Review* 66

Hunter, Tina, 'Converging Energy Governance in Mature Petroleum Provinces: Political, Legal, and Economic Dimensions in Governing Mature Petroleum Fields in the North Sea' in Slawomir Raszewski (ed.), *The International Political Economy of Oil and Gas* (Springer, 2017) 168

Hunter, Tina, 'Law and Policy Frameworks for Local Content in the Development of Petroleum Resources: Norwegian and Australian Perspectives on Cross-Sectoral Linkages and Economic Diversification' (2014) 14(2–3) *Mineral Economics* 115

Hunter, Tina, 'The Development of Shale Gas and Coal Bed Methane in Australia: Best Practice for International Jurisdictions?' (2016) 38(2) *Houston Journal of International Law* 367

Hunter, Tina, Steven Latta and Greg Gordon, 'Current Practice and Emerging Trends in Regulating Onshore Exploration and Production in Great Britain' in Greg Gordon, John Paterson and Emre Usenmez (eds), *UK Oil and Gas Law: Current Practice and Emerging Trends, Volume I* (EUP, 3rd edn, 2018)

Hunter, Tina, Submission No 9 to the Productivity Commission, Regulatory Burden on the Upstream Petroleum (Oil and Gas) Sector, August 2008

Ingelson, Allan, 'Strategic Planning for Energy Development in Canada' (2015) 6 *Journal of Energy and Environmental Law* 35

International Energy Agency (IEA), Energy Policies of IEA Countries – Australia 2018 Review (IEA, 2018)

International Energy Agency, Energy Policies of IEA Countries: France (2016) <http s://www.iea.org/newsroom/news/2017/january/energy-policies-of-iea-countries-fra nce-2016.html>

International Energy Agency, Energy Policies of IEA Countries: Poland 2016 Review (2016) <http://www.iea.org/publications/freepublications/publication/Energy_Poli cies_of_IEA_Countries_Poland_2016_Review.pdf>

International Energy Agency, Energy Policy in Poland until 2030 <https://www.iea. org/policiesandmeasures/pams/poland/name-24723-en.php>

International Energy Agency, Golden Rules for a Golden Age of Gas. World Energy Outlook Special Report on Gas (2012) <http://www.iea.org/publications/freep ublications/publication/WEO_2012_Special_Report_Golden_Rules_for_a_Golden_ Age_of_Gas.pdf>

International Energy Agency, World Energy Outlook 2017: China (2017) <https://www.iea.org/weo/china/>

International Summit on Science and the Precautionary Principle, Lowell Statement on Science and the Precautionary Principle (17 December 2001)

Jacquet, Jeffrey and Richard C. Stedman, 'Natural Gas Landholder Coalitions in New York State: Emerging Benefits of Collective Natural Resource Management' (2011) 26(1) *Journal of Rural Social Sciences*

Jenkins, Heledd, 'Corporate Social Responsibility and the Mining Industry: Conflicts and Constructs' (2004) 11(1) *Corporate Social Responsibility and Environmental Management* 23

JLCY, Letter to President Trump 4-24-17 (2017) <https://www.jlcny.org/site/index.php/press-room/jlcny-press-releases/2632-letter-to-president-trum-4-24-17>

Johnson, Corey and Tim Boersma, 'Energy (in)Security in Poland: The Case of Shale Gas' (2013) 53 *Energy Policy* 389–399, 394

Jones, Glenys, 'The Adaptive Management System for the Tasmanian Wilderness World Heritage Area – Linking Management Planning with Effectiveness Evaluation' in Catherine Allan and George Stankey (eds), *Adaptive Environmental Management* (Springer, Netherlands, 2009) 227

Jordana, Jacint and David Levi-Faur, *The Politics of Regulation: Institutions and Regulatory Reforms for the Age of Governance* (Edward Elgar, 2004)

Jurca, Stephen, 'What's in a Name?: Geographical Indicators, Legal Protection, and the Vulnerability of Zinfandel' (2013) 20(2) *Indiana Journal of Global Legal Studies* 1445, 1457

Kapelus, Paul, 'Mining, Corporate Social Responsibility and the "Community": The Case of Rio Tinto, Richards Bay Minerals and the Mbonambi' (2002) 39 *Journal of Business Ethics* 275

Keeler, John T.S., 'The Politics of Shale Gas and Anti-Fracking Movements in France and the United Kingdom', Masters Thesis, University of Pittsburgh Graduate School of Public and International Affairs (2014)

Keeler, John, 'The Politics of Shale Gas and Anti-Fracking Movements in France and the UK' in Y. Wang and W. Hefley (eds), *The Global Impact of Unconventional Shale Gas Development* (Springer, 2016)

Kellert, Stephen R., Jai N. Mehta, Syma A. Ebbin and Laly L. Lichtenfeld, 'Community Natural Resource Management: Promise, Rhetoric, and Reality' (2000) 13(8) *Society and Natural Resources* 705

Kelly, Paul, *The End of Certainty: Power, Politics and Business in Australia* (Allen & Unwin, 2008)

Kennedy, Amanda, *Environmental Justice and Land Use Conflict: The Governance Of Mineral and Gas Resource Development* (Routledge, 2017)

Kharaka, Y., et al., 'The Energy–Water Nexus: Potential Groundwater-quality Degradation Associated with Production of Shale Gas' (2013) 7 *Procedia Earth and Planetary Science* 417

King, George E., *Hydraulic Fracturing 101: What Every Representative, Environmentalist, Regulator, Reporter, Investor, University Researcher, Neighbor and Engineer Should Know about Estimating Frac risk and Improving Frac Performance in Unconventional Gas and Oil Wells* (SPE International, 2012)

Klass, Alexandra and Hannah Wiseman, *Energy Law* (Foundation Press, 2016)

Klein, Rudolf and Theodore Marmor, 'Reflections on Policy Analysis: Putting it Together Again' in Robert E. Goodin (ed.), *The Oxford Handbook of Political Science* (OUP, Oxford, 2011)

Kolieb, Johnathon, 'When to Punish, When to Persuade and When to Reward: Strengthening Responsive Regulation with the Regulatory Diamond' (2015) 41(1) *Monash Law Review* 136

Kolleen, Guy, *When Champagne Became French: Wine and the Making of a National Identity* (Johns Hopkins University, 2003)

Kovacevic, Aleksandr, *The Impact of the Russia–Ukraine Gas Crisis in South Eastern Europe* (Oxford Institute of Energy Studies, 2009)

Kriebel, David, et al., 'The Precautionary Principle in Environmental Science' (2001) 109(9) *Environmental Health Perspectives* 871

Lamarquea, Pénélope and Eric F. Lambin, 'The Effectiveness of Marked-based Instruments to Foster the Conservation of Extensive Land Use: The Case of Geographical Indications in the French Alps' (2015) 4 *Land Use Policy* 706

Land Access Implementation Committee, Parliament of Queensland, Land Access Implementation Committee Report (30 August 2013) <http://www.parliament.qld.gov.au/Documents/TableOffice/TabledPapers/2014/5414T5893.pdf>

Land Access Review Panel, Land Access Framework – 12-Month Review: Report of the Land Access Review Panel, February 2012 (2012) <http://www.mellorolsson.com.au/Media/Default/News/News%20Documents/Land_Access_Review_Panel_report.pdf>

Landabaso, M., C. Outhton and K. Morgan, 'Learning Regions in Europe: Theory, Policy and Practice through the RIS Experience' (Paper presented at 3rd International Conference on Technology and Innovation Policy Global Knowledge Partnerships, "Creating Value for the 21st Century", Austin, USAAugust 30 – September 2, 1999)

Larson, Rhett B., 'Reconciling Energy and Food Security' (2014) 48 *University of Richmond Law Review* 929

Lawrence, Geoffrey 'Re-evaluating Food Systems and Food Security: A Global Perspective' (2017) 53(4) *Journal of Sociology* 774

Lee, Kai N., *Compass and Gyroscope: Integrating Science and Politics for the Environment* (Island Press, 1993)

Lefebre, Rene, 'Mechanisms Leading to Potential Impacts of Shale Gas Development on Groundwater Quality' (2017) 4(1) *WIREs Water* e1188 (online)

Levi-Faur, David, *Handbook on the Politics of Regulation* (Edward Elgar Publishing, 2011)

Lewis, Steven W., *Natural Gas in the People's Republic of China* (James A. Baker III Institute for Public Policy, Rice University, 2013)

Lockie, Stewart and Vaughan Higgins, 'Roll-out Neoliberalism and Hybrid Practices of Regulation in Australian Agri-environmental Governance' (2007) 23 *Journal of Rural Studies* 1

Lockwood, Michael, Julie Davidson, Allan Curtis, Elaine Stratford and Rod Griffith, 'Governance Principles for Natural Resource Management' (2010) 23(10) *Society and Natural Resources* 986

Lohmar, B., China's Wheat Economy: Current Trends and Prospects for Imports (2004) WHS 04D-01, Economic Research Service, US Department of Agriculture

Long, Stephen, Deal Finalised for Sale of Cubbie Station (Australian Broadcasting Corporation, 2012), <http://www.abc.net.au/news/2012-10-12/sale-of-cubbie-station-confirmed/4310256>

López, Miguel ÁngelMartin, 'A Study of the Application of Food Sovereignty in International Law' (2016) 4(2) *Groningen Journal of International Law* 14

Lucas, Alastair R. and Constance Hunt, *Oil and Gas Law in Canada* (Carswell, 1990)

Lucas, Alastair R. and Simone Fraser, 'Granting of Shale Gas Licences, Land Access and Property Rights in North America' in Tina Hunter, *Shale Gas Handbook of Law and Policy* (Intersentia, 2016) 137–138

Maddison, Sarah and Richard Dennis, *An Introduction to Australian Public Policy: Theory and Practice* (Cambridge University Press, 2nd edn, 2013)

Malin, Stephanie, 'There's No Real Choice But to Sign: Neoliberalization and Normalization of Hydraulic Fracturing on Pennsylvania Farmland' (2014) 4(1) *Journal of Environmental Studies and Sciences* 17

Marshall, Gordon and John Scott, *A Dictionary of Sociology* (Oxford University Press, 2009)

Martin, Paul, Jacqueline Williams and Amanda Kennedy, 'Creating the Next Generation Rural Landscape Governance: The Challenge for Environmental Law Scholarship' in Paul Martin, Li Zhiping, Qin Tianbao, Anel Du Plessis and Yves Le Bouthillier (eds), *Environmental Governance and Sustainability* (IUCN Academy of Environmental Law, 2012) 50

Massicotte, Marie-Josée, 'La Vía Campesina, Brazilian Peasants, and the Agribusiness Model of Agriculture: Towards an Alternative Model of Agrarian Democratic Governance' (2010) 85 *Studies in Political Economy* 69

Mayere, Severine and Paul A. Donehue, 'Perceptions of Land-use Uncertainty in Queensland's Resource-based Regions' (2014) 51(3) *Australian Planner* 212

McGlade, Christopher, Steve Pye, Paul Ekins, Michael Bradshaw and Jim Watson, 'The Future Role of Natural Gas in the UK: A Bridge to Nowhere?' (2018) 113 *Energy Policy* 454

McManus, Phil and Linda H. Connor, 'What's Mine is Mine(d): Contests over Marginalisation of Rural Life in the Upper Hunter, NSW' (2013) 22(2) *Rural Society* 166

Mechlem, Kerstin, 'Food Security and the Right to Food in the Discourse of the United Nations' (2004) 10 *European Law Journal* 4

Meinzen-Dick, Ruth, Monica Di Gregorio and Nancy McCarthy, 'Methods for Studying Collective Action in Rural Development' (2004) 82(3) *Agricultural Systems* 197

Merriam, Dwight H. and Mary Massaron Ross, *Eminent Domain Use and Abuse: Kelo in Context* (American Bar Association, 2006)

Milczarek-Andrzejewska, Dominika, Katarzyna Zawalińska and Adam Czarnecki, 'Land-use Conflicts and the Common Agricultural Policy: Evidence from Poland' (2018) 73 *Land Use Policy* 423

Moartinov-Bennie, Nonna and Angela Hecimovic, 'Assurance of Australian Natural Resource Management' (2010) 12(4) *Public Management Review* 549

Mohamed Behnassi, Mohamed, Shabbir A. Shahid and Joyce D'Silva, *Sustainable Agricultural Development: Recent Approaches in Resources Management and Environmentally-Balanced Production Enhancement* (Springer, 2011)

Morgan, Bronwen and Karen Yeung, *An Introduction to Law and Regulation: Text and Materials* (Cambridge University Press, 2007)

Nahapiet, Janine and Sumantra Ghoshal, 'Social Capital, Intellectual Capital, and the Organizational Advantage' (1998) 23(2) *The Academy of Management Review* 242

Narula, Smita, 'Reclaiming the Right to Food as a Normative Response to the Global Food Crisis' (2011) 13 *Yale Human Rights and Development Law Journal* 405

National Development and Reform Commission. Interfax Global Energy, Chinese Gas Consumption Hits 237 bcm in 2017 (2017) <http://interfaxenergy.com/gasdaily/article/29373/chinese-gas-consumption-hits-237-bcm-in-2017>

National Farmers Federation, Submission 171 to Select Committee on Unconventional Gas Mining, Inquiry into Unconventional Gas Mining, 14 March 2016

National Research Council, *Adaptive Management for Water Resources Project Planning* (The National Academies Press, 2004)

Natural Resource Governance Institute, Extractive Industries Transparency International (2018) <https://eiti.org/>

Nauwelaers, Claire and Alasdair Reid, *Innovative Regions? A Comparative Review of Methods of Evaluating Regional Innovation Potential* (RIDER, 1995)

Nelsen, Brent F., *The State Offshore: Petroleum, Politics, and State Intervention on the British and Norwegian Conventional Shelves* (Praeger Frederick, 1991)

New York Department of Environmental Conservation, 1992 Findings Statement for Oil and Gas GEIS (1992) <https://www.dec.ny.gov/energy/45912.html>

New York Department of Environmental Conservation, About DEC (2018) <https://www.dec.ny.gov/24.html>

New York Department of Environmental Conservation, Draft SGEIS (2009) <ftp://ftp.dec.state.ny.us/dmn/download/OGdSGEISFull.pdf>.

New York Department of Environmental Conservation, Final Supplemental Generic Environmental Impact Statement on the Oil, Gas and Solution Mining Regulatory Program, *Findings Statement* (2015)

New York State Department of Health, A Public Health Review of High Volume Hydraulic Fracturing for Shale Gas Development (2014) <http://www.health.ny.gov/press/reports/docs/high_volume_hydraulic_fracturing.pdf>

Nliam, Sylvester Oscar, 'International Oil and Gas Environmental Legal Framework and the Precautionary Principle: The Implications for the Niger Delta' (2014) 22(1) *African Journal of International and Comparative Law* 23

North Yorkshire County Council, Kirby Misperton Fracking Operations (NYCC, 2017) <https://www.northyorks.gov.uk/kirby-misperton-fracking-operations>

O'Connor, Jack, 'The Enforceability of Agreements to Negotiate in Good Faith' (2010) 29(2) *University of Tasmania Law Review* 177

Office of the Chief Economist, Review of the Socioeconomic Impacts of Coal Seam Gas in Queensland (Commonwealth of Australia, 2015) <https://industry.gov.au/Office-of-the-Chief-   Economist/Publications/Documents/coal-seam-gas/Socio-economic-impacts-of-coal-seam-gas-in-Queensland.pdf>

Ogus, Anthony, 'Regulation' in Bronwen Morgan and Karen Yeung (eds), *An Introduction to Law and Regulation: Text and Materials* (Cambridge University Press, 2007)

Oil and Gas Authority, OGA Traffic Light Monitoring Scheme to Mitigate Induced Seismicity (OGA) <https://www.ogauthority.co.uk/media/3860/traffic-light-system-doc-for-website_final.pdf>

Ojha, Hemant R., Andy Hall and Rasheed Sulaiman, *Adaptive Collaborative Approaches in Natural Resource Governance: Rethinking Participation, Learning and Innovation* (Routledge, 2012)

Organisation for Economic Co-Operation and Development (OECD), OECD Rural Policy Reviews: Poland 2018 (2018)

Organisation for Economic Co-Operation and Development (OECD), Regulatory Performance: Ex Post Evaluation of Regulatory Tools and Institutions (OECD, 2004)

Osborn, Stephen, Avner Vengosh, Nathaniel Warner and Robert Jackson, 'Methane Contamination of Drinking Water Accompanying Gas-well Drilling and Hydraulic Fracturing' (2011) 108(20) *Proceedings of National Academy of Science* 8172–8176

Osborne, David and Ted Gaebler, *Reinvesting Government* (Addison-Wesley, 1992)

Ostrom, Elinor, 'Self-organization and Social Capital' (1995) 4(1) *Industrial and Corporate Change* 131

Ostrom, Elinor, *Governing the Commons* (Cambridge University Press, 2015)

Ostrom, Elinor, Roy Gardner and James Walker, *Rules, Games, and Common-Pool Resources* (The University of Michigan Press, 1994)

Paterson, John, 'Sustainable Development, Sustainable Decisions and the Precautionary Principle' (2007) 42(3) *Natural Hazards* 515

Paterson, John and Tina Hunter, 'Shale Gas Law and Regulation in the United Kingdom' in Tina Hunter (ed.), *Handbook of Shale Gas Law and Policy* (Intersentia, 2016) 254

Peace River Regional District, Regional Agricultural Plan Background Report (Don Cameron Associates, 2014) <http://prrd.bc.ca/wp-content/uploads/Backgroun d-Report-Final-November-2014.pdf>

Pearse, Guy, *Greenwash: Big Brands and Carbon Scams* (Black Inc, 2012)

Peel, Jacqueline, 'Precaution – A Matter of Principle, Approach or Process?' (2004) 5 *Melbourne Journal of International Law* 483

People's Republic of China, 12th Five Year Plan 2011–2015 (2011) <http://cbi.typepad. com/china_direct/2011/05/chinas-twelfth-five-new-plan-the-full-english-version.html>

People's Republic of China, 13th Five Year Plan 2016–2020 (2016)

Pepper, Rachel, et al., Scientific Inquiry into Hydraulic Fracturing in the Northern Territory: Final Report (Northern Territory Government, 2018) ('Pepper Report')

Perfecto, Ivette, John H. Vandermeer and Angus Lindsay Wright, *Nature's Matrix: Linking Agriculture, Conservation and Food Sovereignty* (Earthscan, 2009)

Pimbert, Michel, *Towards Food Sovereignty: Another World is Possible for Food and Agriculture* (IIED, 2008)

Plumb, James and Andrew Shute, Negotiated Access to Land in Queensland – Is This the End of ADR? (2014) <http://www.carternewell.com/page/Publications/ Archive/Negotiated_access_to_land_in_Queensland_is_this_the_end_of_ADR/>

Polish Geological Institute, First Report: Assessment of shale gas and shale oil resources of the Lower Palaeozoic Baltic-Podlasie-Lublin Basin in Poland (2012) <https:// www.pgi.gov.pl/en/dokumenty-pig-pib-all/aktualnosci-2012/zasoby-gazu/769-rap ort-en/file.html>

Polman, Nico, Krijn Poppe, Jan Wilem van der Schans and Jan-Douwe van der Ploeg, 'Nested Markets with Common Pool Resources in Multifunctional Agriculture' (2010) LXV(2) *Rivista di Economia Agraria* 295

Porter, Michael, *Competitive Advantage, Creating and Sustaining Superior Performance* (Free Press, 1985)

Prno, Jason and Scott Slocombe, 'Exploring the Origins of "Social License to Operate" in the Mining Sector: Perspectives from Governance and Sustainability Theories' (2012) 37(3) *Resources Policy* 346

Productivity Commission, *Inquiry into Mineral and Energy Resource Exploration*, (Commonwealth of Australia, Canberra, 2013)

Prud'homme, Alex, *Hydrofracking: What Everyone Needs to Know* (OUPUSA, 2014)

Putnam, Robert, *Bowling Alone: The Collapse and Revival of American Community* (Simon and Schuster, 2000)

Putnam, Robert, *Making Democracy Work: Civic Tradition in Modern Italy* (Princeton University Press, 1993)

Queensland Competition Authority (Qld), Final Report: Coal Seam Gas Review (January 2014) <http://www.qca.org.au/getattachment/aaaeab4b-519f-4a95-8a 65-911bc46cc1d3/CSG-investigation.aspx>

Queensland Department of Infrastructure, Local Government and Planning, RPIA Statutory Guideline 04/14 <https://www.dilgp.qld.gov.au/resources/planning/pla nning/statutory-guideline-04-14.pdf>

Queensland Department of Natural Resources and Mines, Queensland Gas Supply and Demand Action Plan Discussion Paper (2016)

Queensland Department of State Development, Government Response to the Independent Review of the Gasfields Commission (2016) <https://www.statedevelopment.qld.gov.au/resources/report/government-response-to-the-independent-review.pdf>

Queensland Government Department of State Development, Royalties for the Regions (2015) <http://www.drd.wa.gov.au/rfr/whatisrfr/Pages/default.aspx>

Queensland Government, Adaptive Management (Queensland Department of Heritage Protection, 2017) <https://www.ehp.qld.gov.au/management/non-mining/adaptive-management.html>

Queensland Government, Annual Outlook 2015 Queensland Regional Natural Resource Management Investment Program 2013–2018 (Department of Natural Resources and Mines, 2016) <https://www.dnrm.qld.gov.au/__data/assets/pdf_file/0006/281355/nrm-annual-outlook.pdf>

Queensland Government, Darling Downs Regional Plan (Queensland Department of State Development, Infrastructure and Planning, 2013) <http://www.statedevelopment.qld.gov.au/resources/plan/darling-downs/darling-downs-regional-plan.pdf>

Queensland Government, Department of Environment and Heritage Protection, Underground Water (2017) <https://www.ehp.qld.gov.au/management/non-mining/groundwater.html>

Queensland Government, Department of Infrastructure, Local Government and Planning, RPIA Statutory Guideline 04/14 (2017)

Queensland Government, Department of Infrastructure, Local Government and Planning, RPIA Statutory Guideline 02/14 (2017)

Queensland Government, Department of Infrastructure, Local Government and Planning, RPIA Statuary Guideline 03/14 (2017)

Queensland Government, Department of Infrastructure, Local Government and Planning, RPIA Statuary Guideline 09/14 (2017)

Queensland Government, Department of Natural Resources and Mines, Land Access Code (2016)

Queensland Government, Department of Natural Resources and Mines, Queensland Gas Supply and Demand Action Plan Discussion Paper (November 2016)

Queensland Government, Department of Natural Resources and Mines, Strategic Cropping Land Zone Map (2017) <https://www.dnrm.qld.gov.au/__data/assets/pdf_file/0006/171564/scl-zone-map.pdf>

Queensland Government, Department of Natural Resources and Mines, Compliance and Enforcement (2017) <https://www.ehp.qld.gov.au/management/non-mining/enforcement-compliance.html>

Queensland Government, Department of Natural Resources and Mines, CSG Net and CSG Online: Monitoring CSG Impacts on Groundwater (2017) <https://agforceprojects.org.au/file.php?id=319&open=yes>

Queensland Parliamentary Committee, Mineral and Energy Resources (Common Provisions) Bill 2014 (Report No. 46 Agriculture, Resources and Environment Committee)

Queensland Resources Council, Submission No 13 to Productivity Commission, Inquiry into the Non-financial Barriers to Mineral and Energy Resource Exploration (2013) 3

Rankin, Murray, Sandy Carpenter, Patricia Burchmore and Christopher Jones, 'Regulatory Reform in the British Columbia Petroleum Industry: The Oil and Gas Commission' (2000) 38 *Alberta Law Review* 143

Resnik, D., 'Is the Precautionary Principle Unscientific?' (2003) 34 *Studies in History and Philosophy of Science* 329

Reuters, China to Complete Russia Oil, Gas Pipeline Sections by End-2018: Vice Governor (2017) <https://www.reuters.com/article/us-china-silkroad-russia-pip elines-idUSKBN18819I>

Reuters, Poland to Decide Later this Year on Building Nuclear Plant (2018) <https:// www.reuters.com/article/us-poland-nuclear/poland-to-decide-later-this-yea r-on-building-nuclear-plant-idUSKBN1FI1Q8>

Reuters, Shale Gas Will Not Cut EU Import Dependence: Study (2012) <https:// www.reuters.com/article/us-eu-shale-gas/shale-gas-will-not-cut-eu-import-dep endence-study-idUSBRE8860W220120907>

Rimmer, Stephen, 'Best Practice Regulations and Licensing as a Form of Regulation' (2005) 65(2) *Public Administration* 3

Baldwin, Robert, Martin Cave and Martin Lodge, *Understanding Regulation: Theory, Strategy and Practice* (Oxford University Press, 2nd edn, 2011) 243

Rose, Julian and ICPPC, Polish Government Backs Small Farmers' and Food Sover- eignty (2016) <https://theecologist.org/2016/jan/25/polish-government-backs-sma ll-farmers-and-food-sovereignty>

Royal Society and Royal Academy of Engineering, Shale Gas Extraction in the UK: A Review of Hydraulic Fracturing (Royal Society, 2012) <https://www.raeng.org. uk/publications/reports/shale-gas-extraction-in-the-uk>

Rutovitz, Jay StephenHarris, Natasha Kuruppu and Chris Dunstan, 'Drilling Down. Coal Seam Gas: A Background Paper' Prepared by Institute for Sustainable Futures UTS for the City of Sydney Council, November 2011

Salih, M.A.Mohamed, 'Governance of Food Security in the 21st Century' in Hans Günter Brauchet al. (eds), *Facing Global Environmental Change: Environmental, Human, Energy, Food, Health and Water Security Concepts* (Springer, 2009) 34

Sandiglow, David, Jingchao Wu Qing Yang, Anders Hobe and Junda Lin, *Meeting China's Shale Gas Goals* (2014) Columbia, SIPA (working draft)

Schanbacher, William, *The Politics of Food: The Global Conflict Between Food Security and Food Sovereignty* (ABC-CLIO, 2010)

Schieck, Valente, Flavio Luiz and Ana Maria Suarez Franco, 'Human Rights and the Struggle Against Hunger: Laws, Institutions and Instruments in the Fight to Realise the Right to Adequate Food' (2010) 2 *Yale Human Rights and Development Law Journal* 422

Schott, Stephan and Graham Campbell, 'National Energy Strategies of Major Indus- trialized Countries' in Hugh Dyer and Maria Julia Trombetta (eds), *International Handbook of Energy Security* (Edward Elgar, 2013)

Schuck, Peter, *Why Government Fails So Often: And How It Can Do Better* (Princeton University Press, 2014) 405

Scott, Robert P., Independent Review of the Gasfields Commission Queensland and Asso- ciated Matters (Department of State Development (Qld), July 2016) <https://www.sta tedevelopment.qld.gov.au/resources/report/gasfields-commission-review-report.pdf>

Scottish Government, Onshore Oil and Gas (2017) <http://www.gov.scot/Topics/ Business-Industry/Energy/onshoreoilandgas>

Scottish Government, Unconventional Oil and Gas – Statement (Minister for Business, Innovation and Energy, Scottish Government, 2017) <https://news.gov.scot/sp eeches-and-briefings/unconventional-oil-and-gas-statement>

Second International Conference on the Protection of the North Sea. London, 24–25 November, 1987, Ministerial Declaration

Selznick, Phillip, 'Focusing Organisational Research on Regulation' in Roger Noll (ed.), *Regulatory Policy and the Social Sciences* (University of California Press, 1985)

Senate Select Committee on Unconventional Gas, Parliament of Australia, *Inquiry into Unconventional Gas Interim Report* (2016)

Sexton, Richard, 'Market Power, Misconceptions, and Modern Agricultural Markets' (2012) 95(2) *American Journal of Agricultural Economics* 209

Shankwitz, Frank, *The Five Principles of Collaboration: Applying Trust, Respect, Willingness, Empowerment and Effective Communication to Human Relationships* (Ibeh Agbanyim, 2015)

Shell, Ormen Lange (Shell Global, 2018) <https://www.shell.com/about-us/major-p rojects/ormen-lange.html>

Sheppard, David and Andrew Ward, 'Siberian Gas Delivery to UK Offers Relief after Cold Blast: LNG Arrival Highlights Fragile State of British Energy Security', 3 March 2018, *Financial Times* <https://www.ft.com/content/31e076e2-1e28-11e8-956a -43db76e69936>

Sher, Chloe and Cary Wu, 'Fracking in China: Community Impacts and Public Support of Shale Gas Development' (2018) 27 *Journal of Contemporary China* 626

Singleton, Sara, *Constructing Cooperation: The Evolution of Institutions of Co-management* (University of Michigan Press, 1998)

Skogly, Sigrun, 'Right to Adequate Food: National Implementation and Extraterritorial Obligations' (2007) 11 *Max Planck Yearbook of United Nations Law* 340

Smith, Barry, A Work in Progress – The British Columbia Farmland Preservation Program (2012) <http://www.alc.gov.bc.ca/assets/alc/assets/library/archived-publica tions/alr-history/a_work_in_progress_-_farmland_preservation_b_smith_2012.pdf>

Snir, Reut, 'Trends in Global Nanotechnology Regulation: The Public–Private Interplay' (2014) 17 *Vanderbilt Journal of Entertainment & Technology Law* 1

Special Rapporteur on the Right to Food, The New International Economic Order and the Promotion of Human Rights (ECOSOC, 1987), U.N. Doc. E/CN.4/Sub.2/1987/23

Sprankling, John, 'Owning the Centre of the Earth' (2008) 55 *UCLA Law Review* 979

Stamford, Laurence and Adisa Azapagic, 'Life Cycle Environmental Impacts of UK Shale Gas' (2014) 134 *Applied Energy* 506

Standing Council on Energy and Resources, Multiple Land Use Framework (Council of Australian Governments, 2013) <http://www.coagenergycouncil.gov.au/sites/prod. energycouncil/files/publications/documents/Multiple%20Land%20Use%20Fram ework%20-%20Dec%202013.pdf>

State Council of the Peoples Republic of China, White Paper – The Grain Issue in China (1996)

Statistics Canada, 2011 Census of Agriculture (SC, 2011) <https://www.statcan.gc.ca/ eng/ca2011/index>

Steuben County Landowners Coalition (SCLC), Homepage (2010) <http://mysite.ver izon.net/reszcmsk/>

Stobbe, Tracy E., Alison J. Eagle, Geerte Cotteleer and G. Cornelis van Kooten, 'Farmland Preservation Verdicts—Rezoning Agricultural Land in British Columbia' (2011) 59 *Canadian Journal of Agricultural Economics* 555

Storting White Paper 76(1970–71) <https://www.regjeringen.no/globalassets/upload/ kilde/oed/bro/2002/0006/ddd/pdfv/152184-facts_20.pdf>

Styles, Peter, 'Shale, Shale Gas and Hydraulic Fracturing' in Tina Hunter (ed.), *Handbook of Shale Gas Law and Policy: Economics: Access, Law and Regulation in Key Jurisdictions* (Intersentia, 2016)

Sunstein, Cass R., 'Beyond the Precautionary Principle', John M. Olin Law & Economics Working Paper No. 149 (2D SERIES) (2003)

Svantesson, Dan, *Svantesson on the Law of Obligations* (Centre for Commercial Law, 3rd edn, 2012)

Swayne, Nicola, 'Regulating Coal Seam Gas in Queensland: Lessons in an Adaptive Environ Manag Approach?' (2012) 29(2) *Environmental and Planning Law Journal* 163

Tan, Poh-Ling, David George and Maria Comino, 'Cumulative Risk Management, Coal Seam Gas, Sustainable Water, and Agriculture in Australia' (2015) 31 *International Journal of Water Resources Development* 682

Task Force on Shale Gas, Final Conclusions and Recommendations (TFSG, 2015) <https://darkroom.taskforceonshalegas.uk/original/d6f5f84dbfec be9c22bddbc7f93d31bc:cb2ee01d6a9d7a96cd7d10262971d586/task-force-on-sha le-gas-final-conclusions-and-recommendations.pdf>

Taylor, Madeline and Susanne Taylor, 'Agriculture in a Gas Era: A Comparative Analysis of Queensland and British Columbia's Agricultural Land Protection and Unconventional Gas Regimes' (2016) 22(3) *Australian Journal of Regional Studies* 459

Taylor, Susanne and Madeline Taylor, 'The Aroma of Opportunity: The Potential of Wine Geographical Indications in the Comprehensive Economic Cooperation Agreement' in William van Caenegem and Jen Cleary (eds), *The Importance of Place: Geographical Indications as a Tool for Local and Regional Development* (Springer International, 2017) 81–110

The Australian Macquarie Dictionary (2018, online)

The New York State Senate, The 2018–2019 State Budget <https://www.nysenate. gov/issues/2018-19-budget>

Thomas, Derek, *Placemaking: An Urban Design Methodology* (Routledge, 2016)

Thompson, Mark and Martin George, *Thompson's Modern Land Law* (Oxford University Press, 2017)

Timmens, Christopher and Ashley Vissing, Shale Gas Leases: Is Bargaining Efficient and What Are the Implications for Homeowners If It Is Not? (2015) <http://public.econ. duke.edu/~timmins/Timmins_Vissing_11_15.pdf>

Tioga County Landowners Group (TCLG), Homepage (2016) <http://www.tiogaga slease.org/>

Trauger, Amy, *Food Sovereignty in International Context: Discourse, Politics and Practice of Place* (Routledge, 2015)

Tuihedur Rahman, H.M., Gordon M. Hickey and Swapan Kumar Sarker, 'A Framework for Evaluating Collective Action and Informal Institutional Dynamics under a Resource Management Policy of Decentralization' (2012) 83 *Ecological Economics* 32

UK Biodiversity Action Plan, Department of the Environment (1994) (para. 6.8) <http://jncc.defra.gov.uk/PDF/UKBAP_Action-Plan-1994.pdf>

UK Government, Regulatory Roadmap – Onshore Oil and Gas Exploration in the UK: Regulation and Best Practice (United Kingdom Government, 2018) <https:// www.gov.uk/government/publications/regulatory-roadmap-onshore-oil-and-ga s-exploration-in-the-uk-regulation-and-best-practice>

United Nations, International Covenant on Economic, Social and Cultural Rights General Comment No 12 (CESCR, 2009) <http://www.unhcr.org/refworld/docid/ 4a60961f2.html>

United Nations, The Right to Adequate Food Fact Sheet: 34 (OHCHR, 2000)

United States Department of Agriculture (USDA), Adaptive Management of Natural Resources: Theory, Concepts, and Management Institutions (USDA, 2005) <https:// www.wrrb.ca/sites/default/files/18.%20Stankey%20Adaptive%20Management% 20PNW.pdf>

United States Environmental Protection Agency, Hydraulic Fracturing for Oil and Gas: Impacts from the Hydraulic Fracturing Water Cycle on Drinking Water Resources in the United States (2016)

United States Environmental Protection Agency, Investigation of Ground Water Contamination near Pavilion, Wyoming: Draft Report (2011) <https://www.epa.gov/sites/production/files/documents/EPA_ReportOnPavillion_Dec-8-2011.pdf>

US Energy Information Administration, Annual Energy Outlook 2018 with Projections to 2050 (EIA, 2018) <https://www.eia.gov/outlooks/aeo/pdf/AEO2018.pdf>

US Energy Information Administration, How Much Shale Gas is Produced in the United States? (EIA, 2018) <https://www.eia.gov/tools/faqs/faq.php?id=907&t=8>

US Energy Information Administration, Technically Recoverable Shale Oil and Shale Gas Resources: An Assessment of 137 Shale Formations in 41 Countries Outside the United States (2013)

US Energy Information Administration, Technically Recoverable Shale Oil and Shale Gas Resources: China (2015), XX-2

US Energy Information Administration, Technically Recoverable Shale Oil and Shale Gas Resources: Other Western Europe (2015) <https://www.eia.gov/analysis/studies/worldshalegas/pdf/Northern_Western_Europe_2013.pdf>

US Energy Information Administration, Technically Recoverable Shale Oil and Shale Gas Resources: Poland (2015) <https://www.eia.gov/analysis/studies/worldshalegas/pdf/Poland_Lithuania_Kaliningrad_2013.pdf>

US Energy Information Administration, Technically Recoverable Shale Oil and Gas Resources: United Kingdom (United States Government, 2015), XI-3

US Energy Information Administration, *World Shale Resource Assessments* (United States Government, 2015) <https://www.eia.gov/analysis/studies/worldshalegas/>

USGS, *Assessment of Undiscovered Oil and Gas Resources of the Paris Basin, France, 2015* (2015) <https://pubs.usgs.gov/fs/2015/3016/pdf/fs2015-3016.pdf>

van Caenegem, William. Madeline Taylor, Jen Cleary and Brenda Marshall, *Collective Bargaining in the Agricultural Sector* (2015) <https://rirdc.infoservices.com.au/downloads/15-055>

van der Elst, Nicholas, *et al.*, 'Enhanced Remote Earthquake Triggering at Fluid-Injection Sites in the Midwestern United States' (2013) 341(6142) *Science* 164

van der Meulen, Bernd M.J. and Menno van der Velde, *European Food Law Handbook* (Wageningen Academic Publishers, 2008)

van Laerhove, Frank, 'Traditions and Trends in the Study of the Commons' (2007) *International Journal of the Commons* 13

Vanni, Francesco, *Agriculture and Public Foods: The Role of Collective Action* (Springer, 2014)

Vaughan, Adam, 'UK Fracking Backlash: Seven of Eight Plans Rejected in 2018' (8 March 2018) *The Guardian* (online) <https://www.theguardian.com/environment/2018/mar/08/uk-fracking-backlash-seven-out-of-eight-plans-rejected-in-2018>

Vivoda, Vlado, 'State–Market Interaction in Hydrocarbon Sector: The Cases of Australia and Japan' in Andrei Belyi and Kim Talus (eds), *States and Markets in Hydrocarbon Sectors* (Palgrave Macmillan, 2015)

Vogler, John and Hannes R. Stephan, 'Governance Dimensions of Climate and Energy Security' in Julia Maria Trombetta and Julia Dyer (eds), *International Handbook of Energy Security* (Edward Edgar, 2013) 297

Wade, Robert, *Village Republics: Economics Conditions for Collective Action in South India* (Cambridge University Press, 1988)

Walters, Carl and Ray Hilborn, 'Ecological Optimization and Adaptive Management' (1987) 9 *Annual Review of Ecology and Systematics* 157

Walters, Carl J. and Crawford Stanley Holling, 'Large-scale Management Experiments and Learning by Doing' (1990) 71(6) *Ecology* 2060

Walters, Carl, *Adaptive Management of Renewable Resources* (The Blackburn Press, 2002)

Warner, Barbara and Jennifer Shapiro, 'Fractured, Fragmented Federalism: A Study in Fracking Regulatory Policy' 43(3) *Publius The Journal of Federalism* 474

Weatherill, Stephen, *Better Regulation* (Bloomsbury Publishing, 2007)

Weir, Michael and Tina Hunter, Property Rights, and Coal Seam Gas Extraction: The Modern Property Right Conundrum (2014) 2 *Property Law Review* 71

West Sussex County Council, Application No WSCC/040/17/BA (WSCC, 2017) <http://buildings.westsussex.gov.uk/ePlanningOPS/loadFullDetails.do?aplId=2178>

Whitton, John, *et al.*, 'Shale Gas Governance in the UK and US: Opportunities for Public Participation and the Implications for Social Justice' (2017) 26 *Energy Research and Social Science* 11

Wiener, Antje, 'A Theory of Contestation – A Concise Summary of its Argument and Concepts' 49(1) *Polity* 110

Wiercinski, Andrej, Jakub Jedrzejak and Maciej Kazmarek, 'Poland: A (Legal) Step-ahead: New Licensing Procedures for Exploration and Production of Hydrocarbons: New Provisions Dedicated to Shale Gas' in Cecile Musialski, Mattias Altmann and Stefan Lechtenbohmer (eds), *Shale Gas in Europe: A Multidisciplinary Analysis with a Focus on European Specificities* (Claees and Casteels, 2013)

Wilber, Tom, *Under the Surface: Fracking, Fortunes, and the Fate of the Marcellus Shale, Updated Edition* (Cornell University Press, 2015)

Williams, Byron and Fred Johnson, 'Confronting Dynamics and Uncertainty in Optimal Decision Making for Conservation' (2013) 8(2) *Environmental Research Letters* 1

Williams, Byron, 'Adaptive Management of Natural Resources – Framework and Issues' (2011) 92 *Journal of Environmental Management* 1346, 1356

Williams, Byron and Eleanor Brown, 'Adaptive Management: From More Talk to Real Action' (2014) 53(2) *Environmental Management* 465

Williams, John, An Analysis of Coal Seam Gas Production and Natural Resources Management in Australia Issues and Ways Forward (2012) <http://www.aie.org.au/AIE/Documents/Oil_Gas_121114.pdf>

Williams, John, Ann Milligan and Tim Stubbs, 'Whole of Landscape Assessment and Planning in the Management of Unconventional Gas Exploration and Production in Australia' in R. Quentin Grafton, Ian G. Cronshaw and Michal C. Moore (eds), *Risks, Rewards and Regulation of Unconventional Gas: A Global Perspective* (Cambridge University Press, 2016) 427

Wilmsen, Brook, 'Expanding Capitalism in Rural China Through Land Acquisition and Land Reforms' (2016) 25(101) *Journal of Contemporary China* 701

Windfuhr, Michael and Jennie Jonsén, *Food Sovereignty: Towards Democracy in Localized Food Systems* (ITDG Publications, 2005)

Winston, Clifford, Government Failure Versus Market Failure: Microeconomics Policy Research and Government Performance (AEI-Brookings Joint Center for Regulatory Studies, 2006)

Wiseman, Hannah, 'Regulatory Islands' (2014) 89 *New York University Law Review* 1697

Wiseman, Hannah, 'Risk and Response in Fracturing Policy' (2013) 84 *University of Colorado Law Review* 730

Wiseman, Hannah, *Hydraulic Fracturing and Legal Frameworks* (Oxford Handbooks Online, 2017)

World Bank Food and Agricultural Organisation, Cereal Yield (Per Hectare) (2017) <https://data.worldbank.org/indicator/AG.YLD.CREL.KG>

World Bank, *China 2020: Development Challenges in the New Century* (Washington, 1997)

World Bank, Electrical Production from Coal Sources (2016) <https://data.worldbank.org/indicator/EG.ELC.COAL.ZS?locations=PL>

World Bank, World Development Report 2010: development and climate change (2010) <http://siteresources.worldbank.org/INTWDR2010/Resources/5287678-1226014527953/WDR10-Full-Text.pdf>

World Trade Organisation (WTO), Agreement on Trade-Related Aspects of Intellectual Property Rights: Module IV Geographic Indications (2010)

World Trade Organization, Food Security (WTO, 2010) <http://www.who.int/trade/glossary/story028/en/>

Worldometers, China Population (2018) <http://www.worldometers.info/world-population/china-population/>

Wu, Yangfeng, 'Overweight and Obesity in China' (2006) 333 *British Medical Journal* 362

Xu, Muyo and Josephine Mason, China's Energy Demand to Peak in 2040 as Transportation Demand Grows: CNPC (2017) <https://www.reuters.com/article/us-china-cnpc-outlook/chinas-energy-demand-to-peak-in-2040-as-transportation-demand-grows-cnpc-idUSKCN1AW0DF>

Yergen, Daniel, 'Yergin on the Next Energy Revolution' (2014) McKinsey Quarterly, April 2014 <https://www.mckinsey.com/business-functions/sustainability-and-resource-productivity/our-insights/daniel-yergin-on-the-next-energy-revolution>

Yergin, Daniel, *The Prize: The Epic Quest for Oil, Money and Power* (Simon & Schuster, 1991) 358

Yergin, Daniel, *The Quest: Energy, Security and the Remaking of the Modern World* (2012)

Zou, C., D. Dong, S. Wang, J. Li, X. Li, Y. Wang, D. Li and K. Cheng, 'Formation Mechanism, Geologic Properties and Resource Potential of Shale Gas in China' (2010) 37(6) *Petroleum Exploration and Development* 641–653

Cases, legislation and treaties

Cases

ACCC, *Chevron Australia Pty Ltd & Ors A91139 & A91140 & A91160 & A91161*, Draft Determination
Acker v Guinn, 464 SW 2d 348, 352 (Tex, 1971)
Acton v. Blundell (1843) 12 M & W 324, 152 ER 1223
AG v Bronw (1947) Legge 312
Amarillo Oil Co v Energy-Agri Products Inc (1990) 794 S W 2d 20 (Tx)
Australia Pacific LNG Pty Ltd v Golden & Ors [2013] QCA 366.
Benge v Scharbauer (1953) 259 S W 2d 166 (Tx)
Calder v Bull, 3 US 386, 388 (1798)
Case of Mines (1568) 1 Plowd 310
Cooperstown Holstein Corp. v Town of Middlefield 106 AD3d 1170, 964 NYS2d 431 (N.Y. 2013)
Del Monte Mining & Milling Co. v. Last Chance Mining & Milling Co., 171 U.S. 55, 60 (1898)
Eacham Abrasive Blasting Pty Ltd v Gundersen & Anor [2014] QLC 38
Edwards v National Coal Board [1949] 1 All E. R. 743
Fitzgerald & Anor v Struber & Anor [2009] QLC 0076.
International News Service v. Associated Press, (1918) 248 U.S. 215, 248.
Matter of Frew Run Gravel Prods. v Town of Carroll 524 NYS2d 25 [1987]
Matter of Gernatt Asphalt Prods. v Town of Sardinia 642 NYS2d 164 [1996]
Nothdurft v QGC Pty Ltd [2017] QLC 41
Peabody West Burton Pty Ltd v Mason [2012] QLC 0023
Pryce v Stuber & Anor [2016] QLC 1
R (Frack Free Balcombe Residents Association) v West Sussex County Council [2014] EWHC 4108 (Admin) [131].
R v Earl of Northumberland (1568) 1 Plowden 310
Star Energy Weald Basin Ltd & Anor v Bocardo SA [2010] UKSC 35
TEC Desert Pty Ltd v. Commissioner of State Revenue [2010] HCA 49
Udell v Haas, 288 NYS2d 888 [1968]
United Group Rail Services v Rail Corporation of New South Wales [2009] NSWCA 177

Legislation and treaties

Farm Practices Protection (Right to Farm) Act, RSBC 1996, c 131

Foreign Acquisitions and Takeovers Act 1975 (Cth)

Foreign Acquisitions and Takeovers Fees Imposition Act 2015 (Cth)

Forest Act, RSBC 1996, c 157

Forest Practices Code of British Columbia Act, RSBC 1996, c 159

French National Assembly, *Bill No. 155 to End Research and Exploitation of Conventional and Unconventional Hydrocarbons and to introduce Various Provisions Concerning Energy and The Environment*

French National Assembly, *Law No. 2017–1839 of December 30 2017 Act to End Research and Exploitation of Conventional and Unconventional Hydrocarbons and to Introduce Various Provisions Concerning Energy and the Environment.*

Gasfields Commission Act 2013 (Qld) (GCA)

Geological and Mining Law of 2011 (Poland)

Geothermal Energy Act 2010 (Qld)

Greenhouse Gas Storage Act 2009 (Qld)

Heritage Conservation Act, RSBC 1996, c 187

Industries Assistance Omission Act 1973 (Cth).

Infrastructure Act 2015 (UK)

Internal Review Code of 1986, 26 USC

International Covenant on Economic, Social and Cultural Rights, adopted and opened for signature, ratification and accession by General Assembly resolution 2200A (XXI) of 16 December 1966 entry into force 3 January 1976, in accordance with article 27

Judicial Committee Act 1833 (IMP)

Land Access Code 2016 (Qld)

Land Access Code 2010 (Qld)

Land Act RSBC 1996 c 245

Land Administration Law 1999 (Republic of China)

Land Reform (Scotland) Act 2004 (Scot)

Mineral and Energy Resources (Common Provisions) Act 2014 (Qld)

Mineral Resources Act 1989 (Qld) (MRA)

Mineral Resources Law of the People's Republic of China

Mines (Working Facilities and Support) Act 1966 (UK)

Montreal Protocol on Substances that Deplete the Ozone Layer, opened for signature 19 September 1987, 1522 UNTS 29 (entered into force 1 January 1989)

National Energy Board Act, RSC 1985

National Assembly of France, *Charter for the Environment*

National Assembly of France, *Law No. 2011–835 of 13 July 2011 to Ban the Exploration and Exploitation of Oil and Gas by Hydraulic Fracturing and to Repeal the Exclusive Licences of Projects Using This Technique for Mining*

New York Consolidated Laws, Environmental Conservation Law - ENV § 23–0303

New York Constitution

NY *Municipal Home Rule Law*

NY *Town Law*

Oil and Gas Activities Act, SBC 2008, c 36

Oil, Gas and Solution Mining Law Declaration of Policy – *ECL Article 23 regulating oil and gas activities in New York*

Permanent Sovereignty over Natural Resources, GA Res 1803 (XVII), UN GAOR, 17th sess, 1194th plen mtg, UN Doc A/RES/1803(XVII) (14 December 1962)

Petroleum (Onshore) Act 1991 (NSW)

Petroleum Act 1923 (Qld)

Petroleum Act 1934 (Qld)

Petroleum Act 1998 (UK)

Petroleum Activities Act 1996 (Norway)

Petroleum and Gas (Production and Safety Act) 2004 (Qld)

Petroleum and Gas (Production and Safety) Regulation 2007

Petroleum and Geothermal Energy Resources Act 1967 (WA)

Petroleum and Natural Gas Act, RSBC 1996, c 361

Planning Act 2016 (Qld)

PRC Property Law

Protocol to the 1979 Convention on Long Range Trans boundary Air Pollution on Further Reduction of Sulphur Emissions, opened for signature 14 June 1994, UN Doc GE.94.31969 (entered into force 2 September 1987)

Regional Planning Interests Act 2014 (Qld)

Regional Planning Interests Regulation 2014 (Qld)

Register of Foreign Ownership of Water or Agricultural Land Act 2015 (Cth)

Rome Declaration on World Food Security, opened for signature 13 November 1996, E/CN.4/RES/1998/23

Safe Water Drinking Act 1974 (NY)

Sustainable Planning Act 2016 (Qld)

The Constitution of Queensland 2001 (Qld)

The Directive on Common Rules for the Internal Market in Gas (2009/73/EC)

The Directive on Common Rules for the Internal Market in Electricity (2009/72/EC)

The French Constitutional Act of 1 March 2005, relating to the 2004 Environmental Charter Art 5. Loi constitutionnelle 2005–205, 1 March 2005 (Loi constitutionnelle relative à la Charte de l'environnement (1)), JORF 2 March 2005

Town and Country Planning Act 1990 (UK)

Treaty on the Functioning of the European Union, opened for signature 7 February 1992, [2009] OJ C 115/13 (entered into force 1 November 1993) ('TFEU')

United Nations Framework Convention on Climate Change, *Adoption of the Paris Agreement, 21st Conference of the Parties*, Paris: United Nations, U.N. Doc. A/CONF. 541/13 (2015)

United Nations General Assembly, *United Nations Framework Convention on Climate Change 1992*, UN Doc FCCC/INFORMAL/84

United Nations General Assembly, *World Charter for Nature*, GA Res 37/7, 48th sess

United Nations, *General Comment 12: The Right to Adequate Food: U.N. Doc. E/C.12/1999/5* (CESCR, 1999)

Competition and Consumer Regulations 2010 (Cth)

United Nations, *General Comment 15: The Right to Water: U.N. Doc. E/C.12/ 2002/11* (CESCR, 2002)

United States Constitution

Universal Declaration of Human Rights, GA Res 217A, 3rd sess, 183rd plen mtg, UN Doc A/810 at 71 (1948) (*Universal Declaration of Human Rights*)

Water Act 2000 (Cth)

Water Act 2000 (Qld)

Water Act 2014 (UK)

Water Supply (Safety and Reliability Act) 2008 (Qld)

Water Sustainability Act, SBC 2014, c 15

Index